Modeling and Simulation in Engineering

Modeling and Simulation in Engineering

Edited by **Tommy Haynes**

Willford Press

New York

Published by Willford Press,
118-35 Queens Blvd., Suite 400,
Forest Hills, NY 11375, USA
www.willfordpress.com

Modeling and Simulation in Engineering
Edited by Tommy Haynes

International Standard Book Number: 978-1-68285-075-6 (Hardback)

Printed in the United States of America.

Contents

Preface

Modeling and simulation is a well-established branch in engineering which uses various numerical and computational models to compute and collect data as a basis for planning, designing and analysing. Simulation technology is a great aid for all the engineers from diverse domains. It has become crucial for the development and evaluation of state-of-the-art technologies in various engineering disciplines. This book includes system simulation theory, efficient optimisation and embedded simulation technology. It provides an insight into simulation supporting platforms, finite element methods, and computational electrodynamics. Students, researchers and developers will find this book helpful.

This book unites the global concepts and researches in an organized manner for a comprehensive understanding of the subject. It is a ripe text for all researchers, students, scientists or anyone else who is interested in acquiring a better knowledge of this dynamic field.

I extend my sincere thanks to the contributors for such eloquent research chapters. Finally, I thank my family for being a source of support and help.

Editor

1

A posteriori error analysis of stabilised FEM for degenerate convex minimisation problems under weak regularity assumptions

Wolfgang Boiger[1,2] and Carsten Carstensen[1,2*]

*Correspondence:
cc@math.hu-berlin.de
[1] Department of Mathematics, Humboldt-Universität zu Berlin, Unter den Linden 6, 10099, Berlin, Germany
[2] Department of Computational Science and Engineering, Yonsei University, Unter den Linden 6, 120-749, Seoul, Korea

Abstract

Background: The discretisation of degenerate convex minimisation problems experiences numerical difficulties with a singular or nearly singular Hessian matrix.

Methods: Some discrete analog of the surface energy in microstrucures is added to the energy functional to define a stabilisation technique.

Results: This paper proves (a) strong convergence of the stress even without any smoothness assumption for a class of stabilised degenerate convex minimisation problems. Given the limitted a priori error control in those cases, the sharp a posteriori error control is of even higher relevance. This paper derives (b) guaranteed a posteriori error control via some equilibration technique which does not rely on the strict Galerkin orthogonality of the unperturbed problem. In the presence of L^2 control in the original minimisation problem, some realistic model scenario with piecewise smooth exact solution allows for strong convergence of the gradients plus refined a posteriori error estimates. This paper presents (c) an improved a posteriori error control in this interface problem and so narrows the efficiency reliability gap.

Conclusions: Numerical experiments illustrate the theoretical convergence rates for uniform and adaptive mesh-refinements and the improved a posteriori error control for four benchmark examples in the computational microstructures.

Keywords: Adaptive finite element method; Relaxation; Convexification; Calculus of variations; Degenerate convex problems; Energy reduction; Nonconvex minimisation; Partial differential equation; Stabilisation; Strong convergence; A posteriori error estimate; Reliability-efficiency gap; Euler-Lagrange equation; Guaranteed upper bound

Background

Infimising sequences of variational problems with non-quasiconvex energy densities, in general, develop finer and finer oscillations with no classical limit in Sobolev function spaces called microstructure [1-6]. Those oscillations cause difficulty to numerical methods because fine grids are necessary to resolve such oscillations which results in ineffective and tricky mesh-depending computations. Strong convergence of gradients of infimising sequences of the non-quasiconvex problem is impossible.

Relaxation techniques replace the nonconvex energy density by its (semi-)convex hull and lead to a macroscopic model. Since the convexified energy density obtained by this method, in general, lacks strict convexity, numerical algorithms might encounter situations where the Hessian matrix is singular. For instance, the Newton minimisation algorithm fails on the convexified three-well problem of Subsection 'Three-well benchmark' below. Applications of relaxation techniques include models in computational microstructure [5-7], some optimal design problems [8,9], the nonlinear Laplacian [10] (where the Hessian can become arbitrarily ill-conditioned in spite of its strict convexity) and elastoplasticity [1].

Stabilisation techniques regularise the energy term by an additional positive semidefinite stabilisation function. The paper [11] discusses several choices of such stabilisation functions for P_1 conforming finite elements and quasiuniform meshes. It turns out that stabilisation can ensure strong convergence of the strain approximations under particular circumstances. A particular stabilisation in [12] leads to strong convergence even on unstructured grids but is still restricted to unrealistically smooth solutions. This paper studies the stabilisation technique of [12] and addresses the question of convergence (i) without extra regularity assumptions, (ii) in a realistic scenario called model interface problem, and (iii) establishes an a posteriori error control.

The stabilisation leads to improved condition numbers of the Hessian matrix and to reduced errors if the numerical solvers fail without stabilisation. Figure 1 shows the convergence of the discrete stress σ_ℓ of the three-well benchmark corresponding to the discrete minimisers of the energy $E_\ell(v_\ell) = E(v_\ell) + C/2|||v_\ell|||_\ell^2$. The errors are plotted for computations with uniform mesh refinements with various solver tolerances in the discrete minimisation procedure at a fixed triangulation and values of C, cf. Section 'Numerical experiments' for details on the MATLAB implementation. Without stabilisation, the convergence stagnates with a moderate tolerance of 10^{-5} which becomes visible as a "plateau" in Figure 1. The Newton solver even aborts prematurely due to the singular Hessian. In conclusion, stabilisation enables higher accuracies in numerical examples.

For $\beta \geqslant 0$ the convex energy functional assumes the form

$$E(v) := \int_\Omega \big(W(\mathrm{D}v(x)) + \beta|v(x) - g(x)|^2 - f(x) \cdot v(x)\big)\mathrm{d}x. \tag{1.1}$$

Assume that W is convex with quadratic growth so that there exist minimisers $u \in H_0^1(\Omega)$; below p-th order growth is included while $p = 2$ throughout this simplifying

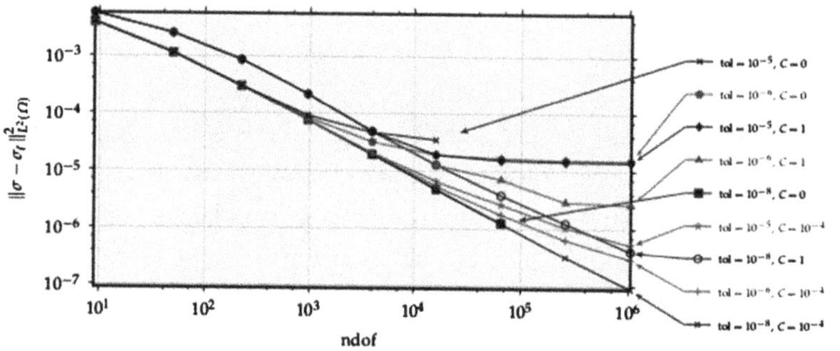

Figure 1 Impact of the stabilisation on the error. Error of the stabilised stress σ_ℓ with coefficients $C = 0, 10^{-4}, 1$ of the stabilisation and various tolerances tol $= 10^{-5}, 10^{-6}, 10^{-8}$ of the Newton solver.

introduction. Given a sequence of shape-regular triangulations $(\mathcal{T}_\ell)_{\ell\in\mathbb{N}_0}$ [13], let u_ℓ minimise the stabilised discrete energy

$$E_\ell(v_\ell) := E(v_\ell) + \frac{1}{2}|||v_\ell|||_\ell^2 \text{ with } |||v_\ell|||_\ell^2 := H_\ell^2 \sum_{F\in\mathcal{F}_\ell(\Omega)} h_F^{-1}||[Dv_\ell]_F||_{L^2(F)}^2$$

amongst all conforming P_1 finite element functions v_ℓ on $\mathcal{T}_\ell$, where $[Dv_\ell]_F$ is the jump of the gradient Dv_ℓ along the interior side F, written $F \in \mathcal{F}_\ell(\Omega)$, and $H_\ell := \max_T h_T$ is the maximal diameter h_T of all simplices $T \in \mathcal{T}_\ell$.

Section 'Global convergence' verifies the strong convergence of the discrete solution u_ℓ and its stress $\sigma_\ell := DW(Du_\ell)$ to their respective continuous conterparts,

$$||\sigma - \sigma_\ell||_{L^2(\Omega)}^2 + \beta||u - u_\ell||_{L^2(\Omega)}^2 + |||u_\ell|||_\ell^2 \to 0 \text{ as } \ell \to \infty.$$

Section 'A posteriori error estimates' presents a novel application of [14-17] to non-linear problems. For the L^2 projection Π_ℓ onto the space of piecewise P_0 functions, any Raviart-Thomas function $\tau_\ell \in RT_0(\mathcal{T}_\ell)$ satisfies

$$\begin{aligned} &||\sigma - \sigma_\ell||_{L^2(\Omega)}^2 \\ &\quad\lesssim \left(||\sigma_\ell - \tau_\ell||_{L^2(\Omega)} + ||\Pi_\ell\Lambda_\ell + \operatorname{div}\tau_\ell||_{L^2(\Omega)} + \operatorname{osc}_{\ell,2}(\Lambda_\ell)\right)||u - u_\ell||_{H^1(\Omega)}. \end{aligned}$$

This error bound holds for any discrete displacement u_ℓ that satisfies the boundary conditions; the point is that inexact solve is included — there is no Galerkin orthogonality required. The drawback is to minimise the expression on the right-hand side with respect to τ_ℓ in order to obtain a sharp error bound. This is a particular selection: degenerate convex minimisation problems do not allow for a control of $||u-u_\ell||_{H^1(\Omega)}$ and may even face multiple exact or discrete solutions while the discrete minimum of E_ℓ is unique. However, in some results of this paper, either W or the lower-order terms lead to some control over $||u - u_\ell||_{H^2(\Omega)}$ and the selection via stabilisation is correct.

Phase transition problems motivate the investigation of scenarios with a smooth solution u up to a one-dimensional interface $\Gamma \subset \overline{\Omega}$ [18]. Section 'Refined analysis for an interface model problem' proves that such problems allow even for strong convergence of the gradients for any unique solution u in $W^{1,\infty}(\Omega)\cap H^2(\Omega\setminus\Gamma)$ [19]. This result also leads to an improvement of the a posteriori error control of the discrete stresses and narrows the efficiency-reliability gap; the efficiency-reliability gap is the difference of the convergence rates of the guaranteed upper a posteriori error bound and the guaranteed lower a posteriori error bound.

Section 'Numerical experiments' complements the theoretical findings with numerical experiments to provide empirical evidence of the improved error control. The stabilisation technique competes in four benchmark examples, with and without known exact solution, for uniform and two different mesh-refining algorithms for the explicit residual-based error estimator of [7] and with an averaging-type error estimator of ([18], (1.11)).

Standard notation on Lebesgue and Sobolev spaces is employed throughout this paper and $a \lesssim b$ abbreviates $a \leqslant Cb$ with some generic constant $0 < C < \infty$ independent of crucial parameters (like the mesh-size on level ℓ); $a \approx b$ means $a \lesssim b \lesssim a$. Furthermore, $A : B$ abbreviates the matrix inner product that corresponds to the Frobenius norm.

Methods: Discretisation and Stabilisation

Based on the convergence results for unstructured grids, this paper will develop reliable error estimators for a class of stabilised convex minimisation problems described in the sequel. Let $\Omega \subset \mathbb{R}^n$ be a bounded Lipshitz domain with polygonal boundary for $n = 2$ or 3. Given a continuous convex energy density $W : \mathbb{R}^{m\times n} \to \mathbb{R}$, $g, f \in L^2(\Omega;\mathbb{R}^m)$, $\beta \geqslant 0$, and $v \in W^{1,p}(\Omega;\mathbb{R}^m)$ with $2 \leqslant p < \infty$ and $m = 1,\ldots,n$, the energy is given by (1.1).

Throughout this paper, the energy density $W \in C^1(\mathbb{R}^{m\times n};\mathbb{R})$ satisfies (2.1)–(2.2) for parameters $1 < r \leqslant 2$, $0 \leqslant s < \infty$ and $s + r + p \leqslant rp$. The *two-sided growth condition* reads

$$|F|^p - 1 \lesssim W(F) \lesssim |F|^p + 1 \quad \text{for all } F \in \mathbb{R}^{m\times n}. \tag{2.1}$$

The *convexity control* assumption reads, for all $F_1, F_2 \in \mathbb{R}^{m\times n}$,

$$|\mathrm{D}\, W(F_1) - \mathrm{D}\, W(F_2)|^r \lesssim \left(1 + |F_1|^s + |F_2|^s\right)(\mathrm{D}\, W(F_1) - \mathrm{D}\, W(F_2)) : (F_1 - F_2). \tag{2.2}$$

The proof of Theorem 2 in [7] shows that (2.2) is crucial for the uniqueness of the stress tensor $\mathrm{D}W(\mathrm{D}u)$.

Given Dirichlet data $u_D \in W^{2,p}(\Omega;\mathbb{R}^m) \cap H^2(\partial\Omega;\mathbb{R}^m)$ for the set of admissible functions $\mathcal{A} := u_D + V := u_D + W_0^{1,p}(\Omega;\mathbb{R}^m)$, the continuous (convex) model problem reads

$$\text{minimise } E(v) \text{ within } v \in \mathcal{A}. \tag{2.3}$$

A finite element approximation of (2.3) is based on a family of regular triangulations $(\mathcal{T}_\ell)_{\ell\in\mathbb{N}_0}$ of the domain Ω into simplices in the sense of Ciarlet [13] (e.g., for $n = 2$, two non-disjoint triangles of $\mathcal{T}_\ell$ share either a common edge or a common node). The set of sides $\mathcal{F}_\ell$ consists of edges (for $n = 2$) or faces (for $n = 3$) of $\mathcal{T}_\ell$ and is split into the union of the sets of all interiour sides $\mathcal{F}_\ell(\Omega)$ and of all boundary sides $\mathcal{F}_\ell(\partial\Omega)$.

For latter reference, define the diameter $h_T := \operatorname{diam} T$ of a triangle (or tetrahedron) $T \in \mathcal{T}_\ell$ and the size $h_F := \operatorname{diam} F$ of a side $F \in \mathcal{F}_\ell$. The *mesh size function* $h_\ell : \Omega \to \mathbb{R}_{>0}$ is given by

$$h_\ell(x) := \begin{cases} h_T & \text{for } x \in \operatorname{int} T \in \mathcal{T}_\ell, \\ \min\{h_F : F \in \mathcal{F}_\ell \text{ and } x \in F\} & \text{otherwise.} \end{cases}$$

The global mesh size will be abbreviated by $H_\ell := \|h_\ell\|_{L^\infty(\Omega)}$. We presume the family $(\mathcal{T}_\ell)_{\ell\in\mathbb{N}_0}$ to be *shape-regular* so that $h_F \approx h_T$ for all $T \in \mathcal{T}_\ell$, $F \in \mathcal{F}_\ell$ and $F \subset T$.

The space of $\mathcal{T}_\ell$-piecewise polynomials of degree $\leqslant k \in \mathbb{N}_0$ is $P_k(\mathcal{T}_\ell)$. The nodal interpolation $I_\ell w \in P_1(\mathcal{T}_\ell) \cap C(\overline{\Omega})$ of $w \in C(\overline{\Omega})$ is given by $I_\ell w(z) = w(z)$ for all nodes z. Let furthermore $\Pi_\ell w$ be the L^2 projection of $w \in L^2(\Omega)$ onto $P_0(\mathcal{T}_\ell)$, and $\operatorname{osc}_{\ell,q}(w) := \|h_\ell(\mathrm{id} - \Pi_\ell)w\|_{L^q(\Omega)}$ be the oscillation of $w \in L^q(\Omega)$ for $2 \leqslant q \leqslant \infty$ with respect to the triangulation $\mathcal{T}_\ell$. The symbol id denotes the identity operator. Let $u_{D,\ell} = I_\ell u_D$, and

$$\mathcal{A}_\ell := u_{D,\ell} + V_\ell \quad \text{with } V_\ell := V \cap P_1(\mathcal{T}_\ell;\mathbb{R}^m) \cap C(\overline{\Omega}).$$

Given a function v on Ω which is possibly discontinuous along some side $F \in \mathcal{F}_\ell(\Omega)$ shared by the two elements $T_\pm$ such that there exist traces from either sides, the *jump* of v along F reads

$$[v]\,(x) = [v]_F\,(x) := \lim_{T_+ \ni y \to x} v(y) - \lim_{T_- \ni y \to x} v(y) \quad \text{for } x \in F.$$

The stabilisation of [12] will be used throughout this paper with $-1 < \gamma < \infty$ and

$$a_\ell(v, w) := \sum_{F \in \mathcal{F}_\ell(\Omega)} \frac{H_\ell^{1+\gamma}}{h_F} \int_F [Dv]_F : [Dw]_F \, ds \quad \text{and} \quad |||v|||_\ell^2 := a_\ell(v, v). \tag{2.4}$$

The stabilised discrete problem reads

$$\text{minimise } E_\ell(v) := E(v) + \frac{1}{2} a_\ell(v, v) \text{ amongst } v \in \mathcal{A}_\ell. \tag{2.5}$$

Convergence of gradients with a guaranteed convergence rate is shown in [12] under unrealistically high regularity assumptions. A comprehensive collection of the results in [12] is summarised in the following theorem.

Theorem 2.1. *([12]) Let $u \in \mathcal{A} \cap H^{3/2+\varepsilon}(\Omega; \mathbb{R}^m)$ be some solution of (2.3) for some $\varepsilon > 0$; let p' and r' be the Hölder conjugate of p and r, $-1 < \gamma < 3$, and set*

$$\zeta := \min\left\{1 + \gamma\,,\, r'\right\} \text{ for } \beta > 0 \text{ and } \zeta := \min\left\{1 + \gamma\,,\, 2\right\} \text{ for } \beta = 0.$$

Then the discrete solution $u_\ell \in \mathcal{A}_\ell$ of (2.5) and the continuous and discrete stress $\sigma := DW(Du) \in L^{p'}(\Omega; \mathbb{R}^{m\times n})$ and $\sigma_\ell := DW(Du_\ell) \in P_0(\mathcal{T}_\ell; \mathbb{R}^{m\times n})$ satisfy

$$\|\sigma - \sigma_\ell\|_{L^{p'}(\Omega)}^{r} + \|u - u_\ell\|_{L^2(\Omega)}^2 + |||u_\ell|||_\ell^2 + H_\ell^{(1+\gamma)/2} \|D(u - u_\ell)\|_{L^2(\Omega)}^2 \lesssim H_\ell^\zeta.$$

Proof. This combines Lemma 3.5 and 4.1–4.2 plus Theorem 3.8 and 4.4 in [12]. □

Global convergence

This section is devoted to the proof of a general convergence result *without* higher regularity assumptions. Let $u \in \mathcal{A}$ and $u_\ell \in \mathcal{A}_\ell$ solve the minimisation problem (2.3) and (2.5) and set $\sigma := DW(Du)$ and $\sigma_\ell := DW(Du_\ell)$. For the unstabilised approximation, the a priori error estimates of [7] plus a density argument prove convergence of

$$\|\sigma - \sigma_\ell\|_{L^{p'}(\Omega)}^{r} + \beta \|u - u_\ell\|_{L^2(\Omega)}^2 \to 0 \quad \text{as } H_\ell \to 0.$$

Note that $\beta = 0$ is permitted. Then, however, uniqueness of u and convergence of $\|u - u_\ell\|_{L^2(\Omega)}^2$ are guaranteed. The point in the following result is that the stabilised approximation converges as well as $|||u_\ell|||_\ell \to 0$ even for non-smooth or non-unique minimisers. Under special circumstances, uniqueness of u and the convergence $\|u - u_\ell\|_{L^2(\Omega)} \to 0$ can be shown even for $\beta = 0$, e.g., in Example 3.3.

Theorem 3.1. *(Global Convergence) Provided $\lim_{\ell\to\infty} H_\ell = 0$ it holds*

$$\|\sigma - \sigma_\ell\|_{L^{p'}(\Omega)}^{r} + \beta \|u - u_\ell\|_{L^2(\Omega)}^2 + |||u_\ell|||_\ell^2 \to 0 \quad \text{as } \ell \to \infty.$$

The proof is based on the following lemma.

Lemma 3.2. *The errors $\delta_\ell := \sigma - \sigma_\ell$ and $e_\ell := u - u_\ell$ satisfy, for all $v_\ell \in V_\ell$, that*

$$\|\delta_\ell\|_{L^{p'}(\Omega)}^{r} + \beta \|e_\ell\|_{L^2(\Omega)}^2 \lesssim |e_\ell - v_\ell|_{W^{1,p}(\Omega)}^{r'} + \beta \|e_\ell - v_\ell\|_{L^2(\Omega)}^2 + a_\ell(u_\ell, v_\ell).$$

Proof. The minimisation problems (2.3) and (2.5) are equivalent to their respective Euler-Lagrange equations, namely for $v \in V$ and $v_\ell \in V_\ell$,

$$\int_\Omega \big(\sigma(x) : Dv(x) + 2\beta(u(x) - g(x)) \cdot v(x) - f(x) \cdot v(x)\big) dx = 0; \tag{3.1}$$

$$\int_\Omega \big(\sigma_\ell(x) : \mathrm{D}v_\ell(x) + 2\beta(u_\ell(x) - g(x)) \cdot v_\ell(x) - f(x) \cdot v_\ell(x)\big)\mathrm{d}x + a_\ell(u_\ell, v_\ell) = 0. \tag{3.2}$$

Algebraic transformations of the difference of these two equations lead to

$$\int_\Omega \delta_\ell : \mathrm{D}e_\ell \mathrm{d}x + 2\beta \|e_\ell\|^2_{L^2(\Omega)} = \int_\Omega (\delta_\ell : \mathrm{D}(e_\ell - v_\ell) + 2\beta e_\ell \cdot (e_\ell - v_\ell))\mathrm{d}x + a_\ell(u_\ell, v_\ell).$$

It is shown in ([12], Lemma 3.5) that

$$\|\delta_\ell\|^r_{L^{p'}(\Omega)} \lesssim \int_\Omega \delta_\ell : \mathrm{D}e_\ell \mathrm{d}x. \tag{3.3}$$

Two Hölder inequalities on the right-hand side and absorbtions of $\|\delta_\ell\|_{L^{p'}(\Omega)}$ and $\|e_\ell\|_{L^2(\Omega)}$ eventually conclude the proof. Further details are dropped for brevity. □

Proof of Theorem 3.1. Given any positive ε, the density of smooth functions in $W_0^{1,p}(\Omega; \mathbb{R}^m)$ leads to some $v_\varepsilon \in \mathcal{D}(\Omega; \mathbb{R}^m)$ such that $\|u - u_D - v_\varepsilon\|_{W^{1,p}(\Omega)} \lesssim \varepsilon$. Hence $v_\ell := I_\ell(v_\varepsilon + u_D) - u_\ell \in V_\ell$ satisfies

$$e_\ell - v_\ell = (u - u_D - v_\varepsilon) + (\mathrm{id} - I_\ell)(v_\varepsilon + u_D).$$

Note that the nodal interpolation $I_\ell(v_\varepsilon + u_D)$ is well-defined since v_ε and u_D are assumed to be smooth. With ([12], Lemma 3.1–3.2) it follows that

$$\|(\mathrm{id} - I_\ell)(v_\varepsilon + u_D)\|_{W^{1,p}(\Omega)} \lesssim H_\ell \to 0 \ \text{ and}$$
$$|||I_\ell(v_\varepsilon + u_D)|||^2_\ell = |||(\mathrm{id} - I_\ell)(v_\varepsilon + u_D)|||^2_\ell \lesssim H_\ell^{1+\gamma} \to 0 \ \text{ as } \ell \to \infty.$$

Since $\|\cdot\|_{L^2(\Omega)} \lesssim \|\cdot\|_{W^{1,p}(\Omega)}$, this yields some $\ell_0 \in \mathbb{N}$ such that

$$|e_\ell - v_\ell|^{r'}_{W^{1,p}(\Omega)} + \beta\|e_\ell - v_\ell\|^2_{L^2}(\Omega) + |||I_\ell(v_\varepsilon + u_D)|||^2_\ell \lesssim \varepsilon \ \text{ for all } \ell \geqslant \ell_0.$$

A Cauchy inequality applied to the stabilisation norm proves

$$a_\ell(u_\ell, v_\ell) = -|||u_\ell|||^2_\ell + a_\ell(u_\ell, I_\ell(v_\varepsilon + u_D)) \leqslant -\frac{1}{2}|||u_\ell|||^2_\ell + \frac{1}{2}|||I_\ell(v_\varepsilon + u_D)|||^2_\ell.$$

Substitute $a_\ell(u_\ell, v_\ell)$ in Lemma 3.2 and add $\frac{1}{2}|||u_\ell|||^2_\ell$ on both sides. This leads to

$$\|\delta_\ell\|^r_{L^{p'}(\Omega)} + \beta\|e_\ell\|^2_{L^2(\Omega)} + |||u_\ell|||^2_\ell \lesssim \varepsilon \ \text{ for all } \ell \geqslant \ell_0.$$

□

Example 3.3. The two-well example from the computational benchmark [18] allows an estimate on $\|e_\ell\|_{L^2(\Omega)}$ even for $\beta = 0$. Let $n = 2$, let $F_1 := -F_2 := (3, 2)/\sqrt{13}$, and let the energy density W be the convex hull of $F \mapsto |F - F_1|^2|F - F_2|^2$. That is

$$W(F) = \left(\max\left\{0, |F|^2 - 1\right\}\right)^2 + 4\left(|F|^2 - (3F(1) + 2F(2))^2/13\right). \tag{3.4}$$

Then ([11], Lemma 9.1) proves, for all $v_\ell \in V_\ell$, that

$$\|e_\ell\|^2_{L^2(\Omega)} \lesssim \int_\Omega \delta_\ell : \mathrm{D}e_\ell \mathrm{d}x + \|e_\ell - v_\ell\|^2_{H^1(\Omega)}.$$

Therefore, the arguments of Lemma 3.2 lead to

$$\|\delta_\ell\|^r_{L^{p'}(\Omega)} + \|e_\ell\|^2_{L^2(\Omega)} \lesssim |e_\ell - v_\ell|^{r'}_{W^{1,p}(\Omega)} + \|e_\ell - v_\ell\|^2_{H^1(\Omega)} + a_\ell(u_\ell, v_\ell).$$

This result can be used in the proof of Theorem 3.1 in order to obtain

$$\|\sigma - \sigma_\ell\|^r_{L^{p'}(\Omega)} + \|u - u_\ell\|^2_{L^2(\Omega)} + |||u_\ell|||^2_\ell \to 0 \ \text{ as } \ell \to \infty.$$

A posteriori error estimates

Beyond the a posteriori error analysis of [7], the additional stabilisation term in the discretisation of this paper causes an additional difficulty in that the Galerkin orthogonality does *not* hold for the natural residual. Inspired from novell developments in the a posteriori error control of elliptic PDEs motivated by inexact solve [14-17], this section presents some guaranteed upper error bound for the discretisation at hand for any approximation u_ℓ which does not necessarily satisfy (3.2) exactly. Thereby inexact solve is included.

Let $u \in \mathcal{A}$ solve (2.3) and let $u_\ell \in \mathcal{A}_\ell$ be arbitrary. It is *not* assumed that u_ℓ solves the discrete problem (2.5); the following theorem holds regardless of this. Recall the definitions of $\mathrm{osc}_{\ell,q}(\cdot)$ and Π_ℓ from Section 'Methods: Discretisation and Stabilisation' and given $\sigma := DW(Du)$ and $\sigma_\ell := DW(Du_\ell)$, abbreviate

$$\Lambda_\ell := -2\beta(u_\ell - g) + f, \quad e_\ell := u - u_\ell \quad \text{and} \quad \delta_\ell := \sigma - \sigma_\ell.$$

Theorem 4.1. *Given any $w_\ell \in W^{1,p}(\Omega;\mathbb{R}^m)$ with $w_\ell = u - u_\ell$ on the boundary $\partial\Omega$, and given any $\tau \in H(div,\Omega;\mathbb{R}^{m\times n})$, it holds, for all $2 \leqslant q \leqslant p$ and for some constant $\varkappa$ known from ([12], Lemma 3.5), that*

$$\varkappa/2\|\delta_\ell\|^r_{L^{p'}(\Omega)} + \beta\|e_\ell\|^2_{L^2(\Omega)} \leqslant (r\varkappa/2)^{1-r'}/r'|w_\ell|^{r'}_{W^{1,p}(\Omega)} + \beta\|w_\ell\|^2_{L^2(\Omega)}$$
$$+ \Big(\|\sigma_\ell - \tau\|_{L^{q'}(\Omega)} + \|\Pi_\ell\Lambda_\ell + div\tau\|_{L^{q'}(\Omega)} + osc_{\ell,q'}(\Lambda_\ell)\Big)\|e_\ell - w_\ell\|_{W^{1,q}(\Omega)}.$$

The constant $\varkappa$ depends on problem-specific data such as $\|u\|_{W^{1,p}(\Omega)}$ and the size of the domain Ω. Refer to the proof of Lemma 3.5 in [12] for details.

Before the proofs conclude this section, some practical choice of τ in Theorem 4.1 is discussed as some Raviart-Thomas finite element functions in

$$RT_0(\mathcal{T}_\ell) := \big\{\tau_{\mathrm{RT}} \in P_1(\mathcal{T}_\ell)\cap H(\mathrm{div},\Omega) : \forall T \in \mathcal{T}_\ell\ \exists a,b,c \in \mathbb{R}\ \forall x \in T, \tau_{\mathrm{RT}}(x) = (a,b)+cx\big\}.$$

We suggest the computation (or an accurate approximation) of

$$\mu_\ell := \min_{\tau\in RT_0(\mathcal{T}_\ell)^m}\Big(\|\sigma_\ell - \tau\|_{L^{q'}(\Omega)} + \|\Pi_\ell\Lambda_\ell + \mathrm{div}\tau\|_{L^{q'}(\Omega)}\Big) \tag{4.1}$$

and emphasise that any upper bound is allowed in Theorem 4.1. This leads to

$$\varkappa/2\|\delta_\ell\|^r_{L^{p'}(\Omega)} + \beta\|e_\ell\|^2_{L^2(\Omega)} \leqslant (r\varkappa/2)^{1-r'}/r'|w_\ell|^{r'}_{W^{1,p}(\Omega)} + \beta\|w_\ell\|^2_{L^2(\Omega)}$$
$$+ (\mu_\ell + \mathrm{osc}_{\ell,q'}(\Lambda_\ell))\|e_\ell - w_\ell\|_{W^{1,q}(\Omega)}.$$

The algorithm of ([20], Prop. 4.1) computes some w_ℓ from $(\mathrm{id} - I_\ell)u_D$ with

$$\|w_\ell\|_{L^q(T)} \approx {h_T}^{1/q}\|(\mathrm{id} - I_\ell)u_D\|_{L^q(\partial T\cap\partial\Omega)} \quad \text{and} \tag{4.2}$$
$$\|Dw_\ell\|_{L^q(T)} \lesssim h_T^{1/q-1}\|(\mathrm{id} - I_\ell)u_D\|_{L^q(\partial T\cap\partial\Omega)} + h_T^{1/q}\|\partial(\mathrm{id} - I_\ell)u_D/\partial s\|_{L^q(\partial T\cap\partial\Omega)}.$$

(The proof of the second assertion is analogous to that of ([20], Prop. 4.1) and the first is an immediate consequence of the design of w_ℓ). This and $\|e_\ell - w_\ell\|_{W^{1,q}(\Omega)} \lesssim 1$ for bounded u_ℓ (i.e. solely $\|u_\ell\|_{W^{1,p}(\Omega)} \lesssim 1$ is assumed) lead to the practical estimate μ_ℓ as a computable guaranteed upper bound of the left-hand side of Theorem 4.1. Since the minimisation of (4.1) is computationally intensive for $q \neq 2$, Section 'Numerical experiments' actually computes an approximation of μ_ℓ, based on $q = 2$.

The choice $\tau = \sigma$ in Theorem 4.1 shows that the right-hand side is in fact optimal up to oscillations. The reliability-efficiency gap of [18] is visible here in that we have no further estimate on $\|u_\ell\|_{W^{1,p}(\Omega)}$ [7,18]. However, additional smoothness assumptions on u may lead to refined estimates on the term $\|e_\ell - w_\ell\|_{W^{1,q}(\Omega)}$ (cf. Section 'Refined analysis for an interface model problem'). The following result indicates that μ_ℓ is sharp in the sense that it converges with the correct convergence rate. This theorem employs the Fortin interpolation operator $I_{F,\ell}$ defined for $\tau \in H(\mathrm{div}, \Omega) \cap L^t(\Omega; \mathbb{R}^n)$ with $t > 2$ by $I_{F,\ell}\tau \in RT_0(\mathcal{T}_\ell)$ and

$$\int_F n_F \cdot (\mathrm{id} - I_{F,\ell})\tau \,\mathrm{d}s = 0 \quad \text{for all } F \in \mathcal{F}_\ell.$$

Here and in the following, n_F denotes a unit normal vector of the side F; the direction of n_F arbitrary, but fixed for a given side F. For the improved regularity of stress in the class of degenerate convex minimisation problems at hand, we refer to [3,21].

Theorem 4.2. (Efficiency) *If the exact stress σ is sufficiently regular such that its Fortin interpolant $\tau_\ell = I_{F,\ell}\sigma \in RT_0(\mathcal{T}_\ell; \mathbb{R}^{m\times n})$ is defined, it holds*

$$\|\sigma_\ell - \tau_\ell\|_{L^{q'}(\Omega)} + \|\Pi_\ell \Lambda_\ell + div\tau_\ell\|_{L^{q'}(\Omega)} \lesssim \|\delta_\ell\|_{L^{q'}\Omega} + 2\beta\|e_\ell\|_{L^{q'}(\Omega)} + \|(id - I_{F,\ell})\sigma\|_{L^{q'}(\Omega)}.$$

It is expected that $\|(\mathrm{id} - I_{F,\ell})\sigma\|_{L^{q'}(\Omega)} \lesssim H_\ell$. This is shown in ([22], Prop. 3.6) for $q' = 2$ and therefore also holds for $q' \leqslant 2$. Hence the right-hand side of the assertion of Theorem 4.2 converges with the (expected) optimal convergence rates.

Proof of Theorem 4.1. Let $\varkappa$ be the reciprocal of c_1 in ([12], Lemma 3.5), which is also the multiplicative constant hidden in (3.3). Recall Young's inequality, which reads $ab \leqslant a^r/r + b^{r'}/r'$ for $a, b > 0$. This, (3.3) and the continuous Euler-Lagrange equation (3.1) show, for $v = e_\ell - w_\ell \in V$, that

$$\begin{aligned}\varkappa\|\delta_\ell\|^r_{L^{p'}(\Omega)} + 2\beta\|e_\ell\|^2_{L^2(\Omega)} &\leqslant \int_\Omega (\delta_\ell : \mathrm{D}v + 2\beta e_\ell \cdot v)\mathrm{d}x \\ &\quad + \int_\Omega (\delta_\ell : \mathrm{D}w_\ell + 2\beta e_\ell \cdot w_\ell)\mathrm{d}x \\ &\leqslant -\int_\Omega (\sigma_\ell : \mathrm{D}v - \Lambda_\ell \cdot v)\mathrm{d}x \\ &\quad + \beta\|e_\ell\|^2_{L^2(\Omega)} + \beta\|w_\ell\|^2_{L^2(\Omega)} \\ &\quad + \varkappa/2\|\delta_\ell\|^r_{L^{p'}(\Omega)} + (r\varkappa/2)^{1-r'}/r'|w_\ell|^{r'}_{W^{1,p}(\Omega)}.\end{aligned}$$

Hence $\mathrm{Res}_\ell(v) := -\int_\Omega(\sigma_\ell : \mathrm{D}v - \Lambda_\ell \cdot v)\mathrm{d}x$ satisfies

$$\varkappa/2\|\delta_\ell\|^r_{L^{p'}(\Omega)} + \beta\|e_\ell\|^2_{L^2(\Omega)} \leqslant \mathrm{Res}_\ell(v) + (r\varkappa/2)^{1-r'}/r'|w_\ell|^{r'}_{W^{1,p}(\Omega)} + \beta\|w_\ell\|^2_{L^2(\Omega)}.$$

Let $C_{q'}$ denote the Poincaré constant of convex domains with respect to the $W^{1,q'}$ norm. The fundamental theorem of calculus on some one-dimensional arc shows that $C_\infty \leqslant 1$. The paper [23] proves $C_1 = 1/2$. Hence, operator-interpolation arguments [24,25] prove $C_{q'} \leqslant (1/2)^{1/q'} \leqslant 1$. The Poincaré inequality shows, for any $2 \leqslant q \leqslant p$, that

$$\begin{aligned}\int_\Omega (\mathrm{id} - \Pi_\ell)\Lambda_\ell \cdot v\mathrm{d}x &= \int_\Omega h_\ell(\mathrm{id} - \Pi_\ell)\Lambda_\ell \cdot \frac{1}{h_\ell}(\mathrm{id} - \Pi_\ell)v\mathrm{d}x \\ &\leqslant \|h_\ell(\mathrm{id} - \Pi_\ell)\Lambda_\ell\|_{L^{q'}(\Omega)}\|\mathrm{D}v\|_{L^q(\Omega)} = \mathrm{osc}_{\ell,q'}(\Lambda_\ell)\|\mathrm{D}v\|_{L^q(\Omega)}.\end{aligned}$$

For any $\tau \in H(\mathrm{div}, \Omega; \mathbb{R}^{m\times n})$, the Hölder and Poincaré inequalities show

$$\begin{aligned}\mathrm{Res}_\ell(v) &= -\int_\Omega \big((\sigma_\ell - \tau) : \mathrm{D}v - (\Pi_\ell\Lambda_\ell + \mathrm{div}\tau) \cdot v - (\mathrm{id} - \Pi_\ell)\Lambda_\ell \cdot v\big)\mathrm{d}x \\ &\leqslant \Big(\|\sigma_\ell - \tau\|_{L^{q'}(\Omega)} + \|\Pi_\ell\Lambda_\ell + \mathrm{div}\tau\|_{L^{q'}(\Omega)} + \mathrm{osc}_{\ell,q'}(\Lambda_\ell)\Big)\|v\|_{W^{1,q}(\Omega)}.\end{aligned}$$

Proof of Theorem 4.2. The triangle inequality yields

$$\|\sigma_\ell - \tau_\ell\|_{L^{q'}(\Omega)} \leqslant \|(\mathrm{id} - I_{F,\ell})\sigma\|_{L^{q'}(\Omega)} + \|\delta_\ell\|_{L^{q'}\Omega}.$$

Since $f = 2\beta(u-g) - \mathrm{div}\sigma$, the commutative property $\mathrm{div} I_{F,\ell} = \Pi_\ell \mathrm{div}$ (cf. ([22], p. 129)) yields

$$\|\Pi_\ell \Lambda_\ell + \mathrm{div}\tau_\ell\|_{L^{q'}(\Omega)} = 2\beta\|\Pi_\ell e_\ell\|_{L^{q'}(\Omega)} \leqslant 2\beta\|e_\ell\|_{L^{q'}(\Omega)}.$$

Refined analysis for an interface model problem

This section is devoted for a model scenario from phase transition problems [18] with some solution u that is smooth outside some one-dimensional interface Γ. Suppose some (possibly non-unique) minimiser u of the continuous problem (2.3) satisfies $u \in W^{1,\infty}(\Omega;\mathbb{R}^m) \cap W^{2,p}(\Omega \setminus \Gamma;\mathbb{R}^m)$ for some finite union Γ of $(n-1)$ dimensional Lipschitz surfaces in $\overline{\Omega}$. Since Ω has a Lipschitz boundary, this implies Lipschitz continuity of u on Ω. We refer to [19] for sufficient conditions for $u \in W^{1,\infty}(\Omega;\mathbb{R}^m)$ and conclude that the remaining assumption $u \in W^{2,p}(\Omega \setminus \Gamma;\mathbb{R}^m)$ is the essential hypothesis expected in many interface problems. Let $u_\ell \in \mathcal{A}_\ell$ be the (unique) minimiser of the discrete stabilised problem (2.5). In the following, also $\Gamma = \emptyset$ is permitted to extend previous results [12] for highly regular minimisers.

The following theorem leads to a priori convergence rates for the interface model problem. Thereby it recovers the results of [12] for problems with piecewise smooth exact solution.

We will abbreviate the set of all triangles that are touched by Γ as $\mathcal{T}_\ell(\Gamma) := \{T \in \mathcal{T}_\ell : \mathrm{dist}(T,\Gamma) = 0\}$, its cardinality as $|\mathcal{T}_\ell(\Gamma)|$, its union as $\Omega_{\Gamma,\ell} := \mathrm{int}(\bigcup \mathcal{T}_\ell(\Gamma))$ with volume $|\Omega_{\Gamma,\ell}|$ and its complement as $\Omega^C_{\Gamma,\ell} := \Omega \setminus \overline{\Omega_{\Gamma,\ell}}$.

Theorem 5.1. *Provided $\beta > 0$, it holds*

$$\|\delta_\ell\|^r_{L^{p'}(\Omega)} + \|e_\ell\|^2_{L^2(\Omega)} + \|\!|u_\ell|\!\|^2_\ell \lesssim H_\ell^{1+\gamma}|u|^2_{H^2(\Omega\setminus\Gamma)} + H_\ell^2|u|^2_{W^{1,\infty}(\Omega)} + H_\ell^{r/(r-1)}|u|^{r/(r-1)}_{W^{2,p}(\Omega^C_{\Gamma,\ell})}$$
$$+ H_\ell^{\gamma+n-1}|u|^2_{W^{1,\infty}(\Omega)}|\mathcal{T}_\ell(\Gamma)| + |u|^{r/(r-1)}_{W^{1,\infty}(\Omega)}|\Omega_{\Gamma,\ell}|^{r/((r-1)p)}.$$

Remark 5.2. In the case of uniform mesh refinements we may expect $|\mathcal{T}_\ell(\Gamma)| \approx H_\ell^{1-n}$ and $|\Omega_{\Gamma,\ell}| \approx H_\ell$ and Theorem 5.1 simplifies to

$$\|\delta_\ell\|^r_{L^{p'}(\Omega)} + \|e_\ell\|^2_{L^2(\Omega)} + \|\!|u_\ell|\!\|^2_\ell \lesssim H_\ell^{\min\{\gamma,2\}}|u|^2_{W^{1,\infty}(\Omega)} + H_\ell^{r/((r-1)p)}|u|^{r/(r-1)}_{W^{1,\infty}(\Omega)}.$$

Proof. With $w_\ell = (\mathrm{id} - I_\ell)e_\ell = (\mathrm{id} - I_\ell)u$, a Young inequality, (3.3) and ([12], Theorem 3.8) yield

$$\|\delta_\ell\|^r_{L^{p'}(\Omega)} + \|e_\ell\|^2_{L^2(\Omega)} + \|\!|u_\ell|\!\|^2_\ell \lesssim |w_\ell|^{r/(r-1)}_{W^{1,p}(\Omega)} + \|w_\ell\|^2_{L^2(\Omega)} + \|\!|I_\ell u|\!\|^2_\ell.$$

Theorem 4.4.4 in [25] shows $\|w_\ell\|_{L^2(\Omega)} \lesssim \|w_\ell\|_{L^\infty(\Omega)} \lesssim H_\ell|u|_{W^{1,\infty}(\Omega)}$ and

$$|w_\ell|^p_{W^{1,p}(\Omega)} = |w_\ell|^p_{W^{1,p}(\Omega_{\Gamma,\ell})} + |w_\ell|^p_{W^{1,p}(\Omega^C_{\Gamma,\ell})}$$
$$\lesssim |u|^p_{W^{1,\infty}(\Omega_{\Gamma,\ell})}|\Omega_{\Gamma,\ell}| + H_\ell^p|u|^p_{W^{2,p}(\Omega^C_{\Gamma,\ell})}.$$

Let $\omega_F = \bigcup_{\substack{T\in\mathcal{T}_\ell \\ F\subset T}} T$ be the patch of a side $F \in \mathcal{F}_\ell$, and set $\mathcal{F}_\ell(\Gamma)= \{F \in \mathcal{F}_\ell(\Omega) : \omega_F \cap \Gamma \neq \emptyset\}$ and $\mathcal{F}_\ell^C(\Gamma) = \mathcal{F}_\ell(\Omega) \setminus \mathcal{F}_\ell(\Gamma)$. Note that $[Du]_F = 0$ for $F \in \mathcal{F}_\ell^C(\Gamma)$. Then

$$|||I_\ell u|||_\ell^2 = H_\ell^{1+\gamma}\left(\sum_{F\in\mathcal{F}_\ell^C(\Gamma)} h_F^{-1}\|[\mathrm{D}w_\ell]_F\|_{L^2(F)}^2 + \sum_{F\in\mathcal{F}_\ell(\Gamma)} h_F^{-1}\|[\mathrm{D}I_\ell u]_F\|_{L^2(F)}^2\right).$$

The first sum can be estimated as in the proof of ([12], Lemma 3.2), the second sum with

$$\|[\mathrm{D}I_\ell u]_F\|_{L^2(F)}^2 \lesssim h_F^{n-1}|I_\ell u|_{W^{1,\infty}(F)}^2 \lesssim h_F^{n-1}|u|_{W^{1,\infty}(F)}^2.$$

The observation $|\mathcal{F}_\ell(\Gamma)| \leqslant (n+1)|\mathcal{T}_\ell(\Gamma)|$ concludes the proof. □

Together with Theorem 5.1, the subsequent result implies strong convergence of the gradients in the model interface problem as $H_\ell \to 0$.

Theorem 5.3. *Under the aforementioned conditions on the (possibly non-unique) exact minimiser $u \in W^{1,\infty}(\Omega;\mathbb{R}^m) \cap W^{2,p}(\Omega \setminus \Gamma;\mathbb{R}^m)$, the error $e_\ell = u - u_\ell$ of the discrete solution $u_\ell \in \mathcal{A}_\ell$ of (2.5) satisfies*

$$\|\mathrm{D}e_\ell\|_{L^2(\Omega)} \lesssim \|e_\ell\|_{L^2(\Omega)}^{1/3} + H_\ell^{5/6}\|\partial^2 u_D/\partial s^2\|_{L^2(\partial\Omega)}^{1/3} + H_\ell^{(1-\gamma)/2}|||u_\ell|||_\ell + H_\ell^{-(1+\gamma)/4}|||u_\ell|||_\ell^{1/2}\left(\|e_\ell\|_{L^2(\Omega)}^{1/2} + H_\ell^{5/4}\|\partial^2 u_D/\partial s^2\|_{L^2(\Omega)}^{1/2}\right).$$

Proof. The basic idea of gradient control is the generalisation of the interpolation estimate $H^1(\Omega) = [L^2(\Omega), H^2(\Omega)]_{1/2}$ for a reduced domain $\Omega\setminus\Gamma$; refer to [24,25] for a detailed analysis of interpolation spaces. Let w_ℓ be the boundary value interpolation of $(\mathrm{id} - I_\ell)u_D$ as described in ([20], Prop. 4.1), such that w_ℓ satisfies the inequalities in (4.2). A piecewise integration by parts shows, for $v := e_\ell - w_\ell \in W_0^{1,p}(\Omega;\mathbb{R}^m)$, that

$$\begin{aligned}\|\mathrm{D}e_\ell\|_{L^2(\Omega)}^2 &= \int_\Omega \mathrm{D}(u-u_\ell) : \mathrm{D}v\,dx + \int_\Omega \mathrm{D}e_\ell : \mathrm{D}w_\ell\,dx \\ &\leqslant \int_\Gamma v\cdot[\mathrm{D}u]_\Gamma n_\Gamma\,ds - \int_{\Omega\setminus\Gamma} v\cdot\Delta u\,dx - \sum_{F\in\mathcal{F}\ell(\Omega)}\int_F v\cdot[\mathrm{D}u_\ell]_F n_F\,ds \\ &\quad + \|\mathrm{D}e_\ell\|_{L^2(\Omega)}\|\mathrm{D}w_\ell\|_{L^2(\Omega)},\end{aligned}$$

where n_Γ is a unit normal vector of the interface Γ. The Lipschitz continuity of u implies $|[Du]_\Gamma n_\Gamma| \lesssim 1$. This and the trace inequality on Γ lead to

$$\int_\Gamma v\cdot[\mathrm{D}u]_\Gamma n_\Gamma\,ds \lesssim \|v\|_{L^2(\Gamma)} \lesssim \|v\|_{L^2(\Omega)} + \|v\|_{L^2(\Omega)}^{1/2}\|\mathrm{D}v\|_{L^2(\Omega)}^{1/2}.$$

The case $\Gamma = \emptyset$ is contained in ([12], Theorem 4.4). The piecewise Laplacian of u is bounded in $L^2(\Omega)$ and so (with the generic constant $C := \|\Delta u\|_{L^2(\Omega\setminus\Gamma)}$ hidden in the notation $C \approx 1$)

$$\int_{\Omega\setminus\Gamma} v\cdot\Delta u\,dx \lesssim \|v\|_{L^2(\Omega)}$$

The elementwise trace inequality ([25], Theorem 1.6.6, p. 39) for an n-dimensional simplex T and one of its sides F, and $f \in W^{1,q}(T;\mathbb{R}^m)$, $1 \leqslant q < \infty$, reads

$$\|f\|_{L^q(F)}^q \lesssim h_T^{-1}\|f\|_{L^q(T)}^q + \|f\|_{L^q(T)}^{q-1}\|\mathrm{D}f\|_{L^q(T)} \lesssim h_T^{-1}\|f\|_{L^q(T)}^q + h_T^{q-1}\|\mathrm{D}f\|_{L^q(T)}^q.$$

The term $\int_F v \cdot [\mathrm{D}u_\ell]_F n_F\, ds$ and the stabilisation $|||u_\ell|||_\ell$ are already analysed in the *Estimate on C* in the proof of ([12], Theorem 4.4). This results in

$$\sum_{F\in\mathcal{F}_\ell(\Omega)} \int_F v \cdot [\mathrm{D}u_\ell]_F n_F\, ds \lesssim |||u_\ell|||_\ell \left(H_\ell^{(1-\gamma)/2}\|\mathrm{D}v\|_{L^2(\Omega)} + H_\ell^{-(1+\gamma)/2}\|v\|_{L^2(\Omega)}\right).$$

The preceding estimates plus the absorbtion of $\|\mathrm{D}e_\ell\|_{L^2(\Omega)}$ lead to

$$\begin{aligned}\|\mathrm{D}e_\ell\|^2_{L^2(\Omega)} &\lesssim \|v\|_{L^2(\Omega)} + \|v\|^{1/2}_{L^2(\Omega)}\|\mathrm{D}v\|^{1/2}_{L^2(\Omega)} + \|\mathrm{D}w_\ell\|^2_{L^2(\Omega)}\\ &\quad + |||u_\ell|||_\ell \left(H_\ell^{(1-\gamma)/2}\|\mathrm{D}v\|_{L^2(\Omega)} + H_\ell^{-(1+\gamma)/2}\|v\|_{L^2(\Omega)}\right).\end{aligned}$$

The triangle inequality applied to $v = e_\ell - w_\ell$ and some careful elementary analysis to absorb $\|\mathrm{D}e_\ell\|^{1/2}_{L^2(\Omega)}$ eventually lead to

$$\begin{aligned}\|\mathrm{D}e_\ell\|_{L^2(\Omega)} &\lesssim \|e_\ell\|^{1/3}_{L^2(\Omega)} + \|w_\ell\|^{1/3}_{L^2(\Omega)} + |w_\ell|_{H^1(\Omega)} + H_\ell^{(1-\gamma)/2}|||u_\ell|||_\ell\\ &\quad + H_\ell^{-(1+\gamma)/4}|||u_\ell|||_\ell^{1/2}\left(\|e_\ell\|_{L^2(\Omega)} + \|w_\ell\|_{L^2(\Omega)}\right)^{1/2}.\end{aligned}$$

The inequalities (4.2), Poincaré and Friedrichs inequalities on sides $F \in \mathcal{F}_\ell(\partial\Omega)$ and removal of higher-order terms in H_ℓ conclude the proof. □

The following theorem is an improved a posteriori estimate based on Theorems 4.1 and 5.3.

Theorem 5.4. *Recall $u \in W^{1,\infty}(\Omega;\mathbb{R}^m) \cap W^{2,p}(\Omega \setminus \Gamma;\mathbb{R}^m)$, the definitions $e_\ell := u - u_\ell$ and $\delta_\ell := \sigma - \sigma_\ell$ for $\sigma := \mathrm{D}W(\mathrm{D}u)$ and $\sigma_\ell := \mathrm{D}W(\mathrm{D}u_\ell)$, and the definition of Λ_ℓ from Section 'A posteriori error estimates'. Set*

$$M(\tau) := \|\sigma_\ell - \tau\|_{L^2(\Omega)} + \|\Pi_\ell\Lambda_\ell + div\tau\|_{L^2(\Omega)} + osc_{\ell,2}(\Lambda_\ell) \qquad \text{for all } \tau \in H(div,\Omega;\mathbb{R}^{m\times n}).$$

Provided $\beta > 0$, it holds

$$\begin{aligned}\|\delta_\ell\|^{r'}_{L^{p'}(\Omega)} + \|e_\ell\|^2_{L^2(\Omega)} &\lesssim M(\tau)^{6/5} + H_\ell^{-(1+\gamma)/3}M(\tau)^{4/3}|||u_\ell|||_\ell^{2/3}\\ &\quad + M(\tau)\left(H_\ell^{(1-\gamma)/2}|||u_\ell|||_\ell + H_\ell^{1-\gamma/4}|||u_\ell|||_\ell^{1/2}\right) + H_\ell^{\min\{5,r'(1+1/p)\}}\end{aligned}$$

and

$$\begin{aligned}\|\mathrm{D}e_\ell\|^2_{L^2(\Omega)} &\lesssim M(\tau)^{2/5} + H_\ell^{-(1+\gamma)/9}M(\tau)^{4/9}|||u_\ell|||_\ell^{2/9} + H_\ell^{\min\{5/3,r'(1+1/p)/3\}}\\ &\quad + M(\tau)^{1/3}\left(H_\ell^{(1-\gamma)/2}|||u_\ell|||_\ell + H_\ell^{1-\gamma/4}|||u_\ell|||_\ell^{1/2}\right)^{1/3} + H_\ell^{1-\gamma}|||u_\ell|||_\ell^2\\ &\quad + H_\ell^{-(1+\gamma)/2}|||u_\ell|||_\ell\left(M(\tau)^{6/5} + H_\ell^{-(1+\gamma)/3}M(\tau)^{4/3}|||u_\ell|||_\ell^{2/3} + H_\ell^{\min\{5,r'(1+1/p)\}}\right)^{1/2}\\ &\quad + H_\ell^{-(1+\gamma)/2}|||u_\ell|||_\ell M(\tau)^{1/2}\left(H_\ell^{(1-\gamma)/2}|||u_\ell|||_\ell + H_\ell^{1-\gamma/4}|||u_\ell|||_\ell^{1/2}\right)^{1/2}\end{aligned}$$

The generic constants in Theorem 5.4 depend on problem-specific data such as the shapes of Ω and Γ as well as the generic constant $\varkappa$ of Theorem 4.1.

Theorem 5.5. *Theorem 5.4 holds verbatim in Example 3.3 and in the modified two-well problem of Subsection 'Modified two-well benchmark', where $\beta = 0$.*

Remark 5.6. The assertion of Theorem 5.4 holds for any discrete $u_\ell \in u_{D,\ell} + V_\ell$ which may approximate the discrete unique exact solution of (2.5). This allows the inexact SOLVE via an iterative procedure.

Proof of Theorem 5.4. Choose w_ℓ as in the proof of Theorem 5.3. Then Theorem 4.1 with $q = 2$ and (4.2) imply

$$\|\delta_\ell\|^r_{L^{p'}(\Omega)} + \|e_\ell\|^2_{L^2(\Omega)} \lesssim M(\tau)\|e_\ell - w_\ell\|_{H^1(\Omega)} + |w_\ell|^{r'}_{W^{1,p}(\Omega)} + \|w_\ell\|^2_{L^2(\Omega)}$$
$$\lesssim M(\tau)\Big(|e_\ell|_{H^1(\Omega)} + \|e_\ell\|_{L^2(\Omega)} + H_\ell^{3/2}\Big) + H_\ell^{\min\{5,r'(1+1/p)\}}.$$

Theorem 5.3 provides an estimate of the semi-norm $|e_\ell|_{H^1(\Omega)}$. A Young inequality shows $H_\ell^{5/6}M(\tau) \lesssim H_\ell^5 + M(\tau)^6$. The absorbtion of $\|e_\ell\|_{L^2(\Omega)}$ then proves the first assertion. The second assertion is an immediate consequence of the first one, Theorem 5.3 and several algebraic transformations.

Numerical experiments

This section illustrates the theoretical estimates and their impact on the reliability-efficiency gap on 2D benchmarks in computational microstructures [18,26].

Numerical algorithms

The adaptive finite element method (AFEM) and algorithmic details on the implementation in MATLAB in the spirit of [27] concern the state-of-the-art AFEM loop

SOLVE $\rightarrow$ ESTIMATE $\rightarrow$ MARK $\rightarrow$ REFINE

and are explained below together with some notation.

Solve

The stabilised discrete problem (2.5) is solved in a nested iteration on a given triangulation $\mathcal{T}_\ell$ with MATLAB's standard-minimiser `fminunc` with default tolerances. Gradient and Hessian of the discrete energy are available and therefore provided to `fminunc`. We set $\gamma = 1$ in the stabilisation term (2.4) in all our experiments. This is motivated by ([12], Theorem 4.4) which suggest that $\gamma = 1$ yields an optimal convergence rate. The discrete solution of the previous AFEM loop iteration serves as a start vector for `fminunc`; for the first iteration, the initial vector is zero everywhere up to the Dirichlet boundary nodes. Since the Galerkin orthogonality is *not* required in Theorem 4.1, the termination of an iterative realisation for SOLVE is *not* a sensitive issue. In the computational PDEs, it is a fundamental issue to involve inexact solve. In this paper, however, the numerical examples are run with the standard settings of MATLAB.

Estimate

The refinement indicator results from the error estimator of Theorem 4.1. As in the work of Repin [28], the computation of the minimiser $\tau \in RT_0(\mathcal{T}_\ell)^m$ of

$$\|\sigma_\ell - \tau\|_{L^2(\Omega)} + \|\Pi_\ell\Lambda_\ell + \mathrm{div}\tau\|_{L^2(\Omega)} \tag{6.1}$$

runs Algorithm 1 based on the formula

$$(a+b)^2 = \min_{s>0}\left((1+s)a^2 + (1+1/s)b^2\right) \quad \text{for } a,b>0$$

The stopping criterion of Algorithm 1 monitors relative changes and avoids degenerate values of s. Undisplayed experiments have conviced us that a maxmium of three iterations and a stopping tolerance of $\varepsilon_{\mathrm{M}}^{0.8}$ (with the machine precision ε_{M}) yield satisfying results.

Algorithm 1 Approximate flux computation

Input: σ_ℓ, $\Pi_\ell \Lambda_\ell$
$s_1 = 1$;
for $k = 1, 2, 3$ **do**
  Compute minimiser τ_k of
  $M(s_k, \tau) = (1 + s_k)\|\sigma_\ell - \tau\|^2_{L^2(\Omega)} + (1 + 1/s_k)\|\Pi_\ell \Lambda_\ell + \operatorname{div}\tau\|^2_{L^2(\Omega)}$;
  if $D^2_\tau(s_k, \tau_k)$ *nearly singular (MATLAB "warning")* **then return** τ_k;
  $s_{k+1} = \|\Pi_\ell \Lambda_\ell + \operatorname{div}\tau_k\|_{L^2(\Omega)} / \|\sigma_\ell - \tau_k\|_{L^2(\Omega)}$;
  if $\max\left\{s_{k+1}, 1/s_{k+1}, \frac{|s_{k+1} - s_k|}{s_{k+1} + s_k}\right\} < \varepsilon_{\mathrm{M}}^{0.8}$ **then return** τ_k;
Output: approximate flux τ

The iteration is stopped whenever s, $1/s$ or the relative change of s drops below this tolerance. As an additional precaution, the iteration also stops if the linear system is deemed "nearly singular" by MATLAB. Our experiments convinced us that ignoring this warning causes a breakdown with NaNs. Note that if $q \neq 2$, we still minimise the L^2 sums in (6.1) to avoid the computational cost of a nonlinear solve. With the computed minimiser τ, Section 'A posteriori error estimates' yields the error estimator

$$\eta_{F,q'} := \|\sigma_\ell - \tau\|_{L^{q'}(\Omega)} + \|\Pi_\ell \Lambda_\ell + \operatorname{div}\tau\|_{L^{q'}(\Omega)} + \operatorname{osc}_{\ell,q'}(\Lambda_\ell).$$

This will be compared with the well-established *residual based a posteriori error estimator* [7]

$$\eta_{R,q'} := \left(\sum_{T \in \mathcal{T}_\ell} h_T^{q'} \|\Lambda_\ell\|^{q'}_{L^{q'}(T)}\right)^{1/q'} + \left(\sum_{F \in \mathcal{F}_\ell(\Omega)} h_F \|[\sigma_\ell]_F \cdot n_F\|^{q'}_{L^{q'}(F)}\right)^{1/q'},$$

which is reliable for the original discretisation without stabilisation. Undisplayed experiments computed the *averaging error estimator* [18], which is founded on the same theoretical background as $\eta_{R,q'}$ and therefore yielded essentially the same convergence rates.

The error estimators in Theorem 5.4 read

$$\begin{aligned}\eta_{L,2} := {}& \eta_{F,2}^{6/5} + H_\ell^{-(1+\gamma)/3} \eta_{F,2}^{4/3} |||u_\ell|||_\ell^{2/3} \\ &+ \eta_{F,2}\left(H_\ell^{(1-\gamma)/2} |||u_\ell|||_\ell + H_\ell^{1-\gamma/4} |||u_\ell|||_\ell^{1/2}\right) + H_\ell^{\min\{5, r'(1+1/p)\}}\end{aligned}$$

$$\begin{aligned}\eta_{H,2} := {}& \eta_{F,2}^{2/5} + H_\ell^{-(1+\gamma)/9} \eta_{F,2}^{4/9} |||u_\ell|||_\ell^{2/9} + H_\ell^{\min\{5/3, r'(1+1/p)/3\}} \\ &+ \eta_{F,2}^{1/3}\left(H_\ell^{(1-\gamma)/2} |||u_\ell|||_\ell + H_\ell^{1-\gamma/4} |||u_\ell|||_\ell^{1/2}\right)^{1/3} + H_\ell^{1-\gamma} |||u_\ell|||_\ell^2 \\ &+ H_\ell^{-(1+\gamma)/2} |||u_\ell|||_\ell \left(\eta_{F,2}^{6/5} + H_\ell^{-(1+\gamma)/3} \eta_{F,2}^{4/3} |||u_\ell|||_\ell^{2/3} + H_\ell^{\min\{5, r'(1+1/p)\}}\right)^{1/2} \\ &+ H_\ell^{-(1+\gamma)/2} |||u_\ell|||_\ell \eta_{F,2}^{1/2}\left(H_\ell^{(1-\gamma)/2} |||u_\ell|||_\ell + H_\ell^{1-\gamma/4} |||u_\ell|||_\ell^{1/2}\right)^{1/2}.\end{aligned}$$

MARK

For any given $T \in \mathcal{T}_\ell$ with its set of faces $\mathcal{F}(T)$, $\partial T = \bigcup \mathcal{F}(T)$, and given τ from (6.1), set

$$\eta_F^{q'}(T) := \|\sigma_\ell - \tau\|_{L^{q'}(T)}^{q'} + \|\Pi_\ell \Lambda_\ell + \operatorname{div}\tau\|_{L^{q'}(T)}^{q'} + h_T{}^{q'} \|(\mathrm{id} - \Pi_\ell)\Lambda_\ell\|_{L^{q'}(T)}^{q'}.$$

$$\eta_R^{q'}(T) := |T|^{q'/n} \|\Lambda_\ell\|_{L^{q'}(T)}^{q'} + |T|^{1/n} \sum_{F \in \mathcal{F}_\ell(\Omega) \cap \mathcal{F}(T)} \|[\sigma_\ell]_F \cdot \nu_F\|_{L^{q'}(F)}^{q'}.$$

Let $\eta^{q'}(T)$ be one of the refinement indicators $\eta_F^{q'}(T)$ and $\eta_R^{q'}(T)$. Some greedy algorithm computes $\mathcal{M}_\ell \subset \mathcal{T}_\ell$ of (almost) minimal cardinality such that

$$\sum_{T \in \mathcal{M}_\ell} \eta^{q'}(T) \geqslant 1/2 \sum_{T \in \mathcal{T}_\ell} \eta^{q'}(T).$$

Refine

This step computes the smallest refinement $\mathcal{T}_{\ell+1}$ of $\mathcal{T}_\ell$ with $\mathcal{M}_\ell \subset \mathcal{T}_\ell \setminus \mathcal{T}_{\ell+1}$ based on the red-green-blue refinement strategy as illustrated in Figure 2. This refinement involves some closure algorithm to avoid hanging nodes.

Two-well benchmark

The computational microstructure benchmark of ([18], Section 2) considers two wells with W from (3.4) in Example 3.3. The energy is given by (1.1) on the domain $\Omega = (0,1) \times (0,3/2) \subset \mathbb{R}^2$ with

$$g(x) := -3t^5/128 - t^3/3 \text{ and } u_D(x) := \begin{cases} g(x) & \text{for } t \leqslant 0, \\ t^3/24 + t & \text{for } t \geqslant 0 \end{cases}$$

for $t := (3(x_1 - 1) + 2x_2)/\sqrt{13}$; $p = q = 4$ and $f \equiv 0$. The unique minimiser u of $\min_{v \in \mathcal{A}} E(v)$ with $\mathcal{A} = u_D + W_0^{1,4}(\Omega)$ reads $u = u_D$ ([18], Theorem 2.1) and $\beta = 1$ allows for Theorems 5.1–5.4 to hold. An initial triangulation $\mathcal{T}_0$ is given by a criss triangulation of $(0,1) \times (0,3/2)$ with 12 congruent triangles and the two interior nodes $(1/2, 1/2)$ and $(1/2, 1)$. The adaptive algorithm of Subsection 'Numerical algorithms' computes a sequence of discrete solutions $(u_\ell)_\ell$ and stresses $(\sigma_\ell)_\ell$, as well as error estimators η_F and η_R with and without stabilisation for uniform and adaptive meshes and led to Figure 3 with overall observations of Section 'Conclusions'. The empirical convergence rates for uniform and R- as well as F-adapted mesh-refining are collected in Table 1. Note that the error estimator η_L performs better than η_F. This is evident from the table for uniform mesh refinements, but a closer look at Figure 3 reveals that even in the adaptive scenarios,

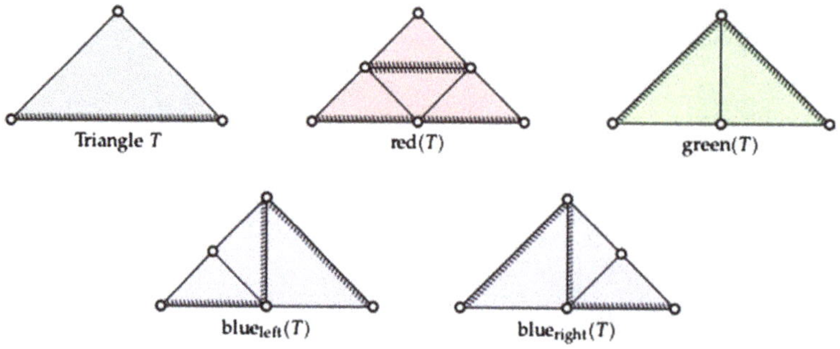

Figure 2 Possible refinements of a triangle T.

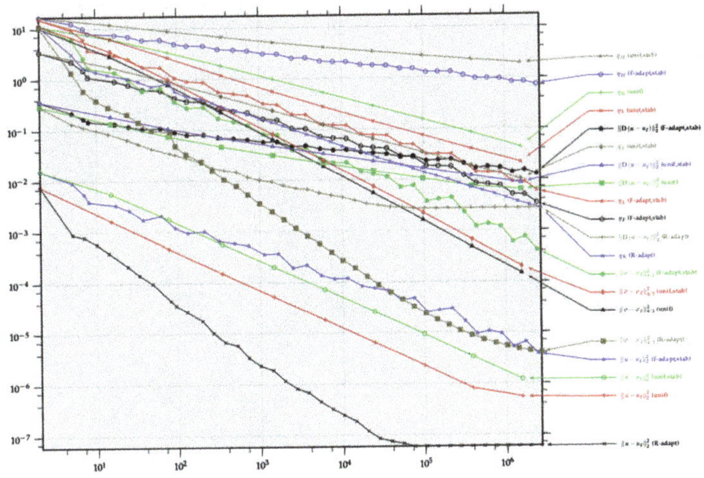

Figure 3 Convergence plot of the two-well benchmark. Errors and error estimators of the two-well benchmark of Subsection 'Two-well benchmark', plotted against the number of degrees of freedom.

η_L converges slightly faster than η_F. This is in accordance to the theory of Section 'Refined analysis for an interface model problem' where η_L is derived from η_F based on additional smoothness assumptions.

Modified two-well benchmark

This subsection concerns a modification of the previous problem with (3.4) and a linear right-hand side for $\beta = 0$ and $f(x) := -\operatorname{div}(DW(Du_D(x)))$ and unique solution $u = u_D$ as before. Note that Example 3.3 applies to this problem, and so the proof of Theorem 3.1 yields

$$\|\sigma - \sigma_\ell\|_{L^{p'}(\Omega)}^{r} + \|u - u_\ell\|_{L^2(\Omega)}^{2} + |||u_\ell|||_\ell^2 \to 0 \quad \text{as } \ell \to \infty$$

and Theorems 5.1–5.4 hold as well. The algorithms of Subsection 'Numerical algorithms' ran with and without stabilisation for uniform and adaptive meshes with the same initial triangulation as in Subsection 'Two-well benchmark' and led to Figure 4 with overall observations of Section 'Conclusions'. The empirical convergence rates for uniform and R- as well as F-adapted mesh-refining are collected in Table 1 for completeness although they are almost identical with those observed in Subsection 'Two-well benchmark'.

Three-well benchmark

The energy density W of ([26], Example 5.9.3, p. 72) is the convex hull of $\min\{|F|^2, |F - (1,0)|^2, |F - (0,1)|^2\}$ with explicit form in ([26], Example 5.6.4, p. 58). Let furthermore $\Omega = (0,1)^2 \subset \mathbb{R}^2$ and $u_D(x_1, x_2) := a(x_1 - 1/4) + a(x_2 - 1/4)$ with $a(t) := t^3/6 + t/8$ for $t \leqslant 0$ and $a(t) := t^5/40 + t^3/8$ for $t \geqslant 0$. Then the energy is given by (1.1) with $\beta = 0$ and $f := -\operatorname{div} DW(Du_D)$. The exact solution $u = u_D$ satisfies the interface condition of Section 'Refined analysis for an interface model problem' and allows Theorem 5.3 to hold. Theorems 5.1 and 5.4 do not apply because $\beta = 0$. We use the grid of Figure 5 as initial triangulation to resolve discontinuities in ∇f.

The algorithms of Subsection 'Numerical algorithms' ran with and without stabilisation for uniform and adaptive meshes and led to Figure 6 with overall observations of

Table 1 Observed convergence rates in Figures 3, 4, 6 and 7 for uniform and adaptive mesh refining

Example of subsection		$\|\sigma - \sigma_\ell\|^2_{L^{p'}(\Omega)}$		$\|u - u_\ell\|^2_{L^2(\Omega)}$		η_R	η_F	η_L	$\|D(u - u_\ell)\|^2_{L^2(\Omega)}$		η_H
		unstab.	**stab.**	**unstab.**	**stab.**	**unstab.**	**stab.**	**stab.**	**unstab.**	**stab.**	**stab.**
'Two-well benchmark'	unif	5/3	5/3	3/2	7/5	4/5	4/5	1	3/5	1/2	1/3
	R-adapt	2	7/5	(5/3)	6/5	1	1	1	(2/3)	2/5	2/5
	F-adapt	2	4/3	(5/3)	6/5	1	1	1	(2/3)	2/5	2/5
'Modified two-well benchmark'	unif	5/3	5/3	3/2	7/5	4/5	4/5	1	3/5	1/2	1/3
	R-adapt	2	7/5	(5/3)	6/5	1	1	1	(2/3)	2/5	2/5
	F-adapt	11/5	4/3	(7/4)	6/5	1	1	1	(2/3)	2/5	2/5
'Three-well benchmark'	unif	(1)	3/2	—	7/5	1	4/5	1	—	1/2	2/5
	R-adapt	2	(1/4)	—	(1/4)	1	—	(1/3)	(1)	(1/5)	—
	F-adapt	9/5	1	—	4/5	1	3/5	4/5	—	1/3	1/3
'An optimal design example'	unif					4/5	4/5	6/5			2/5
	R-adapt					1	4/5	6/5			2/5
	F-adapt					1	4/5	1			2/5

Convergence rates are given as powers of the representative mesh-size $1/\sqrt{\text{ndof}}$ which is proportional to H_ℓ on uniform grids. Unavailable values are left blank, non-continuous rates are put in parantheses, inconclusive convergence behaviour is marked by "—".

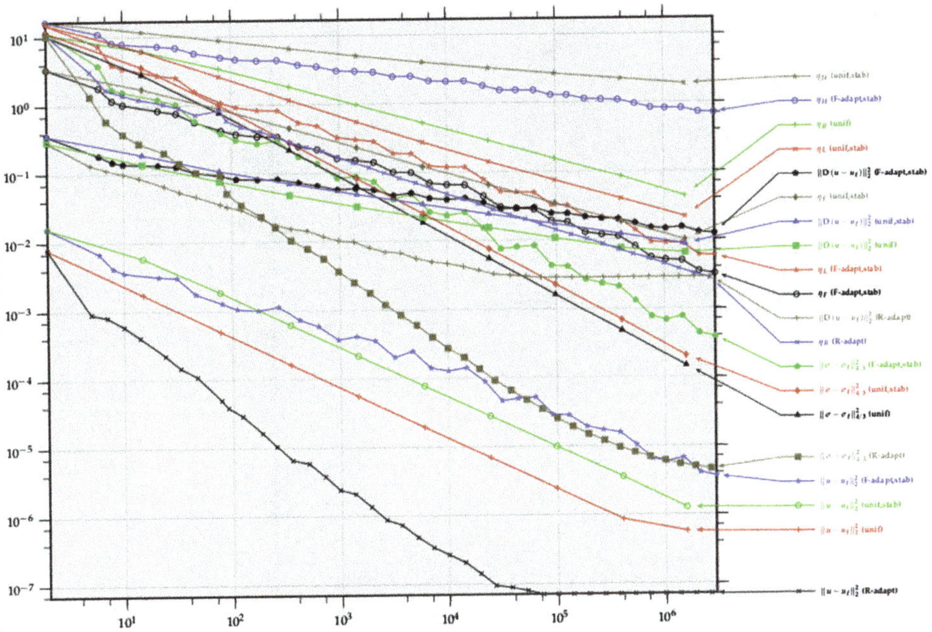

Figure 4 Convergence plot of the modified two-well benchmark. Errors and error estimators of the modified two-well benchmark of Subsection 'Modified two-well benchmark', plotted against the number of degrees of freedom.

Section 'Conclusions'. Beyond those general conclusions, this example demonstrates the difficulties with ill-conditioned Hessians. While the unstabilised method reaches 10^6 degrees of freedom without difficulty on uniform meshes, the adapted algorithms fail without stabilisation beyond 687 324 degrees of freedom (η_F-adaptive) and 33 169 degrees of freedom (η_R-adaptive). MATLAB's error message "Input to EIG must not contain NaN or Inf" indicates that a matrix operation returned non-finite numbers let `fminunc` break down. Undisplayed numerical experiments show condition numbers up to 10^{10} and

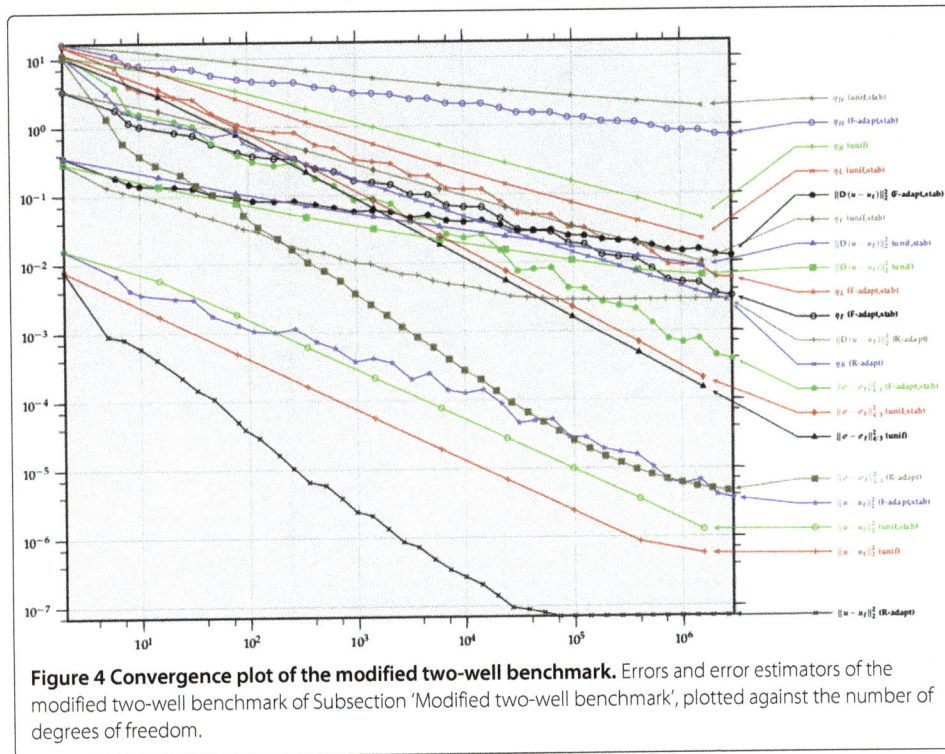

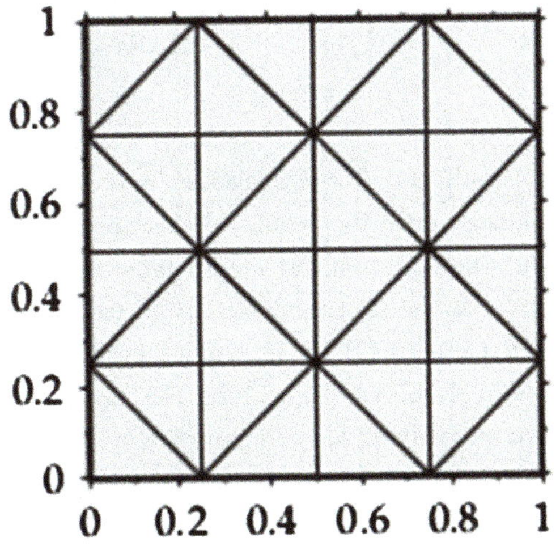

Figure 5 Initial grid for the three-well benchmark. Initial grid for the three-well benchmark of Subsection 'Three-well benchmark'.

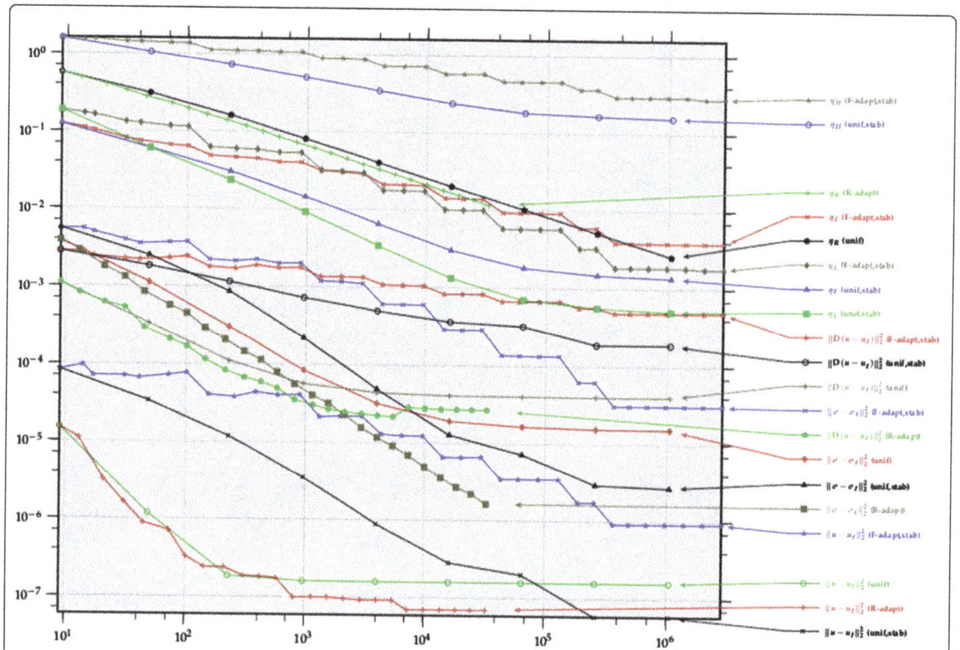

Figure 6 Convergence plot of the three-well benchmark. Errors and error estimators of the three-well benchmark of Subsection 'Three-well benchmark', plotted against the number of degrees of freedom.

beyond. The empirical convergence rates for uniform and R- as well as F-adapted mesh-refining are collected in Table 1. Moreover, Figure 1 in Section 'Background' reveals that stabilisation not only remedies ill-conditioned Hessians but thereby indeed allows for reduced errors in the discrete solution.

An optimal design example

The energy density of the topology optimisation problem of [3,8,29-33] reads

$$W(F) := \phi(|F|) \quad \text{for } F \in \mathbb{R}^2$$

$$\text{with } \phi(t) := \lambda/2 + \begin{cases} t^2 & \text{for } 0 \leqslant t \leqslant \sqrt{\lambda}, \\ 2\sqrt{\lambda}(t - \sqrt{\lambda}/2) & \text{for } \sqrt{\lambda} \leqslant t \leqslant 2\sqrt{\lambda}, \\ t^2/2 + \lambda & \text{for } t \geqslant 2\sqrt{\lambda}. \end{cases}$$

This leads to problem (2.3) with $\beta = 0, \lambda = 0.0084, u_D \equiv 0$ and $f \equiv 1$. Since regularity of the solutions is unclear, only the results of Sections 'Global convergence', 'A posteriori error estimates', 'Refined analysis for an interface model problem' and 'Numerical experiments' apply. As initial triangulation $\mathcal{T}_0$, we use the coarsest cross triangulation $\mathcal{T}_0 = \{\text{conv}\{(0,0),(1,0),(0,1)\}, \text{conv}\{(1,0),(0,1),(1,1)\}\}$ of $\Omega = (0,1)^2$.

The algorithms of Subsection 'Numerical algorithms' ran with and without stabilisation for uniform and adaptive meshes and led to Figure 7 with the overall observations of Section 'Conclusions'. The empirical convergence rates for uniform and R- as well as F-adapted mesh-refining are collected in Table 1. Undocumented experiments with a modified lower-order term f and known exact solution u led to the same convergence rates of the error estimators and confirm their accuracy.

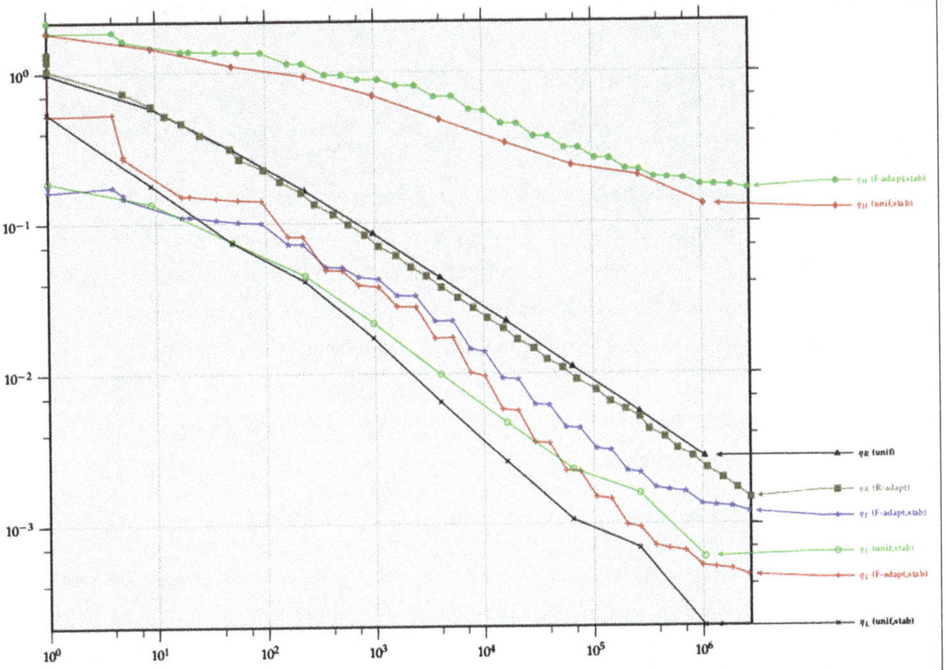

Figure 7 Convergence plot of the optimal design benchmark. Error estimators of the optimal design example of Subsection 'An optimal design example', plotted against the number of degrees of freedom.

Discussion of Empirical Convergence Rates

Global convergence without regularity assumptions

Theorem 3.1 asserts that $\|\sigma - \sigma_\ell\|_{L^{p'}(\Omega)}$, $\beta\|u - u_\ell\|_{L^2(\Omega)}$, and $|||u_\ell|||_\ell$ all tend to zero as $H_\ell \to 0$. The plain convergence result applies to all examples from Subsections 'Two-well benchmark', 'An optimal design example', 'Three-well benchmark', and 'An optimal design example' for the uniform mesh-refinements with $H_{\ell+1} = H_\ell/2$. The numerical experiments, however, show empirical convergence rates displayed in the first columns of Table 1. The adaptive algorithms do not reflect the condition $H_\ell \to 0$ explicitly and hence convergence is not guaranteed a priori. Undisplayed investigations show that indeed in the R-adapted version of the three-well example of Subsection 'Three-well benchmark', this condition $H_\ell \to 0$ does not appear to be true for more than 4 978 degrees of freedom. In all other experiments we observe convergence rates even for unstabilised discretisations.

Empirical convergence rates for interface model problems

Theorem 5.1 provides an a priori error estimate and an estimate of the stabilisation norm. It applies to the benchmark of Subsections 'Two-well benchmark', 'An optimal design example', 'Three-well benchmark', and 'An optimal design example' only, because of $\beta > 0$ and Example 3.3, and the smoothness conditions imposed upon u from Section 'Refined analysis for an interface model problem'. Recall the definitions of $\mathcal{T}_\ell(\Gamma)$, $\Omega_{\Gamma,\ell}$ and $\Omega_{\Gamma,\ell}^C$ from Section 'Refined analysis for an interface model problem' and assume $\|u\|_{L^2(\Omega\setminus\Gamma)} \approx 1 \approx \|u\|_{W^{2,p}(\Omega_{\Gamma,\ell}^C)}$, $|\mathcal{T}_\ell(\Gamma)| \approx H_\ell^{-1}$ and $|\Omega_{\Gamma,\ell}| \approx H_\ell$ in this discussion. This leads to a convergence rate of $H_\ell^{2/p}$ for the right-hand side of Theorem 5.1. The observed convergence rates of $\|\sigma - \sigma_\ell\|_{L^{p'}(\Omega)}$ and $\|u - u_\ell\|_{L^2(\Omega)}$ for the stabilised benchmark examples in Table 1 show convergence rates beyond those guaranteed in Theorem 5.1.

Theorem 5.3 implies, up to perturbations on the boundary,

$$\|D(u-u_\ell)\|_{L^2(\Omega)} \lesssim \|u-u_\ell\|_{L^2(\Omega)}^{1/3} + |||u_\ell|||_\ell + H_\ell^{-1/2}|||u_\ell|||_\ell^{1/2}\|u-u_\ell\|_{L^2(\Omega)}^{1/2}.$$

Since the exact solutions of Subsections 'Two-well benchmark', 'An optimal design example', 'Three-well benchmark', and 'An optimal design example' are all smooth up to a one-dimensional interface line, Theorem 5.3 applies to these examples. The experiments shows that the right-hand side of Theorem 5.3 is dominated by $H_\ell^{-1/2}|||u_\ell|||_\ell^{1/2}\|u-u_\ell\|_{L^2(\Omega)}^{1/2}$ in all examples and that the inequality is satisfied.

Reliability without regularity assumptions

Up to boundary terms, Theorem 4.1 states

$$\|\sigma-\sigma_\ell\|_{L^{p'}(\Omega)}^2 + \beta\|u-u_\ell\|_{L^2(\Omega)}^2 \lesssim \eta_F\|u-u_\ell\|_{W^{1,p}(\Omega)}.$$

The convergence rates confirm this assertion for the general and rough estimate $\|u-u_\ell\|_{W^{1,p}(\Omega)} \lesssim 1$ in the sense that the rates for η_F are worse than or equal to those of $\|\sigma-\sigma_\ell\|_{L^{p'}(\Omega)}^2$ and $\|u-u_\ell\|_{L^2(\Omega)}^2$. In the numerical examples, $\|u-u_\ell\|_{H^1(\Omega)}$ is computed and displayed in Table 1 and the convergence rates of the product $\|u-u_\ell\|_{H^1(\Omega)}\eta_F$ can be compared with those of $\|\sigma-\sigma_\ell\|_{L^{p'}(\Omega)}^2 + \|u-u_\ell\|_{L^2(\Omega)}^2$. This comparison confirms the above a posteriori error estimate. In the examples with $p=2$ (of Subsections 'Two-well benchmark', 'An optimal design example', 'Three-well benchmark', and 'An optimal design example'), there holds even equality of the convergence rates which demonstrates the efficiency of the estimate of Theorem 4.1.

Efficiency without regularity assumptions

Up to oscillations and the (possibly) higher-order term $\|(\mathrm{id}-I_{F,\ell})\sigma\|_{L^{q'}(\Omega)}$, Theorem 4.2 states

$$\eta_F \lesssim \|\sigma-\sigma_\ell\|_{L^{p'}(\Omega)} + \beta\|u-u_\ell\|_{L^{p'}(\Omega)}.$$

The displayed convergence rates of Table 1 confirm this estimate.

Reliability of the refined a posteriori error control

Theorem 5.4 applies to the example of Subsection 'Two-well benchmark' and states

$$\|\sigma-\sigma_\ell\|_{L^{p'}(\Omega)}^2 + \|u-u_\ell\|_{L^2(\Omega)}^2 \lesssim \eta_L \quad\text{and}\quad \|D(u-u_\ell)\|_{L^2(\Omega)}^2 \lesssim \eta_H.$$

Table 1 confirms this estimate and shows that the estimators η_L and η_H accurately predict the convergence rate of the errors, even with equality of the convergence rates in the case of adaptive mesh refinements in the examples of Subsections 'Two-well benchmark', 'An optimal design example', 'Three-well benchmark', and 'An optimal design example'.

All displayed convergence rates of η_L are better or at least equal to those of η_F. For instance, for uniform mesh-refining in Subsections 'Two-well benchmark', 'An optimal design example', 'Three-well benchmark', and 'An optimal design example', the error terms $\|\sigma-\sigma_\ell\|_{L^{p'}(\Omega)}^2 + \|u-u_\ell\|_{L^2(\Omega)}^2$ converge with the empirical convergence rate 7/5 while the upper bound η_F does so with a reduced convergence rate 4/5. The refined error estimator η_L is a guaranteed upper bound (via Theorem 5.4) and converges with an empirical convergence rate 1.

Performance of the minimisation algorithm 1

In all numerical experiments of this paper, Algorithm 1 reaches the maximal number 3 of iterations. While this suggests that the optimal s is *not* found after three iterations, undisplayed experiments with higher iteration counts and hence higher computational efforts result solely in marginal improvements.

Conclusions

Effects of stabilisation

The empirical convergence rates of the error estimators η_F, η_R and the errors $\|u - u_\ell\|_{L^2(\Omega)}$ and $\|\sigma - \sigma_\ell\|_{L^{p'}(\Omega)}$ for uniform mesh-refinement with and without stabilisation coincide. This indicates that the choice $\gamma = 1$ leads to some significant perturbation but maintains the correct convergence rate at the same time. This is different for adaptive mesh refinement with less optimal convergence rates. Our conclusion is that an improved adaptive algorithm has to be developed with balance of local mesh-refinement and global stabilisation parameters in future research. The tested algorithm from Subsection 'Numerical algorithms' does neither reflect the effects of stabilisation nor that of inexact solve.

Another important aspect of the stabilisation is the regularisation of the Hessian in the step SOLVE of Subsection 'Numerical algorithms'. In the three-well problem of Subsection 'Three-well benchmark', the unstabilised adaptive algorithms fail.

Adaptive versus uniform mesh-refinement

The overall empirical convergence rates of the errors and estimators of the unstabilised computation for adaptive mesh-refinements are better than those for uniform mesh-refinements. This is in contrast to the stabilised computation, where the true errors $\|\sigma - \sigma_\ell\|_{L^{p'}(\Omega)}$ and $\|u - u_\ell\|_{L^2(\Omega)}$ behave better for uniform compared with the two adaptive mesh-refinments (with the exception in Subsection 'An optimal design example' where there is equality). It is observed that adaptivity does not necessarily improve the converegnce rates of the error $\|\sigma - \sigma_\ell\|_{L^{p'}(\Omega)}$ and $\|u - u_\ell\|_{L^2(\Omega)}$ in a stabilised computation. Surprisingly, the convergence of the gradient errors $\|\mathrm{D}(u - u_\ell)\|_{L^2(\Omega)}$ are slightly improved in the instabilised calculation by adaptive mesh-refinements. The adaptive mesh-refinement is expected to reduce the a posteriori error estimators in the first place: cf. [1,34] for the estimator reduction property. Indeed, the convergence rates of the a posteriori error estimators η_R, η_F, η_L, η_H are improved (or optimal) for adaptive mesh-refinements (except for the three-well example of Subsection 'Three-well benchmark').

Strong convergence of the gradients

The convergence of the gradient error of the stabilised problem surpasses the expectations of [12] in Subsection 'An optimal design example' but fails to do so in Subsections 'Two-well benchmark', 'An optimal design example', 'Three-well benchmark', and 'An optimal design example'. The improved error estimator η_H shows the same convergence rate as the error of the gradients in Subsections 'Two-well benchmark',

'An optimal design example', 'Three-well benchmark', and 'An optimal design example'. This holds for uniform and for adapted mesh refinements and suggests that η_H is in fact reliable and efficient for $\beta > 0$.

Guaranteed error control

The assertion on η_F in Theorem 4.1 is reflected in the numerical examples in that the stress approximations converge faster than η_F in all cases. This suggests that the estimate $\|u - u_\ell\|_{W^{1,p}(\Omega)} \lesssim 1$ is by far too pessimistic. In fact, the benchmark examples with known exact solution fulfil $\|\sigma - \sigma_\ell\|^2_{L^2(\Omega)} \lesssim \eta_F \|u - u_\ell\|_{H^1(\Omega)}$. Similar affirmative conclusions follow for Theorem 4.2 and 5.4.

Reliability-efficiency gap

In comparison with the residual-based error estimator of [7,18], the new a posteriori error estimators η_L and η_H of Theorem 5.4 lead to refined error control. The improvement is marginal for uniform meshes without stabilisation but significant for adaptive stabilised computations. η_L and η_H match the convergence of the errors and so narrow the reliability-efficiency gap.

Competing interests

The authors declare that they have no competing interests.

Authors' contributions

All authors contributed equally to all parts of this article. All authors read and approved the final manuscript.

References

1. Carstensen C (2008) Convergence of an adaptive fem for a class of degenerate convex minimisation problems. IMA J Numer Anal 28(3): 423–439
2. Dacorogna B (2008) Direct methods in the calculus of variations, 2nd Ed. Applied Mathematical Sciences 78. Springer, Berlin. xii
3. Carstensen C, Müller S (2002) Local stress regularity in scalar non-convex variational problems. SIAM J Math Anal 34(2): 495–509
4. Chipot M (2000) Elements of Nonlinear Analysis. Birkhäuser Advanced Texts. Basel, Birkhäuser. vi
5. Müller S (1999) Variational models for microstructure and phase transisions. In: Hildebrandt S, et al. (eds) Calculus of variations and geometric evolution problems. Lectures given at the 2nd session of the, Centro Internazionale Matematico Estivo (CIME), Cetraro, Italy, June 15–22, 1996, Lect. Notes Math., 1713. Springer, Berlin, pp 85–210
6. Ball JM, James RD (1992) Proposed experimental tests for the theory of fine microstructures and the two-well problem. Phil Trans R Soc Lond A 338: 389–450
7. Carstensen C, Plecháĉ P (1997) Numerical solution of the scalar double-well problem allowing microstructure. Math Comp 66(219): 997–1026
8. Bartels S, Carstensen C (2007) A convergent adaptive finite element method for an optimal design problem. Numer Math 108: 359–385
9. Goodman J, Kohn RV, Reyna L (1986) Numerical study of a relaxed variational problem from optimal design. Comput Methods Appl Mech Eng 57: 107–127
10. Carstensen C, Klose R (2003) Guaranteed a posteriori finite element error control for the p-Laplace problem. SIAM J Sci Comput 25: 792–814
11. Bartels S, Carstensen C, Plecháĉ P, Prohl A (2004) Convergence for stabilisation of degenerate convex minimsation problems. IFB 6(2): 253–269
12. Boiger W, Carstensen C (2010) On the strong convergence of gradients in stabilised degenerate convex minimisation problems. SIAM J Numer Anal 47(6): 4569–4580
13. Ciarlet PG (2002) The finite element method for elliptic problems. Society for Industrial Mathematics, Philadelphia, PA, USA
14. El Alaoui L, Ern A, Vohralík M (2011) Guaranteed and robust a posteriori error estimates and balancing discretization and linearization errors for monotone nonlinear problems. Comp Meth Appl Mech Eng 200(37–40): 2782–2795
15. Ern A, Nicaise S, Vohralík M (2007) An accurate *h*(div) flux reconstruction for discontinuous galerkin approximations of elliptic problems. C R, Math, Acad Sci Paris 345(12): 709–712
16. Luce R, Wohlmuth B (2004) A local a posteriori error estimator based on equilibrated fluxes. SIAM J Numer Anal 42(4): 1394–1414
17. Ainsworth M (2005) A synthesis of a posteriori error estimation techniques for conforming, non-conforming and discontinuous galerkin finite element methods. American Mathematical Society (AMS), Providence

18. Carstensen C, Jochimsen K (2003) Adaptive finite element methods for microstructures? Numerical experiments for a Two-well benchmark. Computing 71: 175–204
19. Chipot M, Evans LC (1986) Linearisation at infinity and Lipschitz estimates for certain problems in the calculus of variations. Proc Roy Soc Edinburgh Sect A 102(3–4): 291–303
20. Bartels S, Carstensen C, Dolzmann G (2004) Inhomogeneous Dirichlet conditions in a priori and a posteriori finite element error anylysis. Numer Math 99(1): 1–24
21. Knees D (2008) Global stress regularity of convex and some nonconvex variational problems. Ann Mat Pura Appl (4) 187(1): 157–184
22. Brezzi F, Fortin M (1991) Mixed and hybrid finite element methods. Springer series in computational mathematics. Springer-Verlag, New York
23. Acosta G, Durán RG (2004) An optimal Poincaré inequality in L^1 for convex domains. Proc Amer Math Soc 132(1): 195–202
24. Bergh J, Löfstrom J (1976) Interpolation spaces. Springer-Verlag, Berlin
25. Brenner SC, Scott LR (2002) The mathematical theory of finite element methods, 2nd Ed. Texts in Applied Mathematics. 15. Springer, Berlin. p361, xv
26. Bartels S (2001) Numerical analysis of some non-convex variational problems. Ph.D. thesis. Christian-Albrechts Universität zu Kiel, Kiel, Germany. [http://eldiss.uni-kiel.de/macau/receive/dissertation_diss_00000519]
27. Alberty J, Carstensen C, Funken SA (1999) Remarks around 50 lines of Matlab: short finite element implementation. Numer. Algorithms 20(2–3): 117–137
28. Repin SI, Sauter S, Smolianski A (2003) A posteriori error estimation for the dirichlet problem with account of the error in the approximation of boundary conditions. Computing 70(3): 205–233
29. Carstensen C, Günther D, Rabus H (2012) Mixed finite element method for a degenerate convex variational problem from topology optimization. SIAM J Math Anal 50(2): 522–543
30. Murat F, Tartar L (1985) Calcul des variations et homogénéisation. In: Bergman D, et al. (eds) Homogenization methods: theory and applications in physics. Collect Dir Études Rech Élec France, vol. 57. Éditions Eyrolles, Paris, France, pp 319–369
31. Kohn RV, Strang G (1986) Optimal design and relaxation of variational problems I–III. Comm Pure Appl Math 39(1–3): 113–137139182353377
32. Kawohl B, Stará J, Wittum G (1991) Analysis and numerical studies of a problem of shape design. Arch Rational Mech Anal 114(4): 349–363
33. Glowinski R, Lions J-L, Trémolières R (1981) Numerical analysis of variational inequalities. Studies in Mathematics and its Applications, vol. 8. North-Holland Publishing Co., Amsterdam. p 776
34. Cascón JM, Kreuzer C, Nochetto RH, Siebert KG (2008) Quasi-optimal convergence rate for an adaptive finite element method. SIAM J Numer Anal 46(5): 2524–2550

2

Weakly periodic boundary conditions for the homogenization of flow in porous media

Carl Sandstöm*, Fredrik Larsson and Kenneth Runesson

*Correspondence:
carl.sandstrom@chalmers.se
Department of Applied Mechanics,
Chalmers University of Technology,
Hörsalsvägen 7, 412 96 Göteborg,
Sweden

Abstract

Background: Seepage in porous media is modeled as a Stokes flow in an open pore system contained in a rigid, impermeable and spatially periodic matrix. By homogenization, the problem is turned into a two-scale problem consisting of a Darcy type problem on the macroscale and a Stokes flow on the subscale.

Methods: The pertinent equations are derived by minimization of a potential and in order to satisfy the Variationally Consistent Macrohomogeneity Condition, Lagrange multipliers are used to impose periodicity on the subscale RVE. Special attention is given to the bounds produced by confining the solutions spaces of the subscale problem.

Results: In the numerical section, we choose to discretize the Lagrange multipliers as global polynomials along the boundary of the computational domain and investigate how the order of the polynomial influence the permeability of the RVE. Furthermore, we investigate how the size of the RVE affect its permeability for two types of domains.

Conclusions: The permeability of the RVE depends highly on the discretization of the Lagrange multipliers. However, the flow quickly converges towards strong periodicity as the multipliers are refined.

Keywords: Multiscale modeling; Computational homogenization; Stokes flow; Weak periodicity; Porous media

Background

We consider the classical problem of flow in porous media. On the macroscale, this phenomenon is often modeled as seepage governed by Darcy's law. Such seepage occur in a vast amount of natural as well as engineered materials, and applications include geomechanics, biomechanics and foam materials designed for energy absorption.

On the subscale, where the details of the pore system are resolved, Stokes' flow is an accurate description of the problem. In order to capture the effective properties of the subscale, the Stokes flow is solved on a Representative Volume Element (RVE) which should be large enough to represent the true subscale yet small enough to be as computationally efficient as possible [1]. In order to allow for the use of RVEs, the microstructure of the pertinent material should be ergodic and statistically spatially homogeneous. For further reading on the size of the RVE for homogenization of Stokes flow, we refer to [2].

Following the work by Sandström et al. [3,4], we consider a two-scale problem where the subscale is represented as a Stokes flow on a strongly heterogeneous domain, consisting of a fluid within an open pore system. By adopting the concept of Variationally Consistent

Homogenization [5], a macroscale representation in the form of a Darcy flow is produced. In the special case of linear flow, the homogenized tangent represents the permeability tensor whereas in the non-linear case it serves as the consistent tangent in the macroscale Newton iterations.

In mechanics, homogenization is used to capture microstructural effects in a material subject to some load by deriving smooth effective properties on the structural scale. The model defined by the RVE can be used as a constitutive relation in itself, in a concurrent manner, or as a tool for calibrating an existing macroscale model. Several types of homogenization exists, such as asymptotic expansion which can be used to determine the macroscale properties in an analytical manner, see e.g [6-8]. In recent years, computational homogenization, where a local boundary value problem is solved on an RVE, has been subject to intense research, see e.g [9-12]. In the context of computational homogenization of porous materials, important areas of application are Resin Transfer Molding (RTM) [13,14], oil geology [15], sintering [16] and transportation of matter [17].

Assuming separation of scales, we may adopt homogenization to derive the problem on two scales: the macroscale, representing the global structure, and the microscale, where the microstructure of the material is resolved. Classical homogenization concerns average theorems for the macroscale (effective) fluxes and primal variables (including possible gradients). Enforcing energy or work equivalence for the formulations on the different scales defines the so-called macrohomogeneity condition, cf. [9].

For single field problems, such as e.g. elasticity or heat conduction, there are three classical boundary conditions that satisfy macrohomogeneity: Dirichlet, Neumann and Periodic. However, in the case of a Stokes flow, it is not obvious how to choose the Dirichlet and Neumann conditions since there exists two primary fields of unknowns, namely velocity and pressure. Suggestions on how to choose Dirichlet and Neumann boundary conditions are given in [18]. It should also be mentioned that, in most cases, the periodic boundary condition performs better than Dirichlet and Neumann boundary conditions in terms of convergence with the size of the RVE [19]. In Sandström and Larsson [3], it was shown that periodic boundary conditions on the subscale (fluctuating) pressure and the (total) velocity defines a prolongation condition that satisfies the generalized macrohomogeneity presented in [5] thus ensuring no energy production on the subscale.

In this paper, we consider homogenization of the saddle-point problem pertinent to the fully resolved Stokes' problem within an open pore system. In contrast to the derivation by Sandström and Larsson [3], we thus carry out computational homogenization on pertinent potentials, rather than balance equations. We shall consider the particular choice of periodic boundary conditions, whereby the end result will be identical to that in [3]. However, we present this alternative derivation with the motivation that the arising subscale potentials will be utilized for computing upper and lower bounds on the effective properties, cf. below.

The classical approach in Finite Element Analysis of RVEs is to enforce periodic boundary conditions by treating two degrees of freedom on opposing sides of a domain as one single degree of freedom. Although computationally effective, this approach calls for a mesh which has identical discretization on either opposite side, which is a severe difficulty in 3D in the case of unstructured meshes.

The purpose of this work is to void the dependence on mesh periodicity for the periodic boundary conditions and instead impose periodicity in a weak sense, cf. [20,21] where the momentum equation has been solved on the subscale for an elasticity problem. We note that in the former paper, the Lagrange multipliers are discretized with piecewise polynomials and in the latter, the displacement is interpolated by polynomials. With a minimization problem as point of departure, constraints pertinent to the boundary condition are added and as a result, Stokes flow with additional terms containing Lagrange multipliers is produced. The Lagrange multipliers can be identified as the required influx and traction necessary to maintain periodicity in pressure and velocity. From the minimization problem, bounds for the effective permeability are produced.

The remainder of this paper is organized as follows: In the Section "Methods", the two-scale formulation of the saddle-point problem pertinent to Stokes flow is derived in detail. Two numerical examples are presented in Section "Results and discussion". The first example concerns an RVE in the form of a unit cell with a non-periodic mesh. In the second example, we investigate how the size of the RVE affects the macroscale permeability. Finally, the conclusions and an outlook to future work are presented in Section "Conclusions".

Methods

The single scale problem

Consider a fully resolved porous domain $\Omega = \Omega^F \cup \Omega^S$, such as the one depicted in Figure 1(a). The domain consists of a topologically periodic substructure where Ω^F is the part of Ω occupied by the fluid phase and Ω^S the part occupied by the solid phase[a]. The

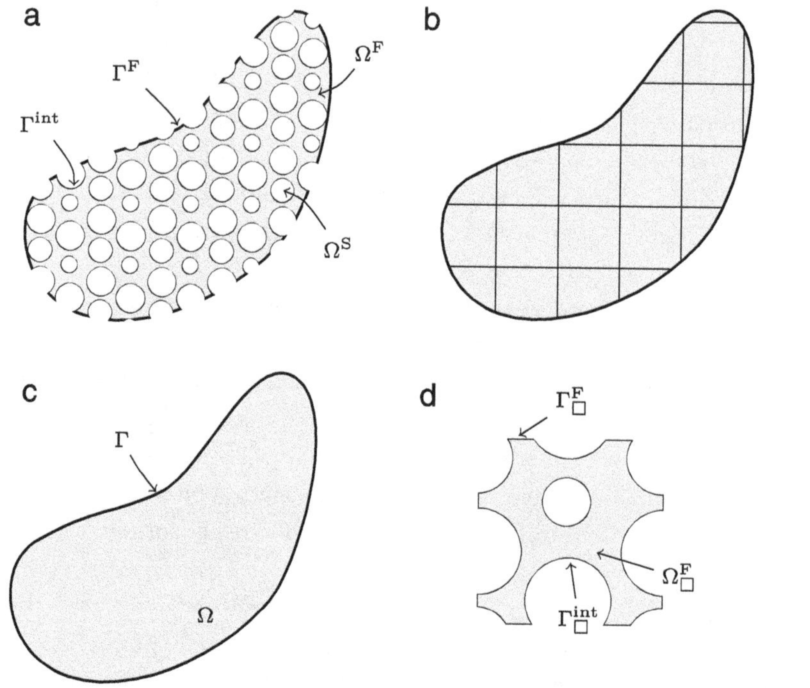

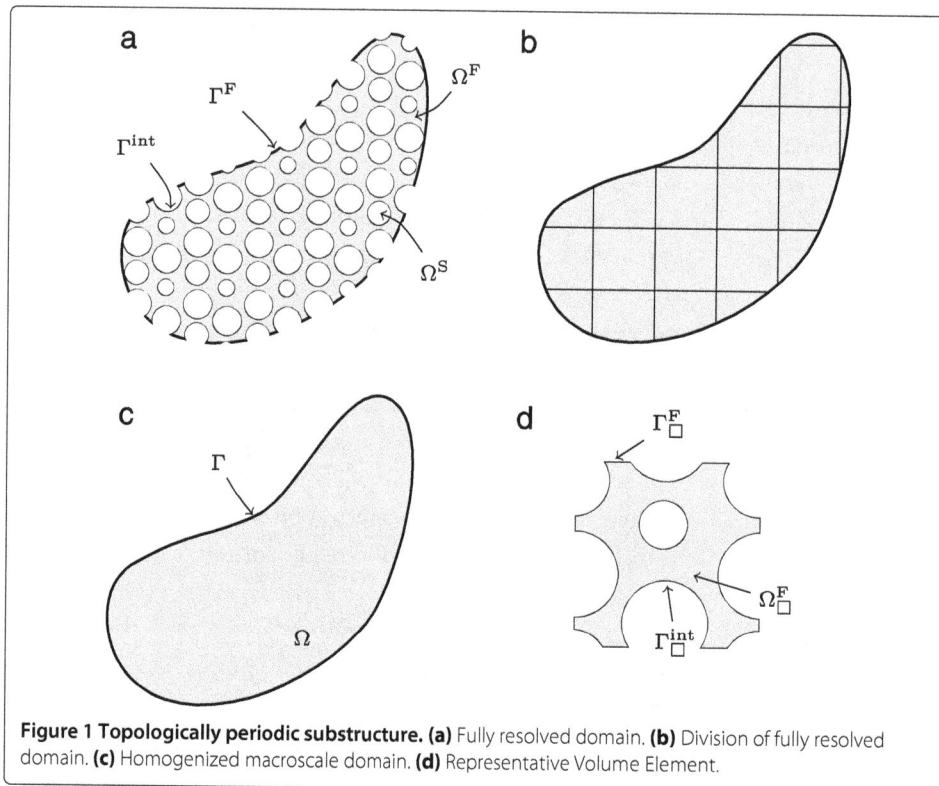

Figure 1 Topologically periodic substructure. (a) Fully resolved domain. **(b)** Division of fully resolved domain. **(c)** Homogenized macroscale domain. **(d)** Representative Volume Element.

interface between the solid and fluid phases is denoted Γ^{int} and the part of Γ where fluid can enter and exit the domain is denoted $\Gamma^{\text{F}} = \partial\Omega^{\text{F}} \setminus \Gamma^{\text{int}} = \Gamma \cap \partial\Omega^{\text{F}}$. The fluid part of the boundary Γ^{F} is further divided into $\Gamma^{\text{F}}_{\text{P}}$ where the pressure p is prescribed and $\Gamma^{\text{F}}_{\text{V}}$ where the velocity $\boldsymbol{v}$ is prescribed. We hereby restrict ourselves to flows with low Reynolds numbers and purely viscous, incompressible fluids, whereby the fluid velocity field $\boldsymbol{v}$ can be found by minimizing the energy potential pertaining to a local viscous potential $\Phi(\boldsymbol{v} \otimes \boldsymbol{\nabla})$, defined such that $\frac{\partial \Phi(\boldsymbol{v}\otimes\boldsymbol{\nabla})}{\partial \boldsymbol{v}\otimes\boldsymbol{\nabla}} = \boldsymbol{\sigma}^{\text{v}}$ where $\boldsymbol{\sigma}^{\text{v}}$ is the deviatoric part of the Cauchy stress [b]. Thus, we seek $\boldsymbol{v} \in \mathcal{V}$ that satisfies the constrained problem

$$\text{minimize} \int_{\Omega^{\text{F}}} \Phi(\boldsymbol{v} \otimes \boldsymbol{\nabla})\, \text{d}V - \int_{\Gamma^{\text{F}}_{\text{P}}} \hat{\boldsymbol{t}} \cdot \boldsymbol{v}\, \text{d}S \tag{1a}$$

$$\text{subject to} : \boldsymbol{\nabla} \cdot \boldsymbol{v} = 0 \text{ on } \Omega^{\text{F}} \tag{1b}$$

where $\hat{\boldsymbol{t}} = -\hat{p}\boldsymbol{n}$ is the prescribed pressure on the boundary $\Gamma^{\text{F}}_{\text{P}} \subset \Gamma^{\text{F}}$ and $\mathcal{V}$ is defined below. This can equivalently be written as the inf-sup problem

$$\inf_{\boldsymbol{v}\in\mathcal{V}} \sup_{p\in\mathcal{P}} \left\{ \int_{\Omega^{\text{F}}} \Phi\left(\boldsymbol{v} \otimes \boldsymbol{\nabla}\right) \text{d}V - \int_{\Omega^{\text{F}}} p(\boldsymbol{\nabla} \cdot \boldsymbol{v})\text{d}V - \int_{\Gamma^{\text{F}}_{\text{P}}} \hat{\boldsymbol{t}} \cdot \boldsymbol{v}\text{d}S \right\} \tag{2}$$

where

$$\mathcal{V} = \left\{ \boldsymbol{v} \in \left[H^1\left(\Omega^{\text{F}}\right)\right]^3 : \boldsymbol{v} = \boldsymbol{0} \text{ on } \Gamma^{\text{int}},\ \boldsymbol{v} = \hat{v}_n \boldsymbol{n} \text{ on } \Gamma^{\text{F}}_{\text{V}} \right\} \tag{3a}$$

$$\mathcal{P} = \left\{ p \in L_2\left(\Omega^{\text{F}}\right) \right\} \tag{3b}$$

and p is a Lagrange multiplier resulting from the continuity condition. Note that due to the fact that $\boldsymbol{\sigma}^{\text{v}}$ is the deviatoric part of the Cauchy stress $\boldsymbol{\sigma}$, p is interpreted as the pressure.

We proceed by splitting the domain into a finite number of n domains $\Omega_{\square,i}$ such that $\Omega = \cup_{i=1}^{n} \Omega_{\square,i}$ and such that each subdomain retains geometric periodicity (cf. Figure 1(a) and the periodic cutout in Figure 1(d)). By the choice of function spaces $\mathcal{V}$ and $\mathcal{P}$, all functions $(\boldsymbol{v}, p) \in \mathcal{V} \times \mathcal{P}$ is continuous on the whole Ω^{F}. Rewriting Equation 2 as the sum of the energy contribution from all subdomains gives

$$\inf_{\boldsymbol{v}\in\mathcal{V}} \sup_{p\in\mathcal{P}} \left\{ \sum_{i=1}^{n} \left(\int_{\Omega^{\text{F}}_{\square,i}} \Phi(\boldsymbol{v} \otimes \boldsymbol{\nabla})\text{d}V - \int_{\Omega^{\text{F}}_{\square,i}} p(\boldsymbol{\nabla} \cdot \boldsymbol{v})\text{d}V \right) - \int_{\Gamma^{\text{F}}_{\text{P}}} \hat{\boldsymbol{t}} \cdot \boldsymbol{v}\text{d}S \right\} \tag{4}$$

In order to separate the macro and subscales features, we split the pressure term p into a smooth part $p^{\text{M}} \in \mathcal{P}^{\text{M}}$ and a fluctuating part $p^{\text{S}} \in \mathcal{P}^{\text{S}}$ such that $p = p^{\text{M}} + p^{\text{S}}$ and $\mathcal{P} = \mathcal{P}^{\text{M}} \oplus \mathcal{P}^{\text{S}}$, $\mathcal{P}^{\text{S}}$ being the hierachial complement to $\mathcal{P}^{\text{M}}$. Integration by parts on p^{M} in the continuity constraint in Equation 4 yields

$$\sum_{i=1}^{n} \int_{\Omega^{\text{F}}_{\square,i}} p^{\text{M}}(\boldsymbol{\nabla} \cdot \boldsymbol{v})\text{d}V = \sum_{i=1}^{n} \left(-\int_{\Omega^{\text{F}}_{\square,i}} \boldsymbol{\nabla} p^{\text{M}} \cdot \boldsymbol{v}\text{d}V + \int_{\Gamma^{\text{F}}_{\square,i}} \boldsymbol{n} \cdot \boldsymbol{v} p^{\text{M}}\text{d}S \right) \tag{5}$$

where $\boldsymbol{n}$ is the outward pointing normal. The boundary integral on the right hand side in Equation 5 vanish on all internal boundaries as $\boldsymbol{v}$ and p^{M} are continuous. Thus, after introducing the split in p, Equation 4 can be restated as

$$\inf_{\boldsymbol{v}\in\mathcal{V}} \sup_{\substack{p^{\mathrm{M}}\in\mathcal{P}^{\mathrm{M}} \\ p^{\mathrm{S}}\in\mathcal{P}^{\mathrm{S}}}} \Bigg\{ \sum_{i=1}^{n}\left(\int_{\Omega^{\mathrm{F}}_{\square,i}} \Phi(\boldsymbol{v}\otimes\nabla)\mathrm{d}V - \int_{\Omega^{\mathrm{F}}_{\square,i}} p^{\mathrm{S}}(\nabla\cdot\boldsymbol{v})\mathrm{d}V + \int_{\Omega^{\mathrm{F}}_{\square,i}} \nabla p^{\mathrm{M}}\cdot\boldsymbol{v}\mathrm{d}V\right) - \int_{\Gamma^{\mathrm{F}}_{\mathrm{P}}} \hat{\boldsymbol{t}}\cdot\boldsymbol{v}\mathrm{d}S - \int_{\Gamma^{\mathrm{F}}} \boldsymbol{n}\cdot\boldsymbol{v}p^{\mathrm{M}}\mathrm{d}S \Bigg\} \tag{6}$$

Furthermore, the last two terms can be rewritten as

$$-\int_{\Gamma^{\mathrm{F}}_{\mathrm{P}}} \hat{\boldsymbol{t}}\cdot\boldsymbol{v}\mathrm{d}S - \int_{\Gamma^{\mathrm{F}}} \boldsymbol{n}\cdot\boldsymbol{v}p^{\mathrm{M}}\mathrm{d}S = -\int_{\Gamma^{\mathrm{F}}_{\mathrm{P}}} \left[\hat{\boldsymbol{t}}\cdot\boldsymbol{v} + \boldsymbol{n}\cdot\boldsymbol{v}p^{\mathrm{M}}\right]\mathrm{d}S - \int_{\Gamma^{\mathrm{F}}_{\mathrm{V}}} \boldsymbol{n}\cdot\boldsymbol{v}p^{\mathrm{M}}\mathrm{d}S \tag{7}$$

where it is noted that the last term contains the prescribed velocity, $\hat{v}_n = \boldsymbol{v}\cdot\boldsymbol{n}$. Under the assumption that $\boldsymbol{t} = -\hat{p}\boldsymbol{n} = -p^{\mathrm{M}}\boldsymbol{n}$ the integral over $\Gamma^{\mathrm{F}}_{\mathrm{P}}$ vanishes and Equation 6 is equivalent to

$$\inf_{\boldsymbol{v}\in\mathcal{V}} \sup_{\substack{p^{\mathrm{M}}\in\mathcal{P}^{\mathrm{M}} \\ p^{\mathrm{S}}\in\mathcal{P}^{\mathrm{S}}}} \Bigg\{ \sum_{i=1}^{n}\left(\int_{\Omega^{\mathrm{F}}_{\square,i}} \Phi(\boldsymbol{v}\otimes\nabla)\mathrm{d}V - \int_{\Omega^{\mathrm{F}}_{\square,i}} p^{\mathrm{S}}(\nabla\cdot\boldsymbol{v})\mathrm{d}V + \int_{\Omega^{\mathrm{F}}_{\square,i}} \nabla p^{\mathrm{M}}\cdot\boldsymbol{v}\mathrm{d}V\right) - \int_{\Gamma^{\mathrm{F}}_{\mathrm{V}}} \hat{v}_n p^{\mathrm{M}}\mathrm{d}S \Bigg\} \tag{8}$$

Computational homogenization

Up to this point, nothing has changed since the original problem except the formulation. To proceed, we now assume separation of scales, i.e. that the subscale feature has a length scale much smaller than that of the macroscale. Furthermore, we also make the assumption that $\boldsymbol{v}$ and p^{S} are periodic over, and continuous inside, each $\Omega^{\mathrm{F}}_{\square}$, thus replacing the condition on continuity over the boundaries $\Gamma^{\mathrm{F}}_{\square}$. As an intermediate step, we note that by removing continuity over $\Gamma^{\mathrm{F}}_{\square}$, reaction forces arise, which eventually will contribute to the subsequent macrohomogeneity condition. In [3] it is shown that periodic boundary conditions satisfy the aforementioned condition. In order to impose periodicity (either in weak or strong form), we start out by following along the lines of [3] and split the subscale boundary $\Gamma_{\square}$ into two parts; $\Gamma_{\square} = \Gamma^{+}_{\square} \cup \Gamma^{-}_{\square}$ where the $+/-$ sign is the sign of the normal to that part of the boundary[c]. Furthermore, we introduce the jump operator

$$[\![f]\!] = f(\boldsymbol{x}) - f(\boldsymbol{x}^{-}(\boldsymbol{x})) \tag{9}$$

where $\boldsymbol{x}$ is a point on $\Gamma^{+}_{\square}$ and $\boldsymbol{x}^{-}(\boldsymbol{x})$ is the corresponding point on the opposite side of the RVE. The conditions for periodicity are given as

$$[\![p^{\mathrm{S}}]\!] = 0 \tag{10a}$$

$$[\![\boldsymbol{v}]\!] = \boldsymbol{0} \tag{10b}$$

$$\boldsymbol{t}^{\mathrm{S}+} + \boldsymbol{t}^{\mathrm{S}-} = \boldsymbol{0} \tag{10c}$$

where $\boldsymbol{t}^{S+}$ and $\boldsymbol{t}^{S-}$ are the subscale tractions along the edges $\Gamma_{\square}^{+}$ and $\Gamma_{\square}^{-}$ respectively. By imposing the periodicity constraints in a weak sense, i.e. introducing the Lagrange multiplier $\boldsymbol{\beta}$ for the constraint $[\![\boldsymbol{v}]\!] = 0$ and γ for the constraint $[\![p^S]\!] = 0$ and allow the constraints to be fulfilled in average, rather than confining the respective solution spaces, we get

$$\inf_{\boldsymbol{v}\in\mathcal{V}_\square} \sup_{\substack{p^M\in\mathcal{P}^M \\ p^S\in\mathcal{P}^S_\square \\ \boldsymbol{\beta}\in\mathcal{B}_\square}} \inf_{\gamma\in\mathcal{G}_\square} \left\{ \sum_{i=1}^{n} \left(\int_{\Omega^F_{\square,i}} \Phi(\boldsymbol{v}\otimes\boldsymbol{\nabla})\mathrm{d}V - \int_{\Omega^F_{\square,i}} p^S(\boldsymbol{\nabla}\cdot\boldsymbol{v})\mathrm{d}V + \int_{\Omega^F_{\square,i}} \boldsymbol{\nabla}p^M\cdot\boldsymbol{v}\mathrm{d}V - \int_{\Gamma^{F+}_{\square,i}} [\![\boldsymbol{v}]\!]\cdot\boldsymbol{\beta} + [\![p^S]\!]\gamma\mathrm{d}S \right) - \int_{\Gamma^F_V} \hat{v}_n p^M \mathrm{d}S \right\} \quad (11)$$

where

$$\mathcal{V}_\square = \left\{\boldsymbol{v}\in\left[H^1\left(\Omega^F\right)\right]^3 : \boldsymbol{v} = \boldsymbol{0} \text{ on } \Gamma^{\text{int}},\ \boldsymbol{v} = \hat{v}_n\boldsymbol{n} \text{ on } \Gamma^F_V\right\} \quad (12a)$$

$$\mathcal{P}^S_\square = \left\{p\in H^1\left(\Omega^F_\square\right)\right\} \quad (12b)$$

$$\mathcal{P}^M = \left\{p\in H^1\left(\Omega^F\right) : p = \hat{p} \text{ on } \Gamma^F_P\right\} \quad (12c)$$

$$\mathcal{B}_\square = \left\{\boldsymbol{\beta}\in\left[L_2\left(\Gamma^F_\square\right)\right]^3\right\} \quad (12d)$$

$$\mathcal{G}_\square = \left\{\gamma\in L_2\left(\Gamma^F_\square\right)\right\} \quad (12e)$$

The Lagrange multiplier $\boldsymbol{\beta}$ can be interpreted as the traction needed to maintain periodicity on $\boldsymbol{v}$ and γ as the flux needed to maintain periodicity on p^S. The infimum on γ is further discussed in Remark 1. In order to allow for strong (essential) boundary conditions on Γ^F_P in the subsequent macroscale problem, the function space $\mathcal{P}^M$ is confined, replacing the former integral formulation of the condition.

Remark 1. *In order to motivate the infimum on γ, consider the supremum of the term containing p^S (which already is a Lagrange multiplier) in Equation 8.*

$$\sup_{p^S\in\mathcal{P}^S_\square} \sum_{i=1}^{n} -\int_{\Omega^F_{\square,i}} p^S\left(\boldsymbol{\nabla}\cdot\boldsymbol{v}\right)\mathrm{d}V \quad (13)$$

which, when adding the constraint $[\![p^S]\!] = 0$ becomes, locally,

$$\underset{p^S\in\mathcal{P}^S_\square}{\text{maximize}} \ -\int_{\Omega^F_{\square,i}} p^S(\boldsymbol{\nabla}\cdot\boldsymbol{v}) \quad (14a)$$

$$\text{subject to} : [\![p^S]\!] = 0 \text{ on } \Gamma^F_{\square,i} \quad (14b)$$

or in weak form

$$\sup_{p^S\in\mathcal{P}^S_\square} \inf_{\gamma\in\mathcal{G}_\square} \sum_{i=1}^{n} \left(-\int_{\Omega^F_{\square,i}} p^S(\boldsymbol{\nabla}\cdot\boldsymbol{v})\mathrm{d}V - \int_{\Gamma^F_{\square,i}} [\![p^S]\!]\gamma\mathrm{d}S \right) \quad (15)$$

We now introduce the total energy potential Π which is split into n RVE potentials Π_i^{int} which, in turn, can be expressed using the RVE mean potential $\pi_{\square,i} = \frac{\Pi_i^{\mathrm{int}}}{|\Omega_{\square,i}|}$ as

$$\begin{aligned}\Pi\left(p^{\mathrm{M}}, \boldsymbol{v}, p^{\mathrm{S}}, \boldsymbol{\beta}, \gamma\right) &= \sum_{i=1}^{n} \Pi_i^{\mathrm{int}}\left(\nabla p^{\mathrm{M}}, \boldsymbol{v}, p^{\mathrm{S}}, \boldsymbol{\beta}, \gamma\right) - \Pi^{\mathrm{ext}}\left(p^{\mathrm{M}}\right) \\ &= \sum_{i=1}^{n} \pi_{\square,i}\left(\nabla p^{\mathrm{M}}, \boldsymbol{v}, p^{\mathrm{S}}, \boldsymbol{\beta}, \gamma\right) \mid \Omega_{\square,i} \mid - \Pi^{\mathrm{ext}}\left(p^{\mathrm{M}}\right)\end{aligned} \tag{16}$$

Here, the RVE potential is given as

$$\begin{aligned}&\Pi_i^{\mathrm{int}}\left(\nabla p^{\mathrm{M}}, \boldsymbol{v}, p^{\mathrm{S}}, \boldsymbol{\beta}, \gamma\right) \\ &= \int_{\Omega^{\mathrm{F}}_{\square,i}} \Phi(\boldsymbol{v} \otimes \nabla)\mathrm{d}V - \int_{\Omega^{\mathrm{F}}_{\square,i}} p^{\mathrm{S}}(\nabla \cdot \boldsymbol{v}) - \nabla p^{\mathrm{M}} \cdot \boldsymbol{v}\mathrm{d}V - \int_{\Gamma^{\mathrm{F}+}_{\square,i}} [\![\boldsymbol{v}]\!] \cdot \boldsymbol{\beta} + [\![p^{\mathrm{S}}]\!] \cdot \gamma \mathrm{d}S\end{aligned} \tag{17}$$

and the external load as

$$\Pi^{\mathrm{ext}}\left(p^{\mathrm{M}}\right) = \int_{\Gamma^{\mathrm{F}}_{\mathrm{V}}} \hat{v}_n p^{\mathrm{M}} \mathrm{d}S \tag{18}$$

Furthermore, we note that by introducing separation of scales, i.e. for each coordinate $\bar{\boldsymbol{x}} \in \Omega$ there exist one RVE, thus, the RVE mean potential functions can be written as

$$\pi_{\square}\left(\nabla p^{\mathrm{M}}, \boldsymbol{v}, p^{\mathrm{S}}, \boldsymbol{\beta}, \gamma, \bar{\boldsymbol{x}}\right) = \pi_{\square,i}\left(\nabla p^{\mathrm{M}}, \boldsymbol{v}, p^{\mathrm{S}}, \boldsymbol{\beta}, \gamma\right) \tag{19}$$

where i is the number of the RVE occupyed by coordinate $\bar{\boldsymbol{x}}$. Here, we define the RVE such that $\bar{\boldsymbol{x}}$ is the centroid of $\Omega_{\square}$. By the assumption that the RVE is small compared to the macroscale, we identify $\mid \Omega_{\square,i} \mid$ as a volume element on the macroscale and rewrite the sum in Equation 16 as an integral. It should be noted that the term $\mid \Omega_{\square,i} \mid$ in the definition of the RVE mean potential is left unchanged during the transition from sum to integral as we are interested in the mean potential in the vicinity of $\bar{\boldsymbol{x}}$. We give the RVE mean potential $\pi_{\square}$ on explicit integral form as

$$\begin{aligned}\pi_{\square}\left(\nabla p^{\mathrm{M}}, \boldsymbol{v}, p^{\mathrm{S}}, \boldsymbol{\beta}, \gamma\right) = \frac{1}{\mid \Omega_{\square} \mid}\Bigg(&\int_{\Omega^{\mathrm{F}}_{\square}} \Phi\left(\boldsymbol{v} \otimes \nabla\right)\mathrm{d}V - \int_{\Omega^{\mathrm{F}}_{\square}} p^{\mathrm{S}}\left(\nabla \cdot \boldsymbol{v}\right)\mathrm{d}V \\ &+ \int_{\Omega^{\mathrm{F}}_{\square}} \nabla p^{\mathrm{M}} \cdot \boldsymbol{v}\mathrm{d}V - \int_{\Gamma^{\mathrm{F}+}_{\square}} [\![\boldsymbol{v}]\!] \cdot \boldsymbol{\beta} + [\![p^{\mathrm{S}}]\!]\gamma \mathrm{d}S\Bigg)\end{aligned} \tag{20}$$

We proceed by writing the total potential on compact form as

$$\Pi\left(p^{\mathrm{M}}, \boldsymbol{v}, p^{\mathrm{S}}, \boldsymbol{\beta}, \gamma\right) = \int_{\Omega} \pi_{\square}\left(\nabla p^{\mathrm{M}}, \boldsymbol{v}, p^{\mathrm{S}}, \boldsymbol{\beta}, \gamma, \bar{\boldsymbol{x}}\right)\mathrm{d}V - \Pi^{\mathrm{ext}}\left(p^{\mathrm{M}}\right) \tag{21}$$

Nested saddle-point formulation

We proceed by introducing the macroscale potential function $\Psi\left(p^{\mathrm{M}}\right)$ as

$$\begin{aligned}&\inf_{\boldsymbol{v} \in \mathcal{V}_{\square}} \sup_{\substack{p^{\mathrm{M}} \in \mathcal{P}^{\mathrm{M}} \\ p^{\mathrm{S}} \in \mathcal{P}^{\mathrm{S}}_{\square} \\ \boldsymbol{\beta} \in \mathcal{B}_{\square}}} \inf_{\gamma \in \mathcal{G}_{\square}} \Pi\left(p^{\mathrm{M}}, \boldsymbol{v}, p^{\mathrm{S}}, \boldsymbol{\beta}, \gamma\right) \\ &= \sup_{p^{\mathrm{M}} \in \mathcal{P}^{\mathrm{M}}} \inf_{\boldsymbol{v} \in \mathcal{V}_{\square}} \sup_{\substack{p^{\mathrm{S}} \in \mathcal{P}^{\mathrm{S}}_{\square} \\ \boldsymbol{\beta} \in \mathcal{B}_{\square}}} \inf_{\gamma \in \mathcal{G}_{\square}} \Pi\left(p^{\mathrm{M}}, \boldsymbol{v}, p^{\mathrm{S}}, \boldsymbol{\beta}, \gamma\right) = \sup_{p^{\mathrm{M}} \in \mathcal{P}^{\mathrm{M}}} \Psi\left\{p^{\mathrm{M}}\right\}\end{aligned} \tag{22}$$

where we refer to Appendix "Commutativity of inf and sup" for a proof on the commutativity of the inf and sup operators. We now introduce the macroscale pressure $\bar{p}$ and the macroscale pressure gradient $\bar{\boldsymbol{g}}$ as

$$\bar{p} \stackrel{\text{def}}{=} \langle p \rangle_{\square} \qquad \bar{\boldsymbol{g}} \stackrel{\text{def}}{=} \boldsymbol{\nabla} \bar{p} \tag{23}$$

and assume 1st order homogenization, i.e. the macroscale pressure p^{M} varies linearily inside the RVE. Thus, we have

$$p^{\mathrm{M}} = \bar{p}(\bar{\boldsymbol{x}}) + \bar{\boldsymbol{g}}\left(\bar{\boldsymbol{x}}\right) \cdot \left[\boldsymbol{x} - \bar{\boldsymbol{x}}^{\mathrm{F}}\right] \tag{24}$$

where $\bar{\boldsymbol{x}}^{\mathrm{F}}$ is the center of mass of $\Omega_{\square}^{\mathrm{F}}$. We can now express $\Psi\left\{p^{\mathrm{M}}\right\}$ in terms of the macroscale pressure $\bar{p}$ as

$$\Psi\left\{\bar{p}\right\} = \int_{\Omega} \psi_{\square}\left\{\bar{\boldsymbol{g}}\right\} \mathrm{d}\Omega - \Pi^{\mathrm{ext}}\left(\bar{p}\right) \tag{25}$$

where we have introduced the local macroscale potential

$$\psi_{\square}\left\{\bar{\boldsymbol{g}}\right\} := \inf_{\boldsymbol{v} \in \mathcal{V}_{\square}} \sup_{\substack{p^{\mathrm{S}} \in \mathcal{P}_{\square}^{\mathrm{S}} \\ \boldsymbol{\beta} \in \mathcal{B}_{\square}}} \inf_{\gamma \in \mathcal{G}_{\square}} \pi_{\square}\left(\bar{\boldsymbol{g}}, \boldsymbol{v}, p^{\mathrm{S}}, \boldsymbol{\beta}, \gamma\right) \tag{26}$$

Weak form of the macroscale problem

Although this paper mainly focuses on the subscale problem, we choose to present the macroscale equation in order to achieve completeness. By taking the directional derivative of the the global macroscale potential, we produce

$$\Psi_{\bar{p}}' = \int_{\Omega} \frac{\partial \psi_{\square}}{\partial \bar{\boldsymbol{g}}} \cdot \delta \bar{\boldsymbol{g}} \mathrm{d}V - \int_{\Gamma_{\mathrm{V}}^{\mathrm{F}}} \hat{v}_n \delta \bar{p} \mathrm{d}S = 0 \tag{27}$$

We now define the macroscale seepage as

$$\bar{\boldsymbol{w}} = \frac{\partial \psi_{\square}}{\partial \bar{\boldsymbol{g}}} = \frac{1}{\mid \Omega_{\square} \mid} \int_{\Omega_{\square}^{\mathrm{F}}} \boldsymbol{v} \mathrm{d}V = \phi \left\langle \boldsymbol{v} \right\rangle_{\square} \tag{28}$$

where ϕ is the porosity defined as $\mid \Omega_{\square}^{\mathrm{F}} \mid / \mid \Omega_{\square} \mid$ and $\langle \bullet \rangle_{\square}$ is the intrinsic averaging operator. We now recognize the weak form of the macroscale problem as that of finding all $\bar{p} \in \mathcal{P}^{\mathrm{M}}$ such that

$$\int_{\Omega} \bar{\boldsymbol{w}} \cdot \delta \bar{\boldsymbol{g}} \mathrm{d}V - \int_{\Gamma_{\mathrm{V}}^{\mathrm{F}}} \hat{v}_n \delta \bar{p} \mathrm{d}S = 0 \qquad \forall \delta \bar{p} \in \mathcal{P}^{\mathrm{M},0} \tag{29}$$

where P^{M} and $P^{M,0}$ are the trial and test spaces respectively; now pertaining to the macroscale pressure $\bar{p}$.

The RVE problem

The local (subscale) problem for a given macroscale pressure gradient $\bar{\boldsymbol{g}}$ is produced by seeking the stationary point for variations of subscale quantities in Equation 20. The problem is stated as: Find $\left(\boldsymbol{v}, p^{\mathrm{S}}, \boldsymbol{\beta}, \gamma\right) \in \mathcal{V}_{\square} \times \mathcal{P}_{\square}^{\mathrm{S}} \times \mathcal{B}_{\square} \times \mathcal{G}_{\square}$ such that

$$a_{\square}\left(\boldsymbol{v}; \delta \boldsymbol{v}\right) - b_{\square}\left(\delta \boldsymbol{v}, p^{\mathrm{S}}\right) - c_{\square}\left(\delta \boldsymbol{v}, \boldsymbol{\beta}\right) = -e_{\square}\left(\delta \boldsymbol{v}, \bar{\boldsymbol{g}}\right) \tag{30a}$$

$$-b_{\square}\left(\boldsymbol{v}, \delta p^{\mathrm{S}}\right) - d_{\square}\left(\delta p^{\mathrm{S}}, \gamma\right) = 0 \tag{30b}$$

$$-c_{\square}\left(\boldsymbol{v}, \delta \boldsymbol{\beta}\right) = 0 \tag{30c}$$

$$-d_{\square}\left(p^{\mathrm{S}}, \delta \gamma\right) = 0 \tag{30d}$$

for all $\delta \boldsymbol{v} \in \mathcal{V}_{\square}$, $\delta p^{\mathrm{S}} \in \mathcal{P}^{\mathrm{S}}_{\square}$, $\delta \boldsymbol{\beta} \in \mathcal{B}_{\square}$ and $\delta \gamma \in \mathcal{G}_{\square}$, where

$$a_{\square}(\boldsymbol{v}; \delta \boldsymbol{v}) = \frac{1}{|\Omega_{\square}|} \int_{\Omega^{\mathrm{F}}_{\square}} \boldsymbol{\sigma}^{\mathrm{v}}(\boldsymbol{v} \otimes \boldsymbol{\nabla}) : [\delta \boldsymbol{v} \otimes \boldsymbol{\nabla}] \, \mathrm{d}V \tag{31a}$$

$$b_{\square}\left(\delta \boldsymbol{v}, p^{\mathrm{S}}\right) = \frac{1}{|\Omega_{\square}|} \int_{\Omega^{\mathrm{F}}_{\square}} [\delta \boldsymbol{v} \cdot \boldsymbol{\nabla}] p^{\mathrm{S}} \mathrm{d}V \tag{31b}$$

$$c_{\square}(\delta \boldsymbol{v}, \boldsymbol{\beta}) = \frac{1}{|\Omega_{\square}|} \int_{\Gamma^{\mathrm{F}}_{\square}} [\![\delta \boldsymbol{v}]\!] \cdot \boldsymbol{\beta} \mathrm{d}S \tag{31c}$$

$$d_{\square}(\delta p, \gamma) = \frac{1}{|\Omega_{\square}|} \int_{\Gamma^{\mathrm{F}}_{\square}} [\![\delta p^{\mathrm{S}}]\!] \gamma \mathrm{d}S \tag{31d}$$

$$e_{\square}\left(\delta \boldsymbol{v}, \bar{\boldsymbol{g}}\right) = \frac{1}{|\Omega_{\square}|} \int_{\Omega^{\mathrm{F}}_{\square}} \delta \boldsymbol{v} \mathrm{d}V \cdot \bar{\boldsymbol{g}} \tag{31e}$$

Homogenization of velocity and the macroscale tangent

From the definition of seepage in Equation 28, we produce the possibly non-linear relation between the seepage and the macroscale pressure gradient by differentiation as

$$\bar{\boldsymbol{w}} = \bar{\boldsymbol{w}}\left\{\bar{\boldsymbol{g}}\right\} \Rightarrow \mathrm{d}\bar{\boldsymbol{w}} = \frac{\mathrm{d}\bar{\boldsymbol{w}}}{\mathrm{d}\bar{\boldsymbol{g}}} \cdot \mathrm{d}\bar{\boldsymbol{g}} = -\bar{\boldsymbol{K}}\left\{\bar{\boldsymbol{g}}\right\} \cdot \mathrm{d}\bar{\boldsymbol{g}} \tag{32}$$

Remark 2. *Note that the minus sign on the positive definite permeability tensor* $\bar{\boldsymbol{K}}$ *in Equation 32 is to ensure positive dissipation due to drag interaction between the solid and fluid phases.*

From Equation 30, we see that the unit sensitivity field is given as

$$\begin{aligned}
a'_{\square}(\boldsymbol{v}; \delta \boldsymbol{v}, \mathrm{d}\boldsymbol{v}) - b_{\square}\left(\delta \boldsymbol{v}, \mathrm{d}p^{\mathrm{S}}\right) \quad - c_{\square}(\delta \boldsymbol{v}, \mathrm{d}\boldsymbol{\beta}) \quad &= - e_{\square}(\delta \boldsymbol{v}, \boldsymbol{e}_i) && \text{(33a)}\\
-b_{\square}\left(\mathrm{d}\boldsymbol{v}, \delta p^{\mathrm{S}}\right) \quad - d_{\square}\left(\delta p^{\mathrm{S}}, \mathrm{d}\gamma\right) &= 0 && \text{(33b)}\\
-c_{\square}(\mathrm{d}\boldsymbol{v}, \delta \boldsymbol{\beta}) &= 0 && \text{(33c)}\\
- d_{\square}\left(\mathrm{d}p^{\mathrm{S}}, \delta \gamma\right) &= 0 && \text{(33d)}
\end{aligned}$$

for all $\delta \boldsymbol{v} \in \mathcal{V}_{\square}$, $\delta p^{\mathrm{S}} \in \mathcal{P}^{\mathrm{S}}_{\square}$, $\delta \boldsymbol{\beta} \in \mathcal{B}_{\square}$ and $\delta \gamma \in \mathcal{G}_{\square}$ where a' is the directional derivative of a. Following [3], we can express an arbitrary unit pressure gradient as

$$\mathrm{d}\bar{\boldsymbol{g}} = \sum_{i=1}^{n_{\mathrm{dim}}} \boldsymbol{e}_i \left[\boldsymbol{e}_i \cdot \mathrm{d}\bar{\boldsymbol{g}}\right] \tag{34}$$

From here, we make an ansatz of the response $\mathrm{d}\boldsymbol{v}$ as

$$\mathrm{d}\boldsymbol{v} = \sum_{i=1}^{n_{\mathrm{dim}}} \boldsymbol{v}^{(i)} \left[\boldsymbol{e}_i \cdot \mathrm{d}\bar{\boldsymbol{g}}\right] = \left(\sum_{i=1}^{n_{\mathrm{dim}}} \boldsymbol{v}^{(i)} \otimes \boldsymbol{e}_i\right) \cdot \mathrm{d}\bar{\boldsymbol{g}} \tag{35}$$

Using the definition of seepage in Equation 28 on the above equation, we produce the relation between seepage and pressure gradient perturbations as

$$\begin{aligned} \mathrm{d}\bar{w} =& \phi \left\langle \mathrm{d}v \right\rangle_{\square} = \phi \left\langle \left(\sum_{i=1}^{n_{\dim}} v^{(i)} \otimes e_i \right) \cdot \mathrm{d}\bar{g} \right\rangle_{\square} \\ =& \phi \left(\sum_{i=1}^{n_{\dim}} \left\langle v^{(i)} \right\rangle_{\square} \otimes e_i \right) \cdot \mathrm{d}\bar{g} = -\bar{K} \cdot \mathrm{d}\bar{g} \end{aligned} \tag{36}$$

whereby the macroscale tangent is identified.

Bounds on effective properties for strong periodicity

According to Equation 26, an upper bound is produced by confining the function spaces $\mathcal{V}_{\square}$ and $\mathcal{G}_{\square}$. Furthermore, by choosing the function space $\mathcal{V}_{\square}$ in such a way that periodicity is always fulfilled, the supremum on $\boldsymbol{\beta}$ is void, producing the following inequality

$$\begin{aligned} \psi_{\square}\left(\bar{g}\right) =& \inf_{v \in \mathcal{V}_{\square}} \sup_{\substack{p^{\mathrm{S}} \in \mathcal{P}_{\square}^{\mathrm{S}} \\ \boldsymbol{\beta} \in \mathcal{B}_{\square}}} \inf_{\gamma \in \mathcal{G}_{\square}} \pi_{\square}\left(\bar{g}, v, p^{\mathrm{S}}, \boldsymbol{\beta}, \gamma\right) \\ \leq& \inf_{v \in \mathcal{V}_{\square}'} \sup_{p^{\mathrm{S}} \in \mathcal{P}_{\square}^{\mathrm{S}}} \inf_{\gamma \in \mathcal{G}_{\square}'} \pi_{\square}\left(\bar{g}, v, p^{\mathrm{S}}, \mathbf{0}, \gamma\right) = \psi_{\square}^{\mathrm{fU}}\left(\bar{g}\right) \end{aligned} \tag{37}$$

Equivalently, a lower bound is produced by confining the spaces $\mathcal{P}_{\square}^{\mathrm{S}}$ and $\mathcal{B}_{\square}$

$$\begin{aligned} \psi_{\square}\left(\bar{g}\right) =& \inf_{v \in \mathcal{V}_{\square}} \sup_{\substack{p^{\mathrm{S}} \in \mathcal{P}_{\square}^{\mathrm{S}} \\ \boldsymbol{\beta} \in \mathcal{B}_{\square}}} \inf_{\gamma \in \mathcal{G}_{\square}} \pi_{\square}\left(\bar{g}, v, p^{\mathrm{S}}, \boldsymbol{\beta}, \gamma\right) \\ \geq& \inf_{v \in \mathcal{V}_{\square}} \sup_{\substack{p^{\mathrm{S}} \in \mathcal{P}_{\square}^{\mathrm{S}\,\prime} \\ \boldsymbol{\beta} \in \mathcal{B}_{\square}'}} \pi_{\square}\left(\bar{g}, v, p^{\mathrm{S}}, \boldsymbol{\beta}, 0\right) = \psi_{\square}^{\mathrm{L}}\left(\bar{g}\right) \end{aligned} \tag{38}$$

By combining Equations 37 and 38, we get

$$\psi_{\square}^{\mathrm{L}}\left(\bar{g}\right) \leq \psi_{\square}\left(\bar{g}\right) \leq \psi_{\square}^{\mathrm{U}}\left(\bar{g}\right) \tag{39}$$

We shall now consider the special case of linear flow, defined by $\Phi(v \otimes \nabla) = \frac{\mu}{2}[v \otimes \nabla]^{\mathrm{sym}} : [v \otimes \nabla]^{\mathrm{sym}}$. Assuming that v, p^{S}, $\boldsymbol{\beta}$ and γ satisfies Equation 30 for some $\bar{g}$, $\psi_{\square}(\bar{g})$ is rendered stationary. Thus, the stationarity condition for Equation 30 is

$$a_{\square}\left(v; \delta v\right) = -e_{\square}\left(\delta v, \bar{g}\right) \qquad \forall \delta v \in \mathcal{V}_{\square} \tag{40}$$

In the case of a linear flow, choosing $\delta v = v$ in the stationarity condition, Equation 40 is given as

$$\int_{\Omega_{\square}^{\mathrm{F}}} \mu \left[v \otimes \nabla\right]^{\mathrm{sym}} : \left[v \otimes \nabla\right]^{\mathrm{sym}} \mathrm{d}V = -\int_{\Omega_{\square}^{\mathrm{F}}} v \mathrm{d}V \cdot \nabla \bar{p} \tag{41}$$

Inserting the stationary point into $\pi_{\square}$ and using 41, we see that the RVE mean potential is given as

$$\begin{aligned} & \frac{1}{\mid \Omega_{\square}^{\mathrm{F}} \mid} \int_{\Omega_{\square}^{\mathrm{F}}} \frac{\mu}{2} \left[v \otimes \nabla\right]^{\mathrm{sym}} : \left[v \otimes \nabla\right]^{\mathrm{sym}} \mathrm{d}V \\ & + \frac{1}{\mid \Omega_{\square}^{\mathrm{F}} \mid} \int_{\Omega_{\square}^{\mathrm{F}}} \bar{g} \cdot v \mathrm{d}V = \frac{1}{2} \bar{w} \cdot \bar{g} = -\frac{1}{2} \bar{g} \cdot \bar{K} \cdot \bar{g} \end{aligned} \tag{42}$$

Thus, by bounding $\psi_\square$, we have also bounded $\bar{\boldsymbol{K}}$. More specifically, we may represent Equation 39, in terms of the permeability tensor as

$$\bar{\boldsymbol{g}} \cdot \bar{\boldsymbol{K}}^{\mathrm{L}} \cdot \bar{\boldsymbol{g}} \leq \bar{\boldsymbol{g}} \cdot \bar{\boldsymbol{K}} \cdot \bar{\boldsymbol{g}} \leq \bar{\boldsymbol{g}} \cdot \bar{\boldsymbol{K}}^{\mathrm{U}} \cdot \bar{\boldsymbol{g}} \tag{43}$$

where

$$\psi_\square^{\mathrm{L}}(\bar{\boldsymbol{g}}) = -\frac{1}{2}\bar{\boldsymbol{g}} \cdot \bar{\boldsymbol{K}}^{\mathrm{U}} \cdot \bar{\boldsymbol{g}} \tag{44a}$$

$$\psi_\square(\bar{\boldsymbol{g}}) = -\frac{1}{2}\bar{\boldsymbol{g}} \cdot \bar{\boldsymbol{K}} \cdot \bar{\boldsymbol{g}} \tag{44b}$$

$$\psi_\square^{\mathrm{U}}(\bar{\boldsymbol{g}}) = -\frac{1}{2}\bar{\boldsymbol{g}} \cdot \bar{\boldsymbol{K}}^{\mathrm{L}} \cdot \bar{\boldsymbol{g}} \tag{44c}$$

Discretization of solutions spaces on the RVE boundary

As to the specific choice of solution spaces for the Lagrange multipliers we note that which is the most efficient depends on both the discretization and the geometry of the subscale domain. One example of a feasible discretization of the Lagrange multipliers is the one presented in [20] where the pertinent unknown functions are discretized on a mesh consisting of the union of all nodes on opposite sides of the domain. Here, however, we choose to discretize the Lagrange multipliers $\boldsymbol{\beta}$ and γ as global polynomials, i.e.

$$\mathcal{B}_\square = \left\{\boldsymbol{\beta} \in \mathbf{R}^2 : \boldsymbol{\beta} = \sum_{i=0}^{n_p} \boldsymbol{b}_i \frac{s^i}{l_\square}\right\}, \quad \mathcal{G}_\square = \left\{\gamma \in \mathbf{R} : \gamma = \sum_{i=0}^{n_p} g_i \frac{s^i}{l_\square}\right\} \tag{45}$$

where n_p are the polynomial order in the respective approximation, s is a parameterized coordinate along $\Gamma_\square^+$, $\boldsymbol{b}_i$ and g_i are the respective coefficients and $l_\square$ is the side length of the RVE.

For an upper bound of the energy, we choose $\mathcal{V}_\square$ such that the velocity is always periodic, removing the supremum on $\boldsymbol{\beta}$.

$$\mathcal{V}'_\square = \left\{\boldsymbol{v} \in \mathcal{V}_\square : \boldsymbol{v} = \sum_{i=1}^{N_v} a_i \boldsymbol{\phi}_i \text{ on } \Gamma_\square^{\mathrm{F}},\ a_i \in \mathbb{R},\ [\![\boldsymbol{\phi}_i]\!] = \mathbf{0}\right\} \subset \mathcal{V}_\square \tag{46}$$

where $\boldsymbol{\phi}_i$ are basis functions for the N_v velocity degrees of freedom a_i. It should be noted that, if $\boldsymbol{\phi}_i$ is represented in polynomial base, the constraint $[\![\boldsymbol{\phi}_i]\!] = 0$ requires approximations of order higher than 1 in the case where obstacles cross the boundary of the RVE. The reason for this is simply that the no slip condition on the obstacle surface implies zero velocity on the RVE boundary if the velocity approximation is constant or linear. For the same reason, the velocity approximation is applied patchwise between obstacles along the boundary. In practice, we use global quadratic 1D element along the boundary as shown in the example in Figure 2 and make all nodes along the boundary hang on the global element. Furthermore, we connect all nodes located on a corner, i.e. N_1 is a master and W_1, W_6, S_1, S_2, E_1 and E_6 its slaves. Finally, we connect opposite sides, i.e W_2 is a slave to E_2, S_2 to N_2 etc.

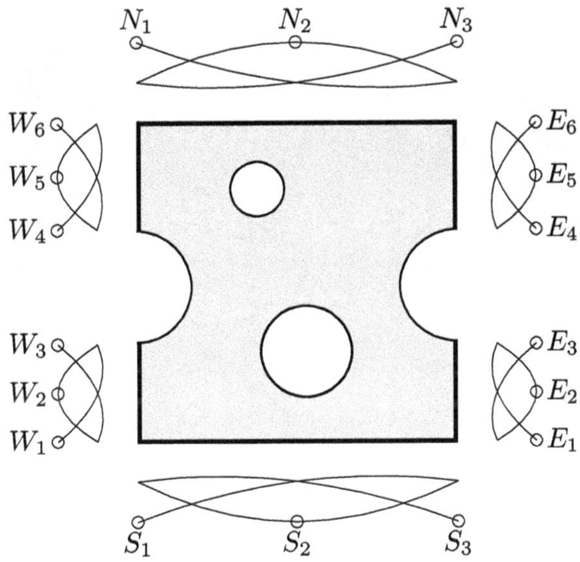

Figure 2 Example of boundary elements approximating the velocity.

For a lower bound on the energy, we choose $\mathcal{P}^{\mathrm{S}}_{\square}$ as

$$\mathcal{P}^{\mathrm{S}\,\prime}_{\square} = \left\{p^{\mathrm{S}} \in \mathcal{P}^{\mathrm{S}}_{\square} : p^{\mathrm{S}} = 0 \text{ on } \Gamma^{\mathrm{F}}_{\square}\right\} \subset \mathcal{P}^{\mathrm{S}}_{\square} \tag{47}$$

Thus, p^{S} is trivially periodic.

Results and discussion

In this section we present two numerical examples. The ambition of the first example is to investigates how the the order of a polynomial approximation of the Lagrange multiplier affect the solution and what order is required to reach convergence in terms of seepage, i.e. when the velocity field has converged to periodicity. This example is performed on a unit cell containing one single, circular, obstacle. The result on a non-periodic mesh is compared the results from the corresponding problem with strong periodicity. Upper and lower bounds for the permeability are also presented. In the second numerical example, a quantitative convergence study is performed on the size effect on seepage of an RVE containing a set of random obstacles or periodic unit cells for a give order.

In all examples, the Stokes flow is solved using the Finite Element Method on triangular Taylor-Hood elements (linear pressure, quadratic velocity). The used fluid model is $\boldsymbol{\sigma}^{\mathrm{v}} = \mu \boldsymbol{l}^{\mathrm{sym}}$ where the viscosity μ is chosen as unity and $\boldsymbol{l}^{\mathrm{sym}}$ is the symmetric velocity gradient.

All numerical simulations are performed using the open source software OOFEM [22].

Influence of polynomial order on permeability

The analysis in this section aim at evaluating how the order of the Lagrange multiplier approximation affect the periodicity of the solution and how the weak periodicity differ from strong periodicity. The simulations are performed on a unit cell containing a circular obstacle with radius 0.25 which is located at (0.26, 0.5) in order to produce a non-periodic mesh (see Figure 3(c)). As to the actual computation of the permeability $\bar{\boldsymbol{K}}$, we

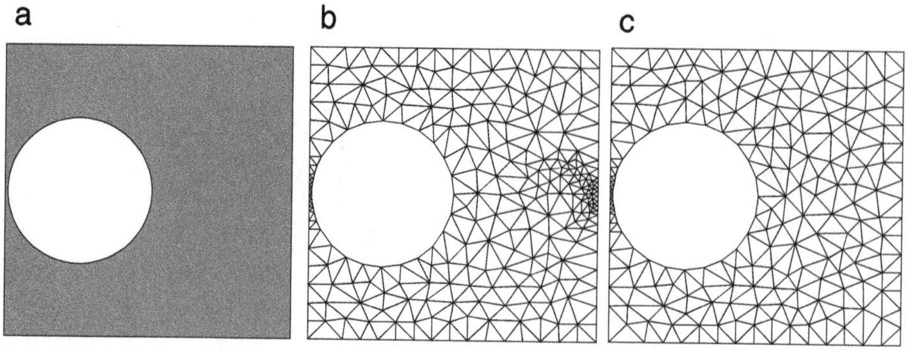

Figure 3 Figures shows the unitcell (a) domain (b) periodic mesh and (c) mesh used during computations.

use the method presented in [4]. In the following example, we compute the permeability on a domain using two different discretizations where each discretization has a set of subproblem as described below.

I Periodic mesh and strong periodicity
II Non-periodic mesh

(a) Upper bound (constant pressure, weakly periodic velocity)
(b) Weak periodicity (weakly periodic pressure and velocity)
(c) Lower bound (weakly periodic pressure, velocity is either constant or quadratic along $\Gamma_{\square}^{\mathrm{F}}$)

Since we aim at replicating the behavior of strong periodicity, I is used as a reference solution.

To compute the respective bounds of the permeability $\bar{\boldsymbol{K}}$, we use the function spaces suggested in Equations 46 and 47 and choose $\mathcal{G}_{\square}$ and $\mathcal{B}_{\square}$ as in Equation 45. Figure 4 shows the first and second eigenvalues K_I and K_{II} ($K_{II} \leq K_I$) of $\bar{\boldsymbol{K}}$ and their respective upper and lower bounds for orders ranging from 0 to 15. As the lower bound pertinent to the quadratic velocity profile gives significanly tighter bounds than the constant velocity profile, the later is omitted in the remaining parts of this paper. Since a unit cell in a periodic pattern is isotropic, K_I and K_{II} should tend to the same value as the solution approaches periodicity [3], which is indeed the case as shown in Figure 4(c). Note that none of the bounds reach isotropy, although the upper bound is significantly closer than the lower bound. According to the results, an order 4 is sufficient.

We choose the discretization of the unknown functions as proposed in Section "Discretization of solutions spaces on the RVE boundary" and note that in practice, the load is applied piecewise and in this case we have two sets of polynomials for each unknown function, one for the horizontal and one for the vertical part of the RVE boundary.

In the case where polynomials are used for the discretization of the Lagrange multipliers, the number of Gauss points, n_G, needed to perform an exact integration of the integrals $d_{\square}(\bullet,\bullet)$ and $c_{\square}(\bullet,\bullet)$ are computed as

$$n_G = \frac{n_p + n_f + 1}{2} \tag{48}$$

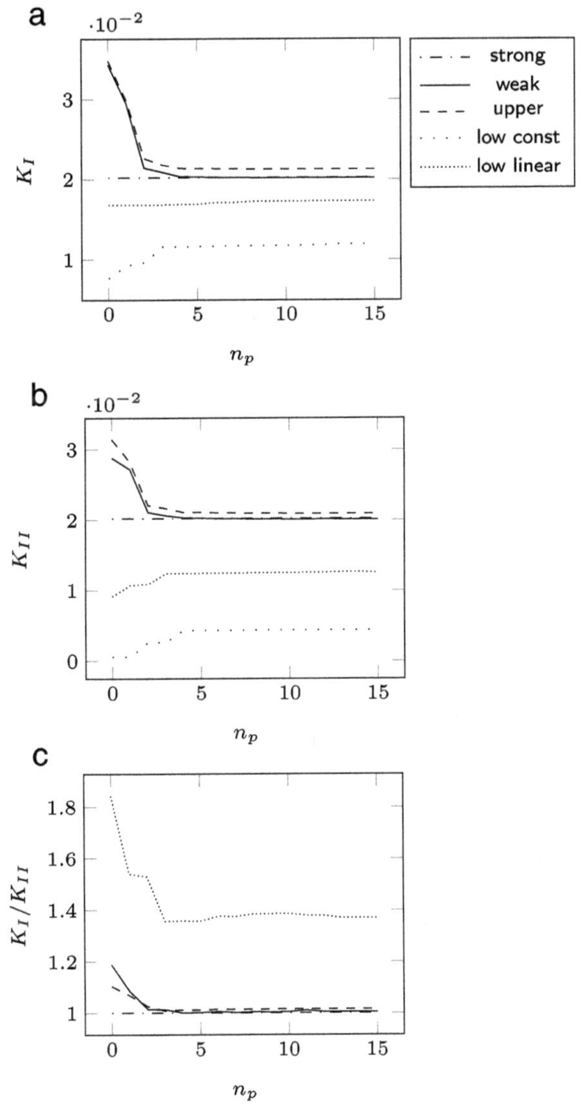

Figure 4 The Figure shows (a) the first eigenvalue of the permeability tensor K_I versus n_p (b) K_{II} versus n_p (c) $\frac{K_I}{K_{II}}$ versus n_p.

where n_f is the order of the approximation of the pertinent field and n_p is the order of the Lagrange multiplier approximation. For a Taylor-Hood element, $n_f = 1$ for the pressure and $n_f = 2$ for the velocity.

As an indicator of how close to strong periodicity the fields are, we compute the L^2 norm of the error as

$$e^2 = \int_{\Gamma_{\square}^{\mathrm{F}+}} [\![\bullet]\!]^2 \mathrm{d}S \tag{49}$$

where $[\![\bullet]\!]$ represents the jump of $\bullet$ on $\Gamma_{\square}^{\mathrm{F}}$. As can be seen in Figure 5, both velocities converge quickly compared to the pressure p^{S}. Indeed, since the pressure is discretized by

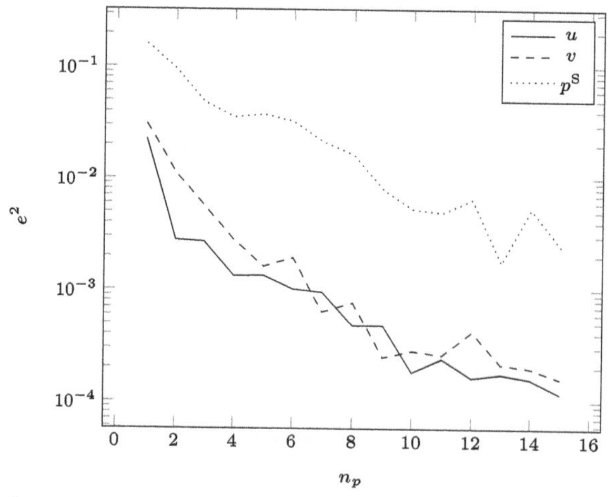

Figure 5 The error e^2 versus n_p.

piecewise linear polynomials, true periodicity can never be achieved on a non-periodic mesh, less in the cases of linear or constant pressure along $\Gamma_{\square}^{F+}$ and $\Gamma_{\square}^{F-}$. The same hold for a quadratic discretization but due to the larger number of degrees of freedom, a solution closer to periodicity is achieved. We would, however, like to point out that as the main interest lies in computing the effective permeability and/or seepage, we consider convergence in terms of the mean velocity.

Figure 6 shows the pressure p^S along the vertical parts of the RVE along with the pertinent Lagrange multiplier for orders 0, 2, 4 and an overkill solution with polynomials of order 15. The effect of linear elements can be seen in these graphs, as even in the overkill solution, relatively large differences between the pressure functions on either side of the RVE are present. However, the large values of the Lagrange multiplier in the corners of the overkill solution suggests that the pressure field is close to periodicity in that area.

Comparing the pressure curves in Figure 6 with the corresponding curves for the velocity u in Figure 7 we again notice that the velocity is closer to periodicity at order 4.

Figure 8 shows the velocity u along the vertical part of the boundary. Notice the even functions in the Lagrange multipliers γ and β_1 and the odd functions in β_2 due to the symmetric shape of the RVE.

Impact of RVE size on permeability

When imitating a material using homogenization on RVEs, two main sources of errors are introduced; boundary conditions and the statistical representation of the microstructure. If the microstructure is truly periodic, the periodic type boundary condition introduces no error, but this is often an approximation of the materials subscale geometry. However, as this paper aims at producing periodicity in a weak sense, we assume that the true solution is indeed periodic. In this case, the error introduced by boundary conditions are the order of the approximation of the Lagrange multipliers as a perfectly periodic solution is not guaranteed. As to the error introduced in terms of statistical representation of the

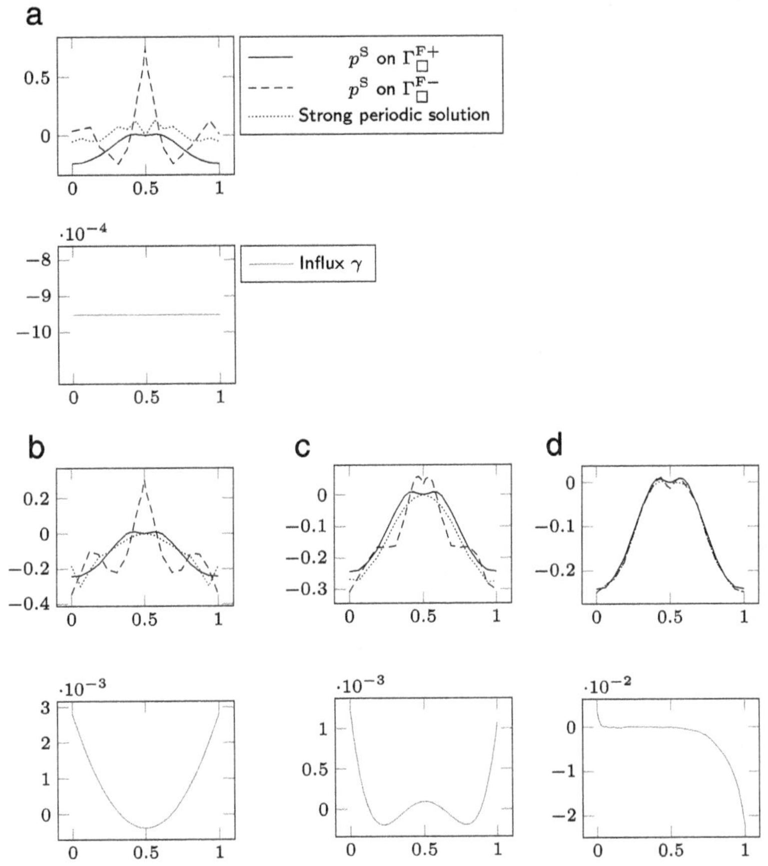

Figure 6 Pressure p^S and flux γ along the vertical parts of $\Gamma_\square^F$ corresponding to Lagrange multipliers of order (a) order 0 (b) order 2 (c) order 4 (d) order 15.

subscale, this can be overcome by either increasing the number of RVEs or increase the size such that all geometrical effects are captured in one RVE. In fact, the relative error introduced by the order of approximation can also be decreased by increasing the size of the RVE.

In order to study the influence of RVE size on the effective permeability, we choose to perform homogenization on two types of domain:

I $\Omega_\square$ contains a perfectly aligned, circular obstacles where no obstacles cross the boundary

   (a) Weak periodicity on pressure and velocity. Order of approximation is 0.

II $\Omega_\square$ contains pseudo-randomly placed, circular obstacles which can cross the boundary

   (a) Weak periodicity on pressure and velocity. Order of approximation is 0.
   (b) Weak periodicity on pressure and velocity. Order of approximation is 4 and the load is applied piecewise.

For type II domains, the RVE is generated according to Algorithm 1.

Algorithm 1: Sketch of algorithm for inserting obstacles into $\Omega_{\square}$

```
1  Initialize i=1, j=1;
2  while i < L□ do
3      while j < L□ do
4          Compute (x_i, y_j) = f(i, j) ;
5          if Distance from circle at (x_i, y_j) and any existing circle < Δ then
6              Go to 4;
7          else
8              Insert circle in Ω□;
9              if Circle at (x_i, y_j) intersect with Γ□ then
10                 Insert the remaining part of the circle at opposite edge;
11             end if
12         end if
13         j=j+1;
14     end while
15     i=i+1;
16 end while
```

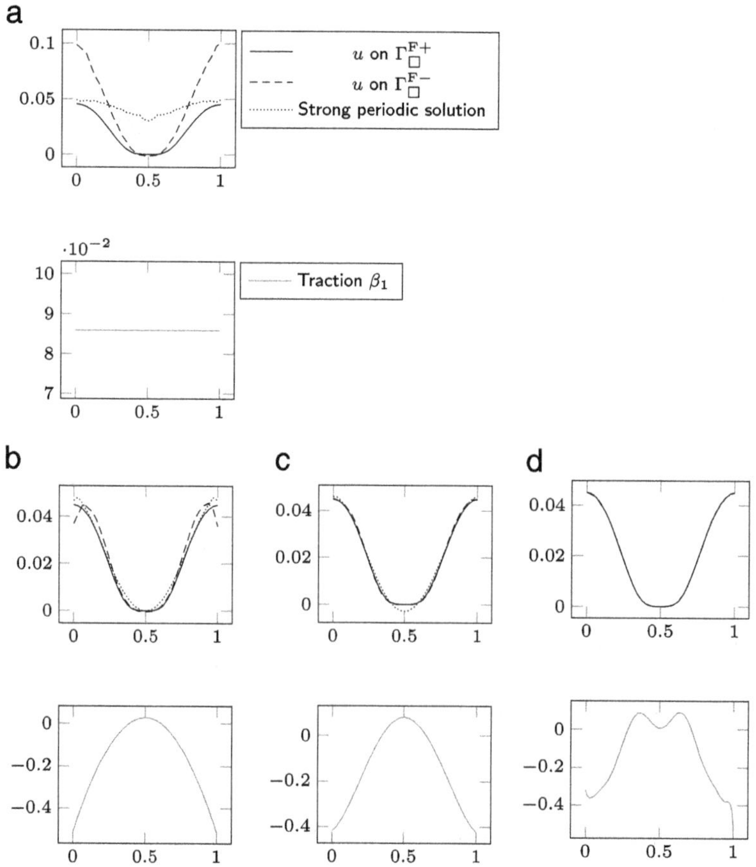

Figure 7 Velocity v and β_1 along the vertical parts of $\Gamma_{\square}^{F}$ corresponding to Lagrange multipliers of order **(a)** order 0 **(b)** order 2 **(c)** order 4 **(d)** order 15.

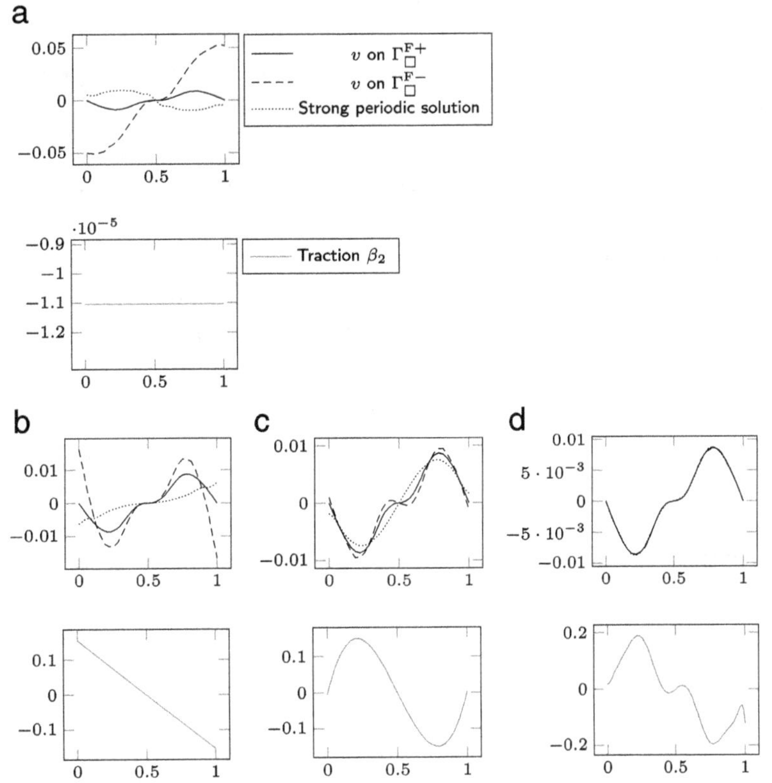

Figure 8 Velocity v and β_2 along the vertical parts of $\Gamma^{\mathrm{F}}_{\square}$ corresponding to Lagrange multipliers of order (a) order 0 (b) order 2 (c) order 4 (d) order 15.

where $L_{\square}$ is the integer length of one side of a rectangular RVE and Δ set to 1% of the radius of the obstacles. Furthermore, we choose the function $f(i,j)$ as

$$f(i,j) = \left(L_{\square}\left(i - \frac{1}{2}\right) + \Phi(\mu,\sigma), L_{\square}\left(j - \frac{1}{2}\right) + \Phi(\mu,\sigma)\right) \tag{50}$$

where Φ is a normally distributed random variable, μ is the mean and σ is the standard deviation. Here, we choose $\mu = 0$ and $\sigma = 0.2$. Note that lines 9-11 in Algorithm 1 implies geometric periodicity and constant porosity.

We use the function spaces suggested in Equations 46 and 47 when computing the bounds of the permeability. In cases where obstacles cross the boundary, the load pertinent to the periodicity condition is applied piecewise, thus, the solution space of the Lagrange multiplier approximation becomes larger.

It should be noted that the highest possible order of the polynomial approximation is limited by the number of boundary elements subject to that constraint. In the case of randomly placed obstacles, it is possible that one element only, separates two obstacles. If that is the case, depending on the discretization of the pertinent function and Lagrange polynomial, the subscale tangent becomes singular. In such cases, a simple rule is used; the number of unknowns in the Lagrange multiplier cannot exceed the number of unknowns belonging to the pertinent function, eg. if only one linear element is present, a linear approximation is used. This is taken into account when producing the RVEs. However, this situation rarely occurs.

In order to produce a reliable estimate for the permeability of an RVE containing pseudo-randomly placed obstacles, a sufficiently large number of RVEs is generated and the permeability computed according to [4]. In the case of an RVE with one periodically repeating unit cell, one realization of each size of the RVE suffices. Examples of meshed RVEs with periodic unit cells and random obstacles are shown in Figure 9.

Figure 10 shows how the permeability changes as the size of the RVE increase for RVE types I (a) and II (a) along with their respective bounds. From Figure 10(a) we can see that the lower bound differs from the weak periodic solution while the upper bound coincides. Comparing to the results in Section "Influence of polynomial order on permeability", this is true for 0th order approximation, but as the order increase, the permeability of the weakly periodic solution decrease. We also emphasize that the error introduced by the 0th order approximation increase, the error in the homogenized result decrease as the RVE grows, since the solution approach the strongly periodic solution. The differences between the solutions are further illustrated in Figure 11 where a pressure gradient in the x direction is imposed on an RVE containing 3×3 periodic unit cells. The image shows the magnitude of the velocity field $\boldsymbol{v}$. The solutions are similar inside the RVE but differs on the boundaries.

Figure 10(b) shows how $\mu\{K_I\}$ behaves as the size of the RVE increases. As in the previous case, the upper bound and the weakly periodic solution produce similar solutions whereas the lower bound yields a significantly lower permeability. By increasing the size of the RVE while keeping the order of the approximation fixed, the error introduced by the approximation increase while the total error decrease. The bump at RVE sizes 2 and 3 are due to two mechanisms; The permeability increase in the direction of the pressure gradient, if the distance between the obstacle orthogonal to the pressure gradient, increase (the volume fluid passing through per second increases quadratically with the distance) and the permeability decrease as the distance parallel to the pressure gradient increase (as this implies a longer distance). The first mechanism is dominant for small RVEs but as the size increase, the two evens out.

In the final example, shown in Figure 12(a), the order of the approximating polynomial has been increased to 4 and the load pertinent to the weak periodic boundary condition is applied patchwise. In order to allow for proper comparison of the two last examples,

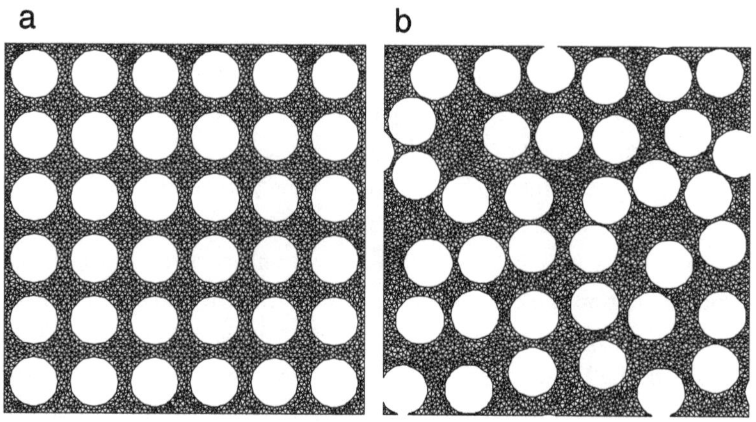

Figure 9 Example meshes for domains consisting of (a) periodic obstacles (b) pseudo-randomly placed obstacles.

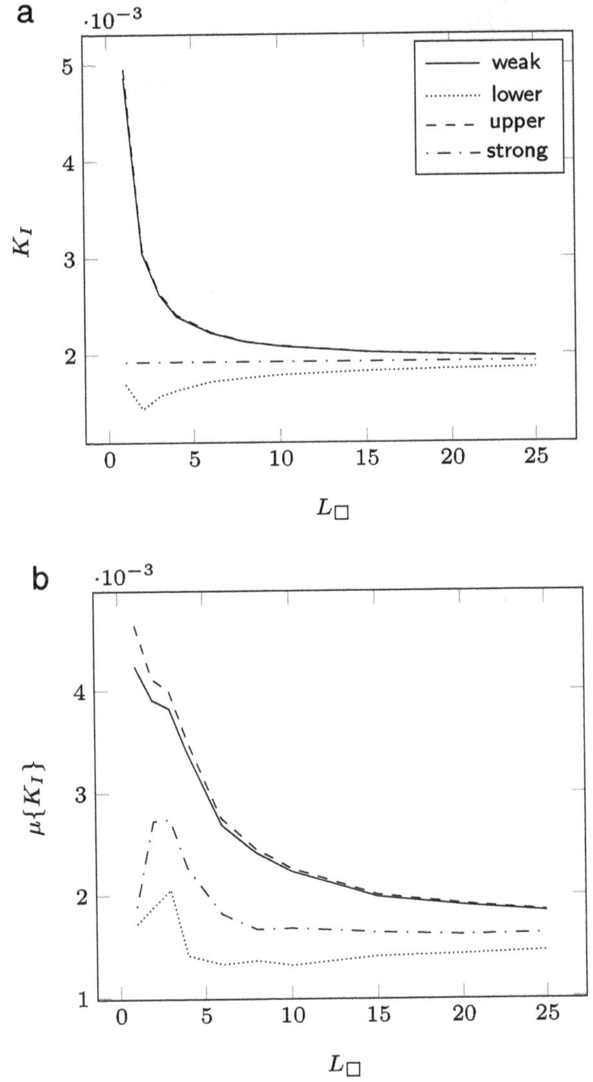

Figure 10 The graphs show **(a)** the first eigenvalue of the permeability tensor for an RVE with periodic unit cells and **(b)** the mean first eigenvalue of the permeability tensor for RVEs with random obstacles and constant Lagrange multipliers.

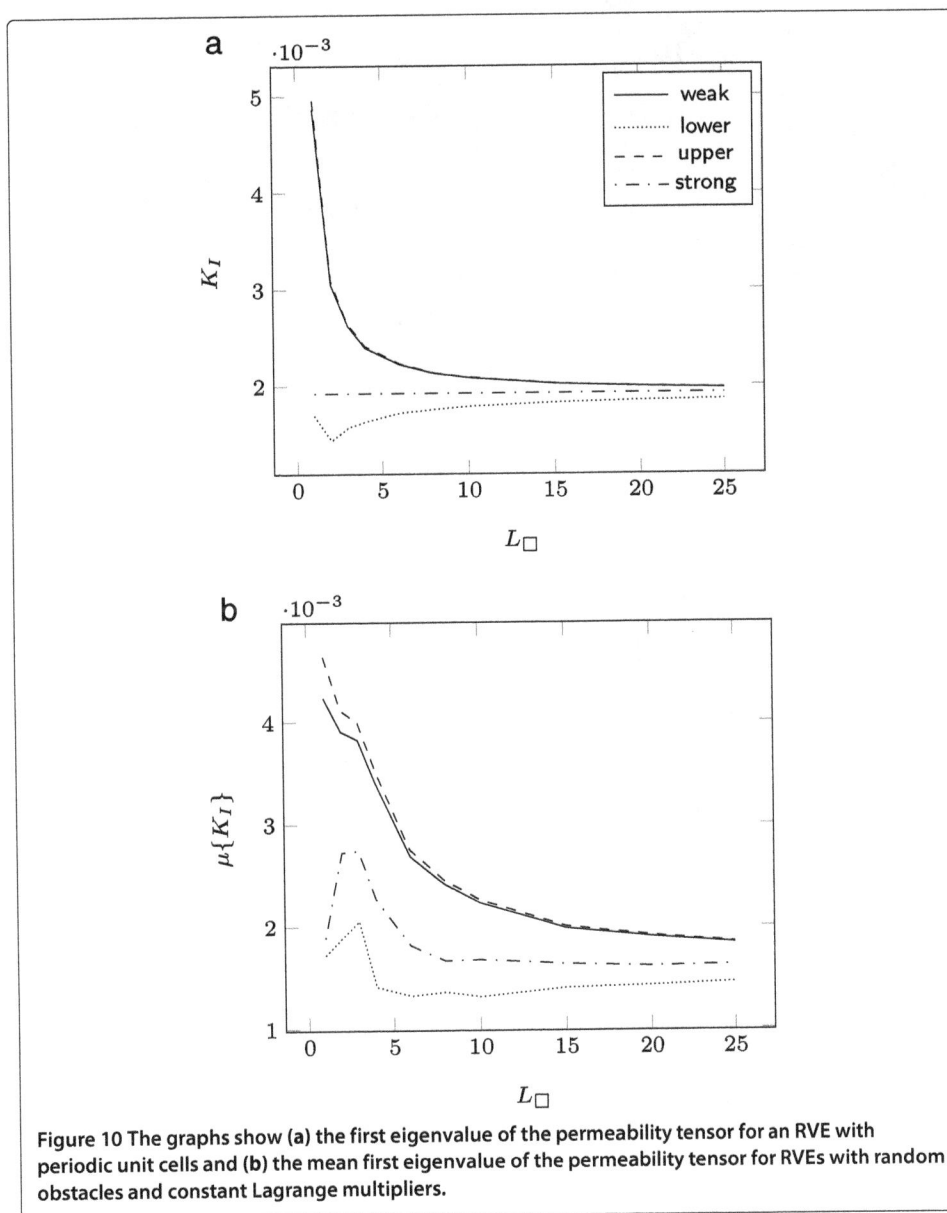

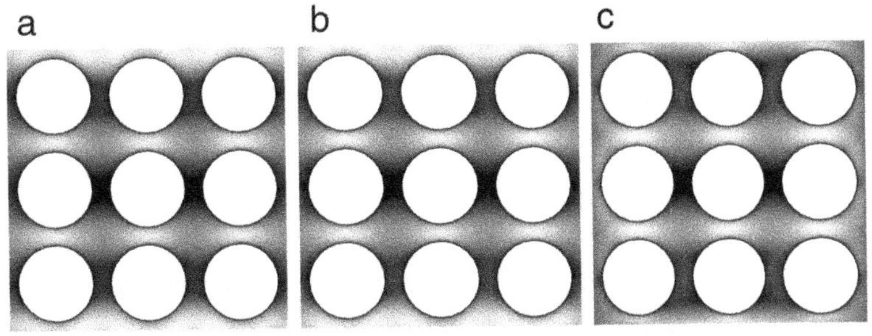

Figure 11 Magnitude of velocity corresponding to **(a)** upper bound, **(b)** weak periodicity and **(c)** lower bound for a pressure gradient $\bar{g} = [1 \;\; 0]$.

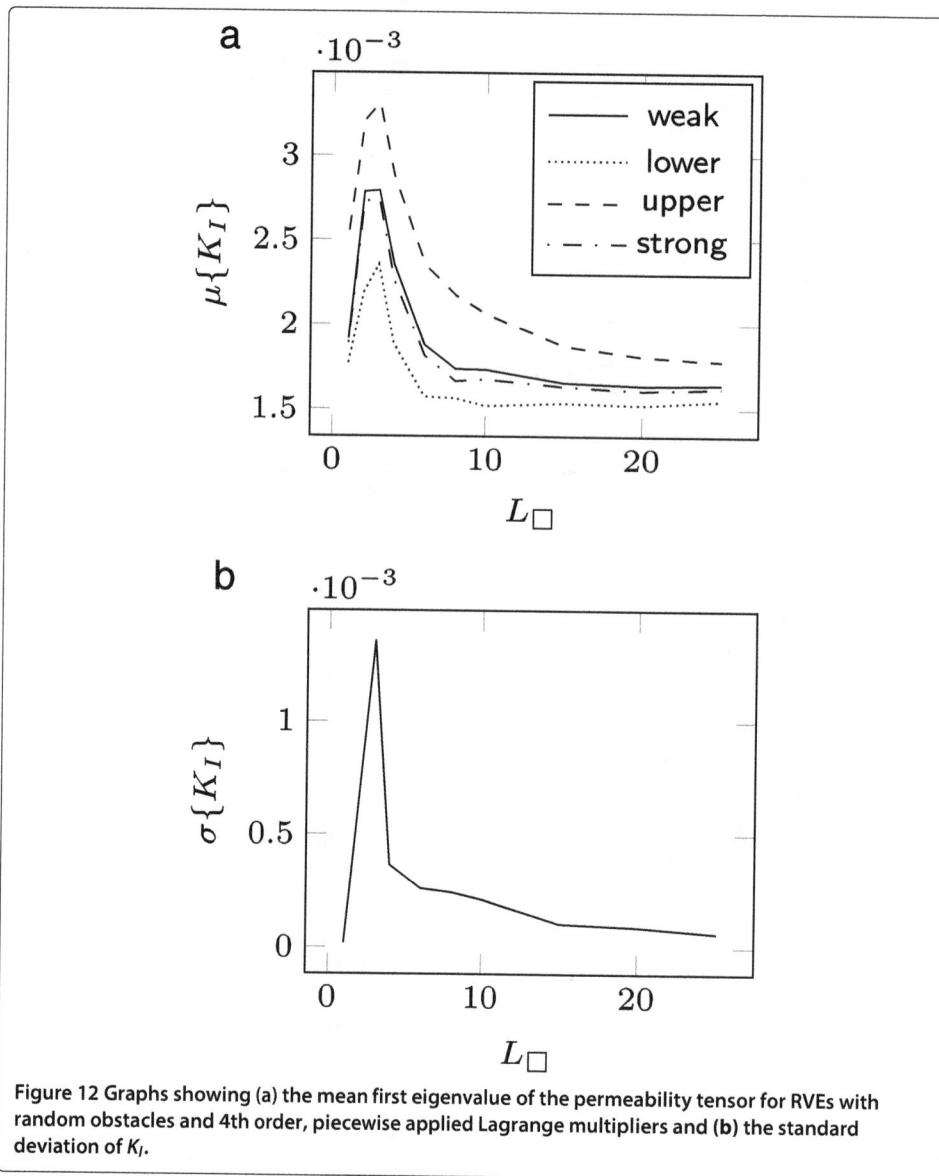

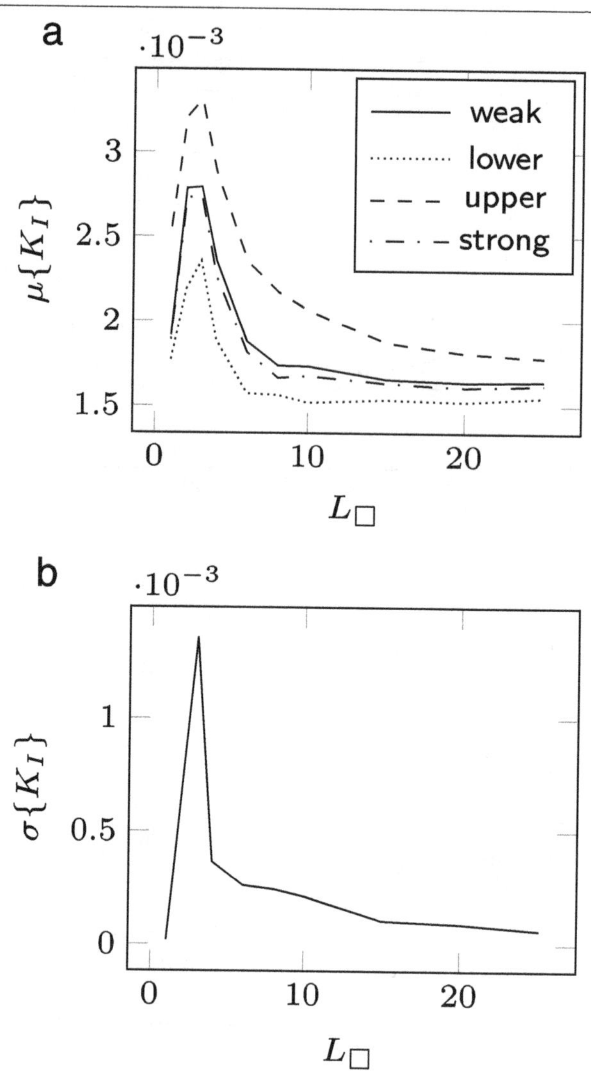

Figure 12 **Graphs showing (a) the mean first eigenvalue of the permeability tensor for RVEs with random obstacles and 4th order, piecewise applied Lagrange multipliers and (b) the standard deviation of** K_I**.**

the pseudo-random domains used in the last example are the same as in the previous. It is apparent that the permeability is dependent on the order of the approximation, even for large RVEs.

A comparison between Figures 13(a) and 14(a) illustrates the impact of additional terms and piecewise application of the loads on the periodicity, especially on the north and south edges of $\Omega_\square$.

To conclude the numerical section, we note that the lower bound is closer to the strongly periodic solution in all examples. Furthermore, we also note that by increasing the resolution of the Lagrange multiplier approximations, the solution approaches strong periodicity fast. The cost of the enriched approximations are the additional degrees of freedom and a stiffness matrix with dense sub matrices pertinent to the boundary integrals in Equation 33.

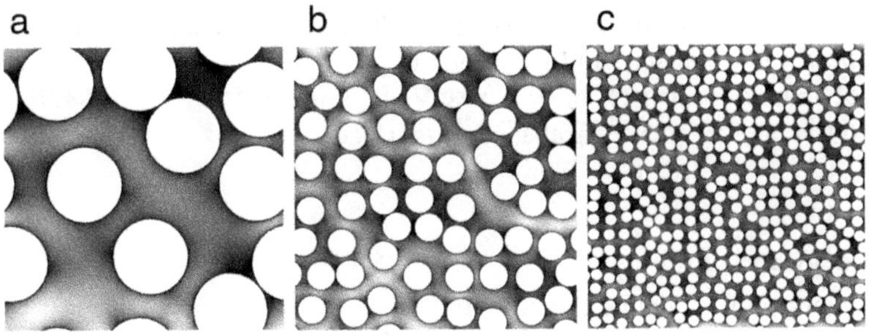

Figure 13 Magnitude of velocity corresponding to a pressure gradient $\bar{g} = [1 \ \ 0]$ on RVEs with weak periodicity of order 0 for sizes (a) 4 × 4 (b) 8 × 8 (c) 20 × 20.

Conclusions

In this paper, we produce a multiscale problem for a Stokes flow by minimization of an energy potential. During the minimization process, a split in the pressure term is introduced, after which the weak form of the problem is introduced by variations of the pertinent quantities. The result is a Stokes flow subscale problem and a Darcy flow macroscale problem.

In order to satisfy the macrohomogeneity condition on a non periodic mesh, weak periodic boundary conditions are imposed using Lagrange multipliers. These are of two types: unknown tractions maintain periodicity on the velocity and unknown fluxes maintain periodicity on the pressure. Due to the saddlepoint-nature of the problem, bounds on the macroscale permeability are produced by confining the pertinent function spaces.

The numerical examples have shown the rapid convergence of periodicity using polynomial approximation of relevant Lagrange multipliers. Furthermore, the expected asymptotic convergence of macroscale permeability due to RVE size has been verified.

Concerning future developments, the primary goal is to be able to couple permeability with deformation of the porous material. Since a 2D representation of an open pore system is unable to carry static load when deformed, it is necessary to extend the study to a more relistic 3D representation.

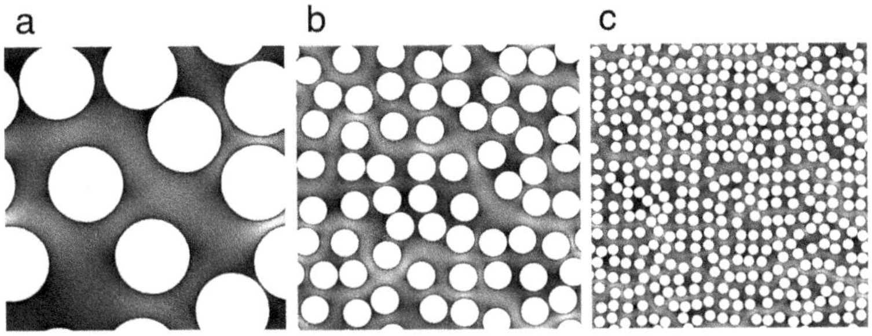

Figure 14 Magnitude of velocity corresponding to a pressure gradient $\nabla p^{\mathrm{M}} = [1 \ \ 0]$ on RVEs with weak periodicity of order 0 for sizes (a) 4 × 4 (b) 8 × 8 (c) 20 × 20.

Endnotes

[a]As this work only concerns the fluid phase, the solid phase is considered rigid and is modeled as impermeable obstacles in the Ω domain.

[b]In the case of linear flow, $\Phi(\boldsymbol{v} \otimes \nabla) = \frac{\mu}{2}\,[\boldsymbol{v} \otimes \nabla]^{\text{sym}} : [\boldsymbol{v} \otimes \nabla]^{\text{sym}}$.

[c]It is assumed that all RVEs have parallel edges/surfaces.

Appendix

Commutativity of inf and sup

For the subsequent proof, we define the potential

$$\hat{\Pi}\left(\boldsymbol{v}, p^{\mathrm{M}}\right) = \sup_{\substack{p^{\mathrm{S}} \in \mathcal{P}_{\square}^{\mathrm{S}} \\ \boldsymbol{\beta} \in \mathcal{B}_{\square}}} \inf_{\gamma \in \mathcal{G}_{\square}} \Pi\left(\boldsymbol{v}, p^{\mathrm{M}}, p^{\mathrm{S}}, \boldsymbol{\beta}, \gamma\right) = \Phi(\boldsymbol{v}) - b\left(\boldsymbol{v}, p^{\mathrm{M}}\right) \tag{51}$$

From [23], we have the max-min inequality as

$$\sup_{p^{\mathrm{M}}} \Theta\left(p^{\mathrm{M}}\right) = \sup_{p^{\mathrm{M}}} \inf_{\boldsymbol{v}} \hat{\Pi}\left(\boldsymbol{v}, p^{\mathrm{M}}\right) \leq \inf_{\boldsymbol{v}} \sup_{p^{\mathrm{M}}} \hat{\Pi}\left(\boldsymbol{v}, p^{\mathrm{M}}\right) \tag{52}$$

which for a saddle point implies equality. What remains to be shown is the validity of the right hand side of the equation, i.e that the term $\Theta(p^{\mathrm{M}})$ is concave in p^{M} for all $\boldsymbol{v}$. For future use, we note that a variation in $\boldsymbol{v}$ yields, by stationarity

$$\Phi'(\boldsymbol{v}; \delta\boldsymbol{v}) - b\left(\delta\boldsymbol{v}, p^{\mathrm{M}}\right) = 0 \tag{53}$$

and a subsequent perturbation in $\boldsymbol{v}$

$$\Phi''(\boldsymbol{v}; \delta\boldsymbol{v}, \mathrm{d}\boldsymbol{v}) - b\left(\delta\boldsymbol{v}, \mathrm{d}p^{\mathrm{M}}\right) = 0 \tag{54}$$

To show concavity of Θ, we make a perturbation in p^{M} and choose $\delta\boldsymbol{v} = \mathrm{d}\boldsymbol{v}$ in Equation 53, which yields

$$\Theta'\left(p^{\mathrm{M}}; \mathrm{d}p^{\mathrm{M}}\right) = \Phi'(\boldsymbol{v}; \mathrm{d}\boldsymbol{v}) - b\left(\mathrm{d}\boldsymbol{v}, p^{\mathrm{M}}\right) - b\left(\boldsymbol{v}, \mathrm{d}p^{\mathrm{M}}\right) = -b\left(\boldsymbol{v}, \mathrm{d}p^{\mathrm{M}}\right) \tag{55}$$

By yet another perturbation and the use of Equation 54

$$\Theta''\left(p^{\mathrm{M}}; \mathrm{d}p^{\mathrm{M}}, \mathrm{d}p^{\mathrm{M}}\right) = -b\left(\mathrm{d}\boldsymbol{v}, \mathrm{d}p^{\mathrm{M}}\right) = -\Phi''\left(\boldsymbol{v}; \mathrm{d}\boldsymbol{v}, \mathrm{d}\boldsymbol{v}\right) < 0 \tag{56}$$

if Φ'' is positive definite and b satisfies the required, classic inf-sup condition

$$\inf_{q} \sup_{\boldsymbol{w}} \frac{b(\boldsymbol{w}, q)}{\| \boldsymbol{w} \| \cdot \| q \|} \geq \gamma_b > 0 \tag{57}$$

Competing interests

The authors declare that they have no competing interests.

Authors' contributions

All authors have carried out the theoretical parts described in this paper. CS has carried out the necessary software development and has had the major responsibility for preparing the manuscript. All authors have read and approved the final manuscript.

Acknowledgements

The project is financed by the Swedish Research Council (www.vr.se) under contract 2011-5388. The authors would also like to thank Håkan Johansson for fruitful discussions.

References

1. Zohdi TI, Wriggers P (2008) An introduction to computational micromechanics. Springer, Berlin Heidelberg
2. Du X, Ostoja-Starzewski M (2006) On the size of representative volume element for Darcy law in random media. Proc R Soc A: Math Phys Eng Sci 462(2074):2949–2963
3. Sandström C, Larsson F (2013) Variationally consistent homogenization of stokes flow in porous media. Int J Multiscale Comput Eng 11(2):117–138

4. Sandström C, Larsson F, Runesson K, Johansson H (2013) A two-scale finite element formulation of Stokes flow in porous media. Comput Methods Appl Mech Eng 261-262:96–104
5. Larsson F, Runesson K, Su F (2009) Variationally consistent computational homogenization of transient heat flow. Int J Numerical Methods Eng 81(13):1659–1686
6. Sánchez-Palencia E (1980) Non-homogeneous Media and Vibration Theory. Springer, Berlin Heidelberg
7. Allaire G (1989) Homogenization of the Stokes flow in a connected porous medium. Asymptotic Anal 2:203–222
8. Hornung U (1997) Homogenization and porous media. Springer, Berlin Heidelberg
9. Nemat-Nasser S, Lori M, Datta SK (1996) Micromechanics: overall properties of heterogeneous materials. J Appl Mech 63(2):561
10. Geers MGD, Kouznetsova VG, Brekelmans WAM (2010) Multi-scale computational homogenization: trends and challenges. J Comput Appl Math 234(7):2175–2182
11. Larsson F, Runesson K (2006) RVE computations with error control and adaptivity: the power of duality. Comput Mech 39(5):647–661
12. Löhnert S, Wriggers P (2003) Homogenisation of microheterogeneous materials considering interfacial delamination at finite strains. Technische Mechanik 23(2–4):167–177
13. Ngo N, Tamma K (2001) Microscale permeability predictions of porous fibrous media. Int J Heat Mass Transf 44:3135–3145
14. Pillai KM, Advani SG (1995) Numerical and analytical study to estimate the effect of two length scales upon the permeability of a fibrous porous medium. Transp Porous Media 21(1):1–17
15. Arbogast T, Lehr H (2006) Homogenization of a Darcy-Stokes system modeling vuggy porous media. Comput Geosci 78712:1–18
16. Ohman M, Runesson K, Larsson F (2011) Computational Mesoscale modeling and homogenization of liquid-phase sintering of particle agglomerates. Technische Mechanik 32:463–483
17. Nilenius F, Larsson F (2012) Macroscopic diffusivity in concrete determined by computational homogenization. Int J Numerical Anal Methods Geomech (May 2012) 37(11):1535–1551
18. Ostoja-Starzewski M (2007) Microstructural randomness and scaling in mechanics of materials. Chapman & Hall/CRC, Boca Raton, FL
19. Yue X (2007) The local microscale problem in the multiscale modeling of strongly heterogeneous media: Effects of boundary conditions and cell size. J Comput Phys 222(2):556–572
20. Larsson F, Runesson K, Saroukhani S, Vafadari R (2011) Computational homogenization based on a weak format of micro-periodicity for RVE-problems. Comput Methods Appl Mech Eng 200(1–4):11–26
21. Nguyen VD, Béchet E, Geuzaine C, Noels L (2011) Imposing periodic boundary condition on arbitrary meshes by polynomial interpolation. Electrical Eng 55(November):390–406
22. Patzák B, Bittnar Z (2001) Design of object oriented finite element code. Adv Eng Softw 32:759–767
23. Boyd S, Vandenberghe L (2010) Convex Optimization. vol. 25. Cambridge University Press. pp 487–487. Chap. 1,10,11.

3

On the sensitivity of the POD technique for a parameterized quasi-nonlinear parabolic equation

Nissrine Akkari[1,2*], Aziz Hamdouni[1*], Erwan Liberge[1] and Mustapha Jazar[2]

*Correspondence:
nissrine.akkari@univ-lr.fr;
aziz.hamdouni@univ-lr.fr
[1] University of La Rochelle, LaSIE - Laboratory of the Engineering Sciences for the Environment, Avenue Michel Crépeau, 17042 La Rochelle Cedex 1, France
[2] Lebanese University, LaMA - Laboratory of Mathematics and Applications, B.P 826 Tripoli, Lebanon

Abstract

Background: In what follows, we consider the Proper Orthogonal Decomposition (POD) technique of model order reduction, for a parameterized quasi-nonlinear parabolic equation.

Methods: A POD basis associated with a set of reference values of the characteristic parameters is considered. From this basis, a parametric reduced order model (ROM) projecting the initial equation is constructed.

Results: A mathematical a priori estimate of the parametric squared L^2-error induced by this projection is developed. This later estimate is based on both, the parametric behavior of the squared L^2-ROM-error thanks to the resolution of a Ricatti differential inequality in the parametric ROM-error, and the convergence rate of the parametric ROM to the full problem, via the augmentation of the basis dimension. Indeed, under restrictive conditions on the solutions regularity of such equations, we are able to precise the slope of the logarithm of the squared L^2-norm of the ROM error, as a function of the logarithm of the basis modes number.
Numerical experiments of our theoretical estimate, are presented for the 2D Navier-Stokes equations in the case of an unsteady and incompressible fluid flow in a channel around a circular cylinder.

Conclusion: A mathematical a priori estimate of the parametric squared L^2-error induced by the model reduction by POD is developped for a parameterized quasi-nonlinear parabolic equation. This estimate is obtained thanks to the resolution of a Ricatti differential inequality.

Keywords: Quasi-nonlinear equations; ROM; POD; Parametric evolution; Sensitivity; A priori error bound; Burgers equation; 2D Navier-Stokes equations

Background

Statement of the problem

High-dimensional Partial Differential Equations (PDE) intervene widely in applications of the field of Mechanics (Fluid Mechanics, Solid Mechanics, etc...). The simulation cost of such equations is in general very high. Model reduction techniques are a very good solution for such problems: They enable an approximation of these equations in subspaces of small dimension. These are applied for Fluid-Structure Interaction problems [1-7], stability study [8], shape optimization problems [9,10], optimal control problems [11,12] etc....

In general, the reduction spaces are obtained from the knowledge of a solution flow. Then, one of the challenges for model reduction, is to enable a good prediction of the solutions behavior in parametric evolution problems: We cite the Greedy algorithm, that enriches a reduction subspace computed for a particular set of parametric values, by evaluating the errors obtained a posteriori by the reduction of the full numerical solutions (obtained for example by finite element dicretisations of the complete model problem) associated with other parametric values. Obviously, this algorithm needs to be accelerated in a way to enable an estimate of the new parameters' values for which the reduced order model (ROM) error is maximal and the computation of a new reduction basis is becoming crucial. At this step only, a complete resolution of the model equations is done for the new parameters' values chosen by the algorithm. Two important questions can be asked:

What is the parametric confidence region of a given reduced order model?

How can we improve the performance of a reduced order model when parameters are variying significantly?

Several techniques of model reduction exist to build a good candidate within the parametric ROMs:

We consider first the reduced basis (RB) method. It is based on showing a given parametric solution as a finite linear combination of solutions associated respectively with particular parameters values. More precisely, we cite the work of Maday et al. [13,14] who developed, for a parameterized elliptic PDE which is symmetric and coercive, an a priori convergence theory of a parametric RB approximation to the parametric full finite element solution. The RB is constructed with association to a sample subset of parametric values which are logarithmically distributed: this allows an a priori convergence of the Greedy algorithm [15].

They were interested also, in showing an *a posteriori* bound of the squared parametric L^2-ROM-RB error, of which computation at each parameter value is less expensive than the parametric ROM-error itself as it is usually done in the original Greedy algorithm. The *a posteriori* character of this error bound is due to its dependency on the squared residual norm of the parametric ROM. We precise also, that the later error bound is obtained by applying classical energetic methods to the discrete equation in the difference between the full numerical solution and the reduced one, so one can bound the ROM-error by the already precised *a posteriori* residual norm. This technique was applied in the context of offline-online procedures, in the following cases: parameterized linear elliptic PDE which is coercive and not symmetric [16]; nonaffine linear elliptic and parabolic equations and nonlinear elliptic and parabolic equations [17]; parameterized linear parabolic equation with a nonaffine source term [18]; the model Burgers equation [19,20].

A very adaptive technique, for building a parametric ROM is the Proper Generalized Decomposition (PGD) method. It is based on building an approximation of the initial PDE as a finite combination of functions of separate variables, including not only the space and time variables, but also all eventual parameters that could be associated with the initial equations. These functions and their coefficients in the later expression are obtained by an iterated algorithm which minimizes the error with respect to the initial problem. This method was introduced by P. Ladevèze in the LATIN method [21,22], where he started by a space-time separation. Then, it was generalised by Chinesta et al. for multidimensional problems [23-28].

We are discussing in this paper, the parametric sensitivity of a reduced order model by the Proper Orthogonal Decomposition (ROM-POD). A detailed literature on this method can be found in [29,30].

It is still an important issue, to determine the confidence interval of a ROM-POD for problems involving parametric evolution. There were few works in this context:

Amsallem et al. [31-33], build a geometric interpolation algorithm of ROM-PODs at different parametric values, in the case of parameterized linear structural dynamics problems. The idea is to interpolate various reduced order, symmetric and positive definite matrices of the complete problem, corresponding each one to a set of characteristic parameters. This interpolation is done in the tangent space to the manifold of symmetric positive definite matrices. Numerical experiments show that the obtained reduced order model, is efficient to represent dynamics asssociated with other parametric sets than the reference ones.

There are also techniques based on an *a posteriori* indicator of the POD modes number we need to add, in order to decrease the ROM-error, when the parameters are varying. Volkwein et al. [34-37] were interested in such estimates for solving linear quadratic elliptic and parabolic optimal control problems. Therefore, they developed an *a posteriori* estimator of the error between the reduced optimal solution and the optimal one. Also, this estimator is based on the residual term we add when solving the reduced adjoint equation. The reduced optimal solution is good when a tolerated error is not trespassed, otherwise we increase the dimension of the reduced optimal control problem.

Besides all these works, we are interested in developing an a priori bound of the squared parametric L^2-ROM-POD error in the case of a parameterized quasi-nonlinear parabolic equation. More precisely, our main issue is to give a mathematical criteria in order to determine a parametric confidence region for a reference POD basis. It appears also the control problem of the dimension of the associated parametric ROM-POD, so we can improve its performance if the parameters are varying considerably with respect to the reference ones.

This type of result will enable us not only to predict the validity extent of a reference reduced order model of fixed dimension with respect to parametric variation, but also to do an enrichement step of the ROM-dimension when necessary, thanks to an a priori indicator of the number of POD modes we need to add. So that, on the one hand the parametric ROM-error is not maximal when the parameters are varying considerably outside the confidence region determined already, and on the other hand the reduction concept is still available.

The result of this paper is an improvement of the one we developed in [38,39]. Where we showed an a priori estimate of the parametric squared L^2-error induced after the model reduction of parameterized semi-discrete quasi-linear parabolic problems, by a reference POD basis associated with a reference solution for a base parameter value. Indeed, the resolution of a Ricatti differential inequality gives sharper a priori upper bounds compared to the ones we developped in [38,39], where we used the Gronwall lemma to solve a first order differential inequality in the parametric ROM-error. Moreover, we present an improvement of this type of results by considering an enriched POD basis associated with a reference solution and its parametric derivative at the same parameter value.

More precisely, the mathematical formulation of the problem is given as follows:

Methods

Mathematical formulation of the problem

Let us consider a general parameterized quasi-nonlinear parabolic PDE: We denote $X = [L^2(\Omega)]^d$, $V = [H^1(\Omega)]^d$. Ω is a bounded open set, connected and lipschitz of $\mathbb{R}^d$, where $d = 1$ or 2. This equation is given by its weak formulation as follows:

$$\begin{cases} \frac{d}{dt}(u_\lambda(t), v)_X + B(u_\lambda(t), u_\lambda(t), v) + a(u_\lambda(t), v; \lambda) \\ = S(u_\lambda(t), v) + l_t(v) \quad \forall v \in V \\ \\ (u_\lambda(0), v)_X = \left(u_\lambda^0, v\right)_X \quad \forall v \in V \end{cases}, \tag{1}$$

where B, a, S and l are given by the following expressions:

- $\forall v_1, v_2, v_3 \in V, B(v_1, v_2, v_3) = \left(f(v_1).g(\nabla v_2), v_3\right)_X$, where f and g are respectively an $\mathbb{R}^d$– valued Lipschitz application defined on $\mathbb{R}^d$ and an $\mathbb{R}^{d^2}$– valued Lipschitz application defined on $\mathbb{R}^{d^2}$.
- $\forall v_1, v_2 \in V, a(v_1, v_2; \lambda)$ is a bilinear symmetric and positive form, which is continuous on $V \times V$ and coercive on $[H_0^1(\Omega)]^d \times [H_0^1(\Omega)]^d$. Moreover, we suppose that a is $\alpha-$ Holder with respect to λ, with Lipschitz constant equal to 1, i.e., $|a(v_1, v_2; \lambda_1) - a(v_1, v_2; \lambda_2)| \leq \|\lambda_1 - \lambda_2\|^\alpha a(v_1, v_2)$.
- $\forall v_1, v_2 \in V, S(v_1, v_2) = (s(v_1), v_2)_X$, where s is an $\mathbb{R}^d$–valued Lipschitz application defined on $\mathbb{R}^d$.
- $\forall v \in V, l_t(v) = (h(t), v)_X$, where h is given in $L^2_{loc}([0, +\infty[, X)$.

(.) denotes the scalar product in the corresponding space. And, $\lambda \in \mathbb{R}^{p,+,*}$ ($p \in \mathbb{N}^*$), denote the parameter vector of these equations.

A solution u_{λ_0} of this equation associated with a reference parameter vector λ_0 is computed once and for all. A POD basis $\Phi^{\lambda_0} = \left(\Phi_n^{\lambda_0}\right)_{n \geq 1}$ in X, associated with u_{λ_0} on a time interval $(0, T)$, will lead to construct a ROM-POD describing the evolution of an approximation $\hat{u}_{\lambda,\lambda_0}$ of u_λ, in a subspace of dimension N very small. We denote $\left(\mu_n^{\lambda_0}\right)_{n \geq 1}$ the POD eigenvalues sequence associated with the POD basis.

We know that:

$$\frac{1}{T}\int_0^T \left(u_{\lambda_0}(t), \Phi_n^{\lambda_0}\right)_X^2 dt = \mu_n^{\lambda_0}.$$

The problem that appears naturally, is to find a way to control the parametric evolution: what is $\frac{1}{T}\int_0^T \left(u_\lambda(t), \Phi_n^{\lambda_0}\right)_X^2 dt$?

More precisely, if we denote $\forall t \in (0, T)$ and $\forall x \in X$:

$$\hat{u}_{\lambda,\lambda_0}(t, x) = \sum_{n=1}^{N} a_n^{\lambda,\lambda_0}(t)\Phi_n^{\lambda_0}(x),$$

where $\forall n = 1, \ldots, N$, $a_n^{\lambda,\lambda_0}(t)$ is the solution of:

$$\begin{cases} \dfrac{da_n^{\lambda,\lambda_0}}{dt} + B\left(\hat{u}_{\lambda,\lambda_0}(t), \hat{u}_{\lambda,\lambda_0}(t), \Phi_n^{\lambda_0}\right)_X + a\left(\hat{u}_{\lambda,\lambda_0}(t), \Phi_n^{\lambda_0}; \lambda\right) \\ = S(\hat{u}_{\lambda,\lambda_0}(t), \Phi_n^{\lambda_0}) + l_t(\Phi_n^{\lambda_0}) \\ \\ a_n^{\lambda,\lambda_0}(0) = \left(u_\lambda^0, \Phi_n^{\lambda_0}\right) \end{cases} \tag{2}$$

Then, the question which arises naturally is the following: to what extent $\hat{u}_{\lambda,\lambda_0}$ remains an accurate approximation to u_λ?

Main results

In whats follows, we present our main results formally, with a lack in rigor:

Reference POD basis associated with a reference parameter vector λ_0

Formal result 1. *There exist two decreasing sequences* $\left(f_1^{\lambda_0}(N)\right)_{N\geq 1}$ *and* $\left(f_2^{\lambda_0}(N)\right)_{N\geq 1}$, *such that:*

$$\left\|u_\lambda - \hat{u}_{\lambda,\lambda_0}\right\|^2_{L^2(0,T;X)} \leq f_1^{\lambda_0}(N) + f_2^{\lambda_0}(N)\frac{\|\lambda - \lambda_0\|^\alpha}{\|\lambda_0\|} \tag{3}$$

$f_1^{\lambda_0}(N)$ is the error induced already by taking $\lambda = \lambda_0$:
$f_1^{\lambda_0}(N) = \left\|u_{\lambda_0} - \hat{u}_{\lambda_0}\right\|^2_{L^2(0,T;X)}$, where $\hat{u}_{\lambda_0} = \hat{u}_{\lambda,\lambda_0}$, for $\lambda = \lambda_0$. For further details concerning the estimate of this term, we refer for instance to [38].

$f_2^{\lambda_0}(N)$is a decreasing sequence, having the same rate of decline as the sequence $\left(\dfrac{1}{N}\right)_{N\geq 1}$.

Precision on result 1

Heuristic result 1. *Under regularity conditions on the solutions difference* $u_\lambda - u_{\lambda_0} \in L^2(0,T;[H^m(\Omega)]^d)$:

$$\left\|u_\lambda - \hat{u}_{\lambda,\lambda_0}\right\|^2_{L^2(0,T;X)} \leq f_1^{\lambda_0}(N) + c(N)\frac{\|\lambda - \lambda_0\|^\alpha}{\|\lambda_0\|}, \tag{4}$$

where $(c(N))_{N\geq 1}$ *is a decreasing sequence, having the same rate of decline as the sequence* $\left(\dfrac{1}{N^m}\right)_{N\geq 1}$.

Improvement of the ROM-POD confidence interval: Enriched POD basis containing the parametric sensitivity of the solution

Formal result 2. *Under the following restrictive conditions:*

- $\nabla_\lambda u_\lambda(\lambda_0) \in L^2(0,T;X)$.
- Φ^{λ_0} *a POD basis associated to snapshots of* $u_{\lambda_0}(t)$ *and* $\nabla_\lambda u_\lambda(\lambda_0)(t)$ *on* $(0,T)$.

We can show that:

$$\left\|u_\lambda - \hat{u}_{\lambda,\lambda_0}\right\|^2_{L^2(0,T;X)} \leq f_1^{\lambda_0}(N) + \sum_{n=N+1}^{\infty} \mu_n^{\lambda_0}\|\lambda - \lambda_0\|^2 + K\|\lambda - \lambda_0\|^4 \tag{5}$$

Result 1 establishes an a priori estimate of the decrease rate of the squared ROM-POD error, especially when the two parameters λ and λ_0 are distant.

Nevertheless, the usefulness of this result is improved thanks to result 1 under regularity conditions on the diffrenece between two solution flows associated respectively with a parameter λ and the base parameter λ_0. Indeed, result 1 shows for a given class of solutions to equation (1), a very good efficiency of a reference reduced order model by a POD basis associated with a base parameter value λ_0.

Also, not to mention supplementary regularity conditions on the solution flows, the estimate (5) improves the validity domain of the reduced order model beside our previous result (estimate (3)) thanks to the term depending on $\|\lambda - \lambda_0\|^4$, and to the remainder of the expanded-POD eigenvalues sum multiplying $\|\lambda - \lambda_0\|^2$, of which decrease rate is optimal thanks to the POD construction.

Organization of the paper

In what follows, we give the proof elements of our principal result. In section "Results and discussion", we prove our formal results 1 and 2 and the heuristic result 1, in the context of multidimensional parametric problems. In section "Numerical experiments: The 2D Navier-Stokes equations", we present numerical experiments of our theoretical estimate for the 2D Navier-Stokes in the case of an unsteady and incompressible fluid flow. In section "Conclusion and prospects", we conclude by giving some prospects to this work.

Results and discussion

Theorems and notations

We recall some theorems and we define some notations, that could be useful to our proof.

Theorem 1. ***Sobolev embeddings***

- *For* $d = 1$, $H^1(\Omega) \subset L^\infty(\Omega)$, *and* $\exists C \in \mathbb{R}^{+*}$ *such that* $\forall v \in H^1(\Omega)$, $\|v\|_{L^\infty(\Omega)} \leq C \|\nabla v\|_{L^2(\Omega)}$.
- *For* $d = 2$, $H^1(\Omega) \subset L^4(\Omega)$, *and* $\exists C \in \mathbb{R}^{+*}$ *such that* $\forall v \in H^1(\Omega)$, $\|v\|_{L^4(\Omega)} \leq C \|\nabla v\|_{[L^2(\Omega)]^2}$.

For more informations about Sobolev embeddings, see for instance [40].

Theorem 2. ***Gagliardo-Nirenberg inequality***

- *For* $d = 1$, $\exists c \in \mathbb{R}^{+*}$ *such that* $\|v\|_{L^\infty(\Omega)} \leq c \|v\|^{1/2}_{L^2(\Omega)} \|\nabla v\|^{1/2}_{L^2(\Omega)}, \forall v \in H^1(\Omega)$.
- *For* $d = 2$, $\exists c \in \mathbb{R}^{+*}$ $\|v\|_{L^4(\Omega)} \leq c \|v\|^{1/2}_{L^2(\Omega)} \|\nabla v\|^{1/2}_{[L^2(\Omega)]^2}, \forall v \in H^1(\Omega)$.

For more informations, see for instance [41].

Notation 1. *C_p^λ denotes the constant relative to the coercivity of the bilinear form a, associated with a given λ. So, $\forall\ v \in [H_0^1(\Omega)]^d$:*

$$a(v, v; \lambda) \geq C_p^\lambda \|\nabla v\|_X^2 .$$

Notation 2. *K_a denotes the constant relative to the continuity of the bilinear form a on the spave V. So, $\forall\ v_1, v_2 \in V$:*

$$a\,(v_1, v_2) \le K_a \left\|\nabla v_1\right\|_X \left\|\nabla v_2\right\|_X.$$

Notation 3. *K_f, K_g and K_s denote respectively the Lipschitz constants of the mappings f, g and s.*

Control of $\left\|u_\lambda - u_{\lambda_0}\right\|^2_{L^2(0,T;X)}$

We recall the following intermediate lemma, based on the resolution of a Ricatti equation of order 3:

Lemma 1. *Let a and b be two strictly positive real numbers, and let z be a time dependent positive quantity verifying the following differential inequality:*

$$\frac{dz}{dt} \le az(t) + bz^3(t).$$

Then,

$$z(t) \le \left(z^{-2}(0) + \frac{b}{a}(exp(-2at) - 1)\right)^{-\frac{1}{2}}.$$

Proof. The proof of this lemma is essentially based on the following key points:

- The change of variable: $y(t) = z^{-2}(t)$.
- $\dfrac{dy}{dt} \ge -2ay(t) - 2b$ (6)
- Application of the Gronwall lemma to inequality (6). □

Thanks to lemma 1, we prove the following proposition:

Proposition 1. *We show the following a priori upper bound of $\left\|u_\lambda - u_{\lambda_0}\right\|^2_{L^2(0,T;X)}$, thanks to a differential Ricatti inequality in the later:*

$$\left\|\left(u_\lambda - u_{\lambda_0}\right)(t)\right\|^2_X \le \left(\left\|u^0_\lambda - u^0_{\lambda_0}\right\|^{-4}_X + \frac{b}{a}(exp(-2at) - 1)\right)^{-\frac{1}{2}} K^2_a \left\|\nabla u_{\lambda_0}(t)\right\|^2_X \left\|\lambda - \lambda_0\right\|^\alpha,$$

where a and b are positive real numbers that will be detailed in what follows:

Proof. We prove this proposition for $d = 1$. The same proof applies directly to the case: $d = 2$.

We denote $w(t) = \left(u_\lambda - u_{\lambda_0}\right)(t)$, which verifies the following weak formulation:

$$\begin{cases} \frac{d}{dt}\left(w(t), v\right)_X + B\left(u_\lambda(t), u_\lambda(t), v\right) - B\left(u_\lambda(t), u_{\lambda_0}(t), v\right) + B\left(u_\lambda(t), u_{\lambda_0}(t), v\right) \\ -B\left(u_{\lambda_0}(t), u_{\lambda_0}(t), v\right) + a(u_\lambda(t), v; \lambda) - a(u_{\lambda_0}(t), v; \lambda_0) \\ = S(u_\lambda(t), v) - S(u_{\lambda_0}(t), v) \quad \forall v \in V \\ \\ (w(0), v)_X = \left(u^0_\lambda - u^0_\lambda, v\right)_X \quad \forall v \in V \end{cases} \tag{7}$$

If we replace v by $w(t)$, then we obtain:

$$\frac{1}{2}\frac{d}{dt}\|w(t)\|_X^2 + \int_\Omega f(u_\lambda(t))\left(g(\nabla u_\lambda(t)) - g(\nabla u_{\lambda_0}(t))\right) w(t)dx =$$
$$-\int_\Omega \left(f(u_\lambda(t)) - f(u_{\lambda_0}(t))\right) g(\nabla u_{\lambda_0}(t)) w(t)dx - a(w(t), w(t); \lambda)$$
$$-a(u_{\lambda_0}(t), w(t); \lambda) + a(u_{\lambda_0}(t), w(t); \lambda_0) + \int_\Omega \left(s(u_\lambda(t)) - s(u_{\lambda_0}(t))\right) w(t)dx.$$

Therefore, by using the Sobolev embeddings mentioned above, we get:

$$\begin{aligned}\frac{1}{2}\frac{d}{dt}\|w(t)\|_X^2 \leq\; & \left\|f(u_\lambda(t))\right\|_{L^\infty(\Omega)} \left\|g(\nabla u_\lambda(t)) - g(\nabla u_{\lambda_0}(t))\right\|_X \|w(t)\|_X \\ & + \left\|f(u_\lambda(t)) - f(u_{\lambda_0}(t))\right\|_{L^\infty(\Omega)} \left\|g(\nabla u_{\lambda_0}(t))\right\|_X \|w(t)\|_X \\ & - a(w(t), w(t); \lambda) - a(u_{\lambda_0}(t), w(t); \lambda) + a(u_{\lambda_0}(t), w(t); \lambda_0) \\ & + \left\|s(u_\lambda(t)) - s(u_{\lambda_0}(t))\right\|_X \|w(t)\|_X\end{aligned}$$

Therefore,

$$\begin{aligned}\frac{1}{2}\frac{d}{dt}\|w(t)\|_X^2 \leq\; & CK_g \left\|f(u_{\lambda_0}(t))\right\|_V \|\nabla w(t)\|_X \|w(t)\|_X \\ & + K_f K_g \|w(t)\|_{L^\infty(\Omega)} \|\nabla w(t)\|_X \|w(t)\|_X \\ & + CK_f \left\|g(\nabla u_{\lambda_0}(t))\right\|_X \|\nabla w(t)\|_X \|w(t)\|_X \\ & - C_p^\lambda \|\nabla w(t)\|_X^2 \\ & + K_a \|\lambda - \lambda_0\|^\alpha \left\|\nabla u_{\lambda_0}(t)\right\|_X \|\nabla w(t)\|_X + K_s \|w(t)\|_X^2\end{aligned}$$

By appyling a Young inequality two times and the Gagliardo-Nirenberg inequality one time, we get:

$$\begin{aligned}\frac{1}{2}\frac{d}{dt}\|w(t)\|_X^2 \leq\; & \left(\frac{C^2 \left\|f(u_{\lambda_0})\right\|_V^2 K_g^2 + C^2 \left\|g(\nabla u_{\lambda_0})\right\|_X^2 K_f^2}{2\beta}\right) \|w(t)\|_X^2 \\ & + \beta \|\nabla w(t)\|_X^2 \\ & + cK_f K_g \|w(t)\|_X^{3/2} \|\nabla w(t)\|_X^{3/2} \\ & - C_p^\lambda \|\nabla w(t)\|_X^2 \\ & + \frac{1}{2\varepsilon_2} \|\lambda - \lambda_0\|^\alpha K_a^2 \left\|\nabla u_{\lambda_0}(t)\right\|_X^2 + \frac{\varepsilon_2}{2} \|\lambda - \lambda_0\|^\alpha \|\nabla w(t)\|_X^2 \\ & + K_s \|w(t)\|_X^2\end{aligned}$$

But $cK_f K_g \|w(t)\|_X^{3/2} \|\nabla w(t)\|_X^{3/2}$ can be upper bounded as follows, thanks to the application of a Young inequality two times:

$$cK_f K_g \|w(t)\|_X^{3/2} \|\nabla w(t)\|_X^{3/2} \leq \|w(t)\|_X \left(\frac{c^2 K_f^2 K_g^2 \delta}{2} \|w(t)\|_X + \frac{1}{2\delta} \|\nabla w(t)\|_X^3\right)$$

If we choose $\delta = \dfrac{\|w(t)\|_X \|\nabla w(t)\|_X}{\varepsilon_3}$, and $\eta = \dfrac{\|w(t)\|_X^2}{\varepsilon_1}$, then we obtain the following:

$$\begin{aligned} cK_fK_g \|w(t)\|_X^{3/2} \|\nabla w(t)\|_X^{3/2} &\le \frac{\varepsilon_3}{2}\|\nabla w(t)\|_X^2 + \frac{c^2K_f^2K_g^2\delta}{2}\|w(t)\|_X^2 \\ &\le \frac{\varepsilon_3}{2}\|\nabla w(t)\|_X^2 + \frac{\delta^2}{8\eta} + \frac{\eta c^4K_f^4K_g^4}{2}\|w(t)\|_X^4 \\ &\le \frac{\varepsilon_3}{2}\|\nabla w(t)\|_X^2 + \frac{\varepsilon_1}{8(\varepsilon_3)^2}\|\nabla w(t)\|_X^2 \\ &\quad + \frac{c^4K_f^4K_g^4}{2\varepsilon_1}\|w(t)\|_X^6 \end{aligned}$$

Therefore,

$$\begin{aligned} \frac{1}{2}\frac{d}{dt}\|w(t)\|_X^2 \le& \frac{c^4K_f^4K_g^4}{2\varepsilon_1}\|w(t)\|_X^6 \\ &+ \left(\frac{C^2\left\|f(u_{\lambda_0})\right\|_V^2 K_g^2 + C^2\left\|g(\nabla u_{\lambda_0})\right\|_X^2 K_f^2}{2\beta} + K_s\right)\|w(t)\|_X^2 \\ &+ \left(\beta + \frac{\varepsilon_2}{2}\|\lambda-\lambda_0\|^\alpha + \frac{\varepsilon_3}{2} + \frac{\varepsilon_1}{8(\varepsilon_3)^2} - C_p^\lambda\right)\|\nabla w(t)\|_X^2 \\ &+ \frac{1}{2\varepsilon_2}\|\lambda-\lambda_0\|^\alpha K_a^2\left\|\nabla u_{\lambda_0}(t)\right\|_X^2 \end{aligned}$$

If we choose β, ε_1, ε_2 and ε_3 such that the term multiplying $\|\nabla w(t)\|_X^2$ is negative, then by analogy to lemma 1 and by taking:

$$a = 2\left(\frac{C^2\left\|f(u_{\lambda_0})\right\|_V^2 K_g^2 + C^2\left\|g(\nabla u_{\lambda_0})\right\|_X^2 K_f^2}{2\beta} + K_s\right).$$

And,

$$b = \frac{c^4K_f^4K_g^4}{\varepsilon_1},$$

we conclude to our result. □

Remark 1. ***On the choice of β, ε_1, ε_2 and ε_3***

Let's suppose for instance, that $p = 1$, $C_p^\lambda = \lambda C_p$ and $\alpha = 1$.

A necessary condition to have $\beta < \lambda C_p - \frac{\varepsilon_2}{2}|\lambda-\lambda_0| - \frac{\varepsilon_3}{2} - \frac{\varepsilon_1}{8(\varepsilon_3)^2}$, is :

$$\lambda C_p - \frac{\varepsilon_2}{2}|\lambda-\lambda_0| - \frac{\varepsilon_3}{2} - \frac{\varepsilon_1}{8(\varepsilon_3)^2} > 0. \tag{8}$$

- *If $\lambda > \lambda_0$, then equation (8) is equivalent to: $\left(C_p - \frac{\varepsilon_2}{2}\right)\lambda + \frac{\varepsilon_2}{2}\lambda_0 - \frac{\varepsilon_3}{2} - \frac{\varepsilon_1}{8(\varepsilon_3)^2} > 0.$*

  Then, we choose ε_1, ε_2 and ε_3 such that: $C_p > \frac{\varepsilon_2}{2}$, and $\frac{\varepsilon_2}{2}\lambda_0 - \frac{\varepsilon_3}{2} - \frac{\varepsilon_1}{8(\varepsilon_3)^2} > 0.$

- *If $\lambda < \lambda_0$, then equation (8) is equivalent to:* $\left(C_p + \frac{\varepsilon_2}{2}\right)\lambda - \frac{\varepsilon_2}{2}\lambda_0 - \frac{\varepsilon_3}{2} - \frac{\varepsilon_1}{8(\varepsilon_3)^2} > 0.$

  Then, $\lambda > \dfrac{\frac{\varepsilon_2}{2}\lambda_0 + \frac{\varepsilon_3}{2} + \frac{\varepsilon_1}{8(\varepsilon_3)^2}}{C_p + \frac{\varepsilon_2}{2}}.$

Choice of $\left(f_2^{\lambda_0}(N)\right)_{N\geq 1}$ and an a priori estimate of its terms

We prove the following proposition, that will be a key point in order to give an a priori estimate of the terms of the sequence $\left(f_2^{\lambda_0}(N)\right)_{N=1,\ldots,M}$.

For instance, we suppose that $d = 2$ and for convenience we take $\Omega = (0,1) \times (0,1)$, without any loss of generality:

Proposition 2. *Let $(\Phi_n)_{n\geq 1}$ be an orthonormal basis of $(V, \|.\|_X)$. We denote $\Phi_n = \left(\Phi_n^1, \Phi_n^2\right)^T$. We define $f_n = \left(f_n^1, f_n^2\right)^T$:*

$$f_n^1(x_1, y_1) = \int_0^{y_1}\int_0^{x_1} \Phi_n^1(x,y)dxdy \quad \text{and } f_n^2(x_1, y_1) = \int_0^{y_1}\int_0^{x_1} \Phi_n^2(x,y)dxdy.$$

Then, $\sum_{n=1}^{\infty} \|f_n\|_X^2 = \frac{1}{2}.$

Proof. For $i = 1, 2$:

$f_n^i(x,y) = \left(\Phi_n^i, \mathbf{1}_{[0,x]\times[0,y]}\right)_X$, then $\|f_n^i\|_X^2 = \int_0^1\int_0^1 \left|\left(\Phi_n^i, \mathbf{1}_{[0,x]\times[0,y]}\right)_X\right|^2 dxdy.$

Therefore, $\sum_{n=1}^{\infty} \|f_n^i\|_X^2 = \int_0^1\int_0^1 \|\mathbf{1}_{[0,x]\times[0,y]}\|_X^2 dxdy.$

Which concludes to the result. □

From now on, we denote: $f_2^{\lambda_0}(N) = \sum_{n=N+1}^{\infty} \|f_n^{\lambda_0}\|_X^2$, where $f_n^{\lambda_0}$ is the primitive function of the POD mode $\Phi_n^{\lambda_0}$.

Completion of the proof of result 1

$$\begin{aligned}\|u_\lambda - \hat{u}_{\lambda,\lambda_0}\|_{L^2(0,T;X)}^2 \leq\ & 2\,\|u_{\lambda_0} - \hat{u}_{\lambda_0}\|_{L^2(0,T;X)}^2 \\ & +2\left\|\left(u_\lambda - u_{\lambda_0}\right) - \Pi_{\Phi^{\lambda_0}}^N\left(u_\lambda - u_{\lambda_0}\right)\right\|_{L^2(0,T;X)}^2 \\ & +2\left\|\Pi_{\Phi^{\lambda_0}}^N\left(u_\lambda - u_{\lambda_0}\right) - \left(\hat{u}_{\lambda,\lambda_0} - \hat{u}_{\lambda_0}\right)\right\|_{L^2(0,T;X)}^2\end{aligned}$$

- $\left\|\left(u_\lambda - u_{\lambda_0}\right) - \Pi_{\Phi^{\lambda_0}}^N\left(u_\lambda - u_{\lambda_0}\right)\right\|_{L^2(0,T;X)}^2 = \sum_{n=N+1}^{\infty} \left\|\left(u_\lambda - u_{\lambda_0}, \Phi_n^{\lambda_0}\right)_X\right\|_{L^2(0,T;\mathbb{R})}^2.$
- The Galerkin error $\left\|\Pi_{\Phi^{\lambda_0}}^N\left(u_\lambda - u_{\lambda_0}\right) - \left(\hat{u}_{\lambda,\lambda_0} - \hat{u}_{\lambda_0}\right)\right\|_{L^2(0,T;X)}^2$ is controled by $\sum_{n=N+1}^{\infty} \left\|\left(u_\lambda - u_{\lambda_0}, \Phi_n^{\lambda_0}\right)_X\right\|_{L^2(0,T;\mathbb{R})}^2$; This is shown easily by reducing the equation describing the evolution of $\left(u_\lambda - u_{\lambda_0}\right)(t)$.

Therefore, the parametric squared POD-Galerkin error is essentially controled by the remainder $\sum_{n=N+1}^{\infty} \left\| \left(u_\lambda^h - u_{\lambda_0}^h, \Phi_n^{\lambda_0} \right)_X \right\|_{L^2(0,T;\mathbb{R})}^2$. Then, based on the previous proposition 2, a way to study the decrease rate of this remainder will be by considering the remainder of the primitives sum of the reference POD modes $\Phi_n^{\lambda_0}$: This is shown simply by applying successively the Green formula and the Cauchy-Schwarz inequality to each one of the orthogonal projection coefficients $\left((u_\lambda - u_{\lambda_0})(t), \Phi_n^{\lambda_0} \right)_X$.

Therefore,

$\sum_{n=N+1}^{\infty} \| (u_\lambda - u_{\lambda_0}, \Phi_n^{\lambda_0})_X \|_{L^2(0,T;\mathbb{R})}^2$ is controled by $f_2^{\lambda_0}(N) \, \|\lambda - \lambda_0\|^\alpha$.

This ends the proof of result 1.

Completion of the proof of heuristic result 1

By applying successively the Green formula and the Cauchy-Schwarz inequality and by repeating this step m-times to each $\left((u_\lambda - u_{\lambda_0})(t), \Phi_n^{\lambda_0} \right)_X$, we prove that $\sum_{n=N+1}^{\infty} \| (u_\lambda - u_{\lambda_0}, \Phi_n^{\lambda_0})_X \|_{L^2(0,T;\mathbb{R})}^2$ is finally controled by the remainder of the m-iterated primitives sum of the reference POD modes.

Which concludes to the result.

Completion of the proof of result 2

Thanks to proposition (1), the first restrictive condition of result 2 is verified for a quasi-nonlinear equation of the form (1). Then, we write the following Taylor expansion of u_λ to the order 1:

$u_\lambda = u_{\lambda_0} + \nabla_\lambda u_\lambda(\lambda_0).(\lambda - \lambda_0) + R_1(\lambda)$, where $\|R_1(\lambda)\|_{L^2(0,T;X)}$ is a function of $\|\lambda - \lambda_0\|^2$.

Now, we impose the second restrictive condition of result 2. Then, we can easily show that:

$$\| (\nabla_\lambda u_\lambda(\lambda_0), \Phi_n^{\lambda_0})_X \|_{L^2(0,T;\mathbb{R})}^2 \le 2T\mu_n^{\lambda_0}.$$

In this case, the remainder $\sum_{n=N+1}^{\infty} \| (u_\lambda - u_{\lambda_0}, \Phi_n^{\lambda_0})_X \|_{L^2(0,T;\mathbb{R})}^2$ is better controled by the POD modes:

$$\sum_{n=N+1}^{M} \| (u_\lambda - u_{\lambda_0}, \Phi_n^{\lambda_0})_X \|_{L^2(0,T;\mathbb{R})}^2 \le \|\lambda - \lambda_0\|^2 \sum_{n=N+1}^{\infty} \| (\nabla_\lambda u_\lambda(\lambda_0), \Phi_n^{\lambda_0})_X \|_{L^2(0,T;\mathbb{R})}^2 + \|R_1(\lambda)\|_{L^2(0,T;X)}^2.$$

Therefore, we deduce the a priori estimate (5).

This ends the proof.

Numerical experiments: The 2D Navier-Stokes equations

We place our problem in the particular case of the Navier-Stokes equations for a 2D incompressible fluid flow. The parameter λ denotes here the viscosity of the flow.

Flow configuration

Our study configuration is for an unsteady and incompressible fluid flow in a channel, around a circular cylinder (see Figure 1). The inlet condition is a uniform fluid flow. We impose an outlet condition, and symmetry conditions on Γ_1 and Γ_2.

The two strategies of POD computations

We present a numerical comparaison of our theoretical estimates (3) and (5), for a fixed POD modes numbers N and by varying only the viscosity λ. For a fixed POD modes number N, we plot the logarithm of parametric ROM-POD error $\left\| u_\lambda^h - \hat{u}_{\lambda,\lambda_0} \right\|_{L^2(0,T;X)}$, as a function of $log(|\lambda - \lambda_0|)$.

Indeed, the reference parameter value $\lambda_0 = 0.001$ corresponds to a Reynolds number equal to 100. And, we varied the viscosity λ in the interval $[5.56 \times 10^{-4}, 1.43 \times 10^{-3}]$ in order to have computations associated with Reynolds numbers varying in the interval $[70, 180]$. We consider two strategies of POD computations. Indeed, two different snapshots sets are considered on the time interval $[0, T]$, where $T = 75s$:

- $\mathcal{S}_1 = \{u_{\lambda_0}(t) \;\; t \in [0, T]\}$.
- $\mathcal{S}_2 = \{v_{\lambda_0}(t) \;\; t \in [0, 2T]\}$. Such that:
  - $v_{\lambda_0}(t) = u_{\lambda_0}(t)$ for $t \in [0, T]$.
  - $v_{\lambda_0}(t) = \dfrac{\partial u}{\partial \lambda}(\lambda_0)(2T - t)$ for $t > T$.

More precisely, if we discretize the time interval to $M = 200$ points, then the snapshots sets are given as follows:

- $\mathcal{S}_1 = \{u_{\lambda_0}(t_i) \;\; i = 1, \ldots, M\}$.
- $\mathcal{S}_2 = \{v_{\lambda_0}(t_i) \;\; \text{for} i = 1, \cdots, 2M\}$

Therefore, the POD eigenvectors $\Phi_n^{\lambda_0}$ associated respectively with these snapshots sets are solutions of the following two eigenvalues problems:

The correlation matrix:

$$C_{ij}^1 = \int_\Omega u_{\lambda_0}(t_i, x, y) \cdot u_{\lambda_0}\left(t_j, x, y\right) d\Omega,$$

of which size is $M \times M$.

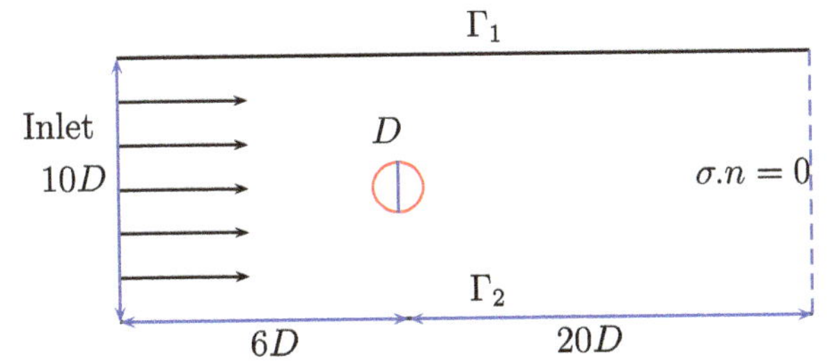

Figure 1 Flow configuration.

We denote by $A_n = (A_{i,n})_{1 \leq i \leq M}$ for $n = 1, \ldots, M$, a set of orthonormal eigenvectors of the matrix C^1. Then, the POD-eigenvectors associated to u_{λ_0}, are given by:

$$\Phi_n^{\lambda_0} = \frac{1}{\sqrt{M}} \sum_{i=1}^{M} A_{i,n} u_{\lambda_0}(t_i).$$

And, the correlation matrix:

$$C_{ij}^2 = \int_\Omega v_{\lambda_0}(t_i, x) \cdot v_{\lambda_0}(t_j, x)\, d\Omega,$$

of which size is $2M \times 2M$.

We denote by $A_n = (A_{i,n})_{1 \leq i \leq 2M}$ for $n = 1, \ldots, 2M$, a set of orthonormal eigenvectors of the matrix C^2. Then, the POD-eigenvectors associated to v_{λ_0}, are given by:

$$\Phi_n^{\lambda_0} = \frac{1}{\sqrt{2M}} \sum_{i=1}^{2M} A_{i,n} v_{\lambda_0}(t_i).$$

We should note that, in the case of $\mathcal{S}_2$, the dimensions of the snapshots are compatible. Indeed, the flow snapshots are normalized by the inlet velocity, and the parametric derivatives snapshots are normalized by their corresponding magnitudes.

Numerical comparaison of the two theoretical estimates (3) and (5)

We compare the two plots of $\log\left(\left\|u_\lambda - \hat{u}_{\lambda,\lambda_0}\right\|_{L^2(0,T;X)}\right)$ as a function of $log(|\lambda - \lambda_0|)$, obtained respectively from the model reduction by the reference POD basis and the enriched one. We get the two plots on (Figure 2).

We retrieve here the power law of the parametric ROM-POD error with respect to $|\lambda - \lambda_0|$, for both strategies of the POD computation.

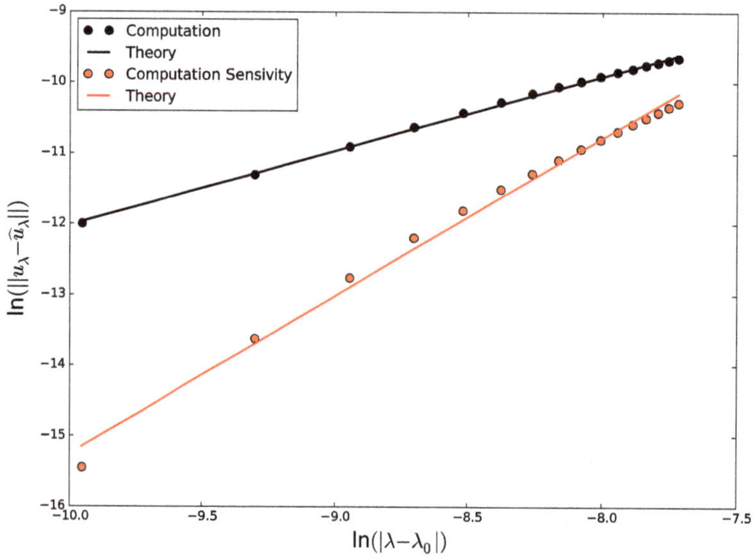

Figure 2 Red line: The logarithm of the error relative to the model reduction by the enriched POD basis. Black line: The logarithm of the error relative to the model reduction by the reference POD basis.

Moreover, (Figure 2) shows that the case for which the parametric sensitivity of the flow is also included in the snapshots set, is better than the one where we had only snapshots associated with a reference solution: The slope of the red line is clearly greater than the black one, and the numerical parametric error in this case is less than the one associated to the reduction by a reference POD basis.

Conclusion and prospects

Conclusion

We were interested in mathematical a priori error bounds of a parametric ROM by a reference POD. A mathematical a priori estimate of the parametric squared L^2-error induced by the corresponding ROM is developped, thanks to the resolution of a Ricatti differential inequality.

This result is an improvement of the one we developed in [38], where we showed an a priori estimate of the parametric squared L^2-error induced after the model reduction of parameterized semi-discrete quasi-linear parabolic problems, by a reference POD basis associated with a reference solution. We obtained here sharper a priori estimates. Moreover, when we considered the enriched POD basis, the parametric ROM-POD error was better controled. Which shows an improvement of the ROM-POD confidence interval with respect to parametric variation.

Prospects

Concerning the prospects of this work, we hope to be able to apply this type of results as an a priori convergence technique of enrichement algorithms as the POD-Greedy one.

It is important also to be able to use these a priori techniques for errors estimate as a sharp a posteriori errors indicator.

Furthermore, it is important to apply these results to interpolation techniques (using the Lie group theory) of reduced order models, available each one in a confidence region around a reference parameter.

Competing interests

The authors declare that they have no competing interests.

Authors' contributions

NA have made substantive intellectual contributions to the mathematical conception of the manuscript and drafted the manuscript. AH have been involved in revising the manuscript critically for important intellectual content and have given final approval of the version to be published. EL have made substantial contributions to the acquisition of the numerical data and to the interpretation of the numerical experiments. MJ have participated to the mathematical content of the manuscript. All authors read and approved the final manuscript.

Acknowledgment

The authors are thankful to FEDER (European Regional Development Fund) for providing the necessary financial facilities for the preparation of the paper.

References

1. Lieu T, Farhat C, Lesoinne M (2006) Reduced-order fluid/structure modeling of a complete aircraft configuration. Comput Methods Appl Mech Engrg 195(41-43):5730–5742
2. Balajewicz M, Farhat C (2013) Model order reduction of embedded boundary models. Bulletin of the American Physical Society 58

3. Liberge E, Hamdouni A (2010) Reducer order modeling method via proper orthogonal decomposition (POD) for flow around an oscillating cylinder. J Fluids Struct 26(2):292–311
4. Liberge E, Benaouicha M, Hamdouni A (2007) Proper Orthogonal Decomposition (POD) investigation in fluid stucture interaction. Eur J Comput Mech 16:401–418
5. Liberge E, Benaouicha M, Hamdouni A (2008) Low order dynamical system for fluid rigid body interaction problem using POD method. Int J Multiphys 2:59–81
6. Liberge E, Pomarède M, Hamdouni A (2010) Reduced-order modeling by POD multiphase approach for fluid-structure interaction. Eur J Comput Mech 19:41–52
7. Lassila T, Quarteroni A, Rozza G (2012) A reduced basis model with parametric coupling for fluid-structure interaction problems. SIAM J Sci Comput 34(2):1187–1213
8. Lassila T, Manzoni A, Rozza G (2012) On the approximation of stability factors for general parametrized partial differential equations with a two-level affine decomposition. ESAIM Math Model Numer Anal 46(6):1555–1576
9. Manzoni A, Quarteroni A, Rozza G (2012) Shape optimization for viscous flows by reduced basis methods and free-form deformation. Int J Numer Methods Fluids 70(5):646–670
10. Lassila T, Rozza G (2010) Parametric free-form shape design with pde models and reduced basis method. Comput Methods Appl Mech Eng 199(23–24):1583–1592
11. Ito K, Ravindran SS (1998) A reduced-order method for simulation and control of fluid flows. J Comp Phys 143:403–425
12. Afanasiev K, Hinze M (2001) Adaptive control of a wake flow using proper orthogonal decomposition. Shape optimization optimal design. Lecture Notes Pure Appl Math:317–332
13. Maday Y, Patera AT, Turinici G (2002) Global a priori convergence theory for reduced-basis approximations of single-parameter symmetric coercive elliptic partial differential equations. C R Acad Sci Paris, Ser I 335:289–294
14. Buffa A, Maday Y, Patera AT, Prud'homme C, Turinici G (2012) A priori convergence theory of the greedy algorithm for the parametrized reduced basis. ESAIM-Math Model Numer Anal 46(3):595–603
15. Chen Y, Hesthaven J-S, Maday Y, Rodriguez J, Zhu X (2012) Certified reduced basis method for electromagnetic scattering and radar cross section estimation. Comput Methods Appl Mech Eng 92(108):233–236
16. Machiels L, Maday Y, Patera AT (2001) Output bounds for reduced-order approximations of elliptic partial differential equations. Comput Methods Appl Mech Eng 190:3413–3426
17. Grepl MA, Maday Y, Nguyen N-C, Patera AT (2007) Efficient reduced-basis treatment of nonaffine and nonlinear partial differential equations. ESAIM: Math Model Numer Anal 41(3):575–605
18. Klindworth D, Grepl MA, Vossen G (2012) Certified reduced basis methods for parametrized parabolic partial differential equations with non-affine source terms. Comput Methods Appl Mech Eng 209(212):144–155
19. Veroy K, Prud'homme C, Patera AT (2003) Reduced-basis approximation of the viscous burgers equation: rigorous a posteriori error bounds. C R Acad Sci Paris, Ser I 337:619–624
20. Nguyen N-C, Rozza G, Patera AT (2009) Reduced basis approximation and a posteriori error estimation for the time-dependent viscous Burgers' equation. Calcolo 46:157–185
21. Ladeveze P, Passieux J-C, Néron D (2010) The LATIN multiscale computational method and the proper generalized decomposition. Comput Methods Appl Mech Eng 199(21-22):1287–1296
22. Ladeveze P, Nouy A (2003) On a multiscale computational strategy with time and space homogenization for structural mechanics. Comput Methods Appl Mech Eng 192(28–30):3061–3087
23. Ammar A, Chinesta F, Diez P, Huerta A (2010) An error estimate for seperated representation of highly multidimensional models. Comput Methods Appl Mech Eng 199:1872–1880
24. Ammar A, Mokdad B, Chinesta F, Keunings R (2006) A new family of solvers for some classes of multidimensionnal partial differential equations encountered in kinetic theory modeling of complex fluids. J Non Newtonian Fluid Mech 139:153–176
25. Ammar A, Normandin M, Chinesta F (2010) Solving parametric complex fluids models in rheometric fows. J Non Newtonian Fluid Mech 165(23–24):1588–1601
26. Chinesta F, Ammar A, Cueto E (2010) Recent advances and new challenges in the use of the proper generalized decomposition for solving multidimensional models. Arch Comput Methods Eng 17(4):327–350
27. Chinesta F, Leygue A, Bordeu F, Aguado JV, Cueto E, Gonzalez D, Alfaro I, Ammar A, Huerta A (2013) Pgd-based computational vademecum for efficient design, optimization and control. Arch Comput Methods Eng 20(1):31–59. doi:10.1007/s11831-013-9080-x
28. Gonzalez D, Masson F, Poulhaon F, Leygue A, Cueto E, Chinesta F (2012) Proper generalized decomposition based dynamic data driven inverse identification. Math Comput Simul 82(9):1677–1695. doi:10.1016/j.matcom.2012.04.001
29. Berkooz G, Holmes P, Lumley JL (1993) The proper orthogonal decomposition in the analysis of turbulent flows. Annu Rev Fluid Mech 25(1):539–575
30. Aubry N, Holmes P, Lumley JL, Stone E (1988) The dynamics of coherent structures in the wall region of a turbulent boundary layer. J Fluid Mech 192:115–173
31. Amsallem D, Cortial J, Carlberg K, Farhat C (2009) A method for interpolation on manifolds structural dynamics reduced-order models. Int J Numer Methods Eng 80:1241–1258
32. Amsallem D, Farhat C (2008) An interpolation method for adapting reduced order models and application to aeroelasticity. Am Inst Aeronautic Astronautics 46(7):1803–1813
33. Amsallem D, Cortial J, Farhat C (2009) On-demand CFD-based aeroelastic predictions using a database of reduced-order bases and models In: 47th AIAA Aerospace Sciences Meeting, vol 18.
34. Hinze M, Volkwein S (2008) Error estimates for abstract linear-quadratic optimal control problems using proper orthogonal decomposition. Comput Optim Appl 39:319–345
35. Troltzsch F, Volkwein S (2009) POD a-posteriori error estimates for linear-quadratic optimal control problems. Comput Optim Appl 44:83–115
36. Tonn T, Urban K, Volkwein S (2011) Comparison of the reduced-basis and POD a posteriori error estimators for an elliptic linear-quadratic optimal control problem. Math Comput Model Dyn Syst 17(4):355–369

37. Vossen G, Volkwein S (2012) Model reduction techniques with a posteriori error analysis for linear quadratic optimal control problems. Math Comput Model Dyn Syst 2(3):465–485
38. Akkari N, Hamdouni A, Liberge E, Jazar M (2013) A mathematical and numerical study of the sensitivity of a reduced order model by POD ROM-POD, for a 2D incompressible fluid flow. J Comput Appl Math 270:522–530
39. Akkari N, Hamdouni A, Jazar M Mathematical and numerical results on the parametric sensitivity of a ROM-POD of the burgers equation. Eur J Comput Mech 23(1-2):78–95
40. Evans L (1998) Partial differential equations. Am Math Soc 662
41. Nirenberg L (1959) On elliptic partial differential equations. Ann Scuola Norm Sup Pisa 13(3):115–162

On a recursive formulation for solving inverse form finding problems in isotropic elastoplasticity

Sandrine Germain*, Philipp Landkammer and Paul Steinmann

*Correspondence:
sandrine.germain@fau.de
University of Erlangen-Nuremberg,
Chair of Applied Mechanics,
Egerlandstrasse 5, 91058 Erlangen,
Germany

Abstract

Background: Inverse form finding methods allow conceiving the design of functional components in less time and at lower costs than with direct experiments. The deformed configuration of the functional component, the applied forces and boundary conditions are given and the undeformed configuration of this component is sought.

Methods: In this paper we present a new recursive formulation for solving inverse form finding problems for isotropic elastoplastic materials, based on an inverse mechanical formulation written in the logarithmic strain space. First, the inverse mechanical formulation is applied to the target deformed configuration of the workpiece with the set of internal variables set to zero. Subsequently a direct mechanical formulation is performed on the resulting undeformed configuration, which will capture the path-dependency in elastoplasticity. The so obtained deformed configuration is furthermore compared with the target deformed configuration of the component. If the difference is negligible, the wanted undeformed configuration of the functional component is obtained. Otherwise the computation of the inverse mechanical formulation is started again with the target deformed configuration and the current state of internal variables obtained at the end of the computed direct formulation. This process is continued until convergence is reached.

Results: In our three numerical examples in isotropic elastoplasticity, the convergence was reached after five, six and nine iterations, respectively, when the set of internal variables is initialised to zero at the beginning of the computation. It was also found that when the initial set of internal variables is initialised to zero at the beginning of the computation the convergence was reached after less iterations and less computational time than with other values. Different starting values for the set of internal variables have no influence on the obtained undeformed configuration, if convergence can be achieved.

Conclusions: With the presented recursive formulation we are able to find an appropriate undeformed configuration for isotropic elastoplastic materials, when only the deformed configuration, the applied forces and boundary conditions are given. An initial homogeneous set of internal variables equal to zero should be considered for such problems.

Keywords: Inverse form finding; Elastoplasticity; Large strain

Background

In this work we present a recursive method for the determination of the undeformed configuration of a functional component, when only the deformed configuration of a workpiece, the applied forces and the boundary conditions are previously known. This is commonly known as an inverse form finding problem, which is inverse to the standard direct kinematic analysis in which the undeformed sheet of metal, the applied forces and boundary conditions are known while the deformed state is sought. Inverse form finding methods are useful because they allow to conceive designs at less time and at lower costs compared to an experimental approach.

Govindjee, 1996 and 1998 [1,2] proposed a numerical method for the determination of the undeformed shape of a continuous body, which is based on the work originally presented in [3]. Their work is limited to isotropic compressible neo-Hookean and incompressible materials. In these contributions it was shown that the weak form of the inverse motion problem based on the Cauchy stress is more efficient and straightforward compared to the weak form based on the Eshelby stress. The governing equation underlying the numerical analysis of the inverse form finding problem is therefore the common weak form of the balance of momentum formulated in terms of the Cauchy stress tensor. The unconventional result lies in the fact that all quantities are parameterised in the spatial coordinates. In [4], temperature changes in the undeformed and deformed configurations have been taken in consideration for orthotropic nonlinear elasticity and axisymmetry using a St.Venant type material, i.e., a material characterised by a quadratic free energy density in terms of the Green-Lagrange strain. Koishi, 2001 [5] used the previous method for the purpose of tire design. Yamada, 1998 [6] proposed another approach as in [1] based on an arbitrary Lagrangian–Eulerian kinematic description. The arbitrary Lagrangian-Eulerian description is approximated by a finite element discretisation. In the last decade, [7,8] extended the method proposed in [1] for the case of anisotropic hyperelasticity for a St.Venant type material. This work is extended in [9,10] to inverse analysis of large-displacement beams in the elastic range. Lu, 2007 [11] proposed a computational method of inverse elastostatics for anisotropic hyperelastic solids in the context of fibrous hyperelastic solids and provide explicit stress function for soft tissue models. In [12] an inverse method for thin-wall structures modelled as geometrically exact stress resultant shells is presented. Germain, 2010 and 2013 [13-15] extended the method originally proposed in [1] to anisotropic hyperelasticity that is based on logarithmic strains. This work was further extended to anisotropic elastoplasticity in [15,16]. The authors demonstrated that the inverse mechanical formulation in elastoplasticity can be used only if the set of internal variables at the deformed state is previously given. However, when dealing with metal forming processes, this set of internal variables is not known at the deformed state. To overcome this problem in anisotropic elastoplasticity, [15,17] proposed a numerical method based on shape optimisation in order to solve inverse form finding problems. A gradient-based shape optimisation is used in the sense of an inverse problem via successive iterations of a direct mechanical problem. The objective function is defined by a least-square minimisation of the difference between the target and the current deformed configuration of the workpiece. The design variables are chosen as the node coordinates stemming from the Finite Element (FE) formulation. A drawback of a node-based shape optimisation is the possible occurrence of mesh distortions. Germain, 2013 and 2012 [15,18,19] proposed a recursive algorithm using an update of the reference configuration. This proposal allows to avoid mesh distortions but leads to large computational costs. Germain, 2011 and 2012 [20,21] compared the inverse mechanical and the shape optimisation formulation in terms of computational costs and accuracy of the obtained undeformed functional component. They have shown that both methods lead to the same results, but the shape optimisation formulation has larger computational costs. In a similar way [22] dealt with the optimal design and optimal control of structures undergoing large rotations and large elastic deformations. Ibrahimbegovic, 2003 [23] introduced shape optimization of elastic structural systems undergoing large rotations. Sousa, 2002 [24] proposed an approach to optimal shape design in forging. In their recursive formulation the inverse problem is formulated as an optimisation problem, where the objective function sensitivity is calculated by the accumulated sensitivities of the nodal coordinates throughout the entire simulation of the process. Ponthot, 2006 [25] presented optimisation methodologies for automatic parameter identification and shape/process optimisation in metal forming simulation. In the sensitivity analysis they used a disturbed balanced configuration, which is updated until the residual equilibrium of the disturbed problem ends under a fixed tolerance. Recently, [26] proposed an inverse-motion-based form finding for electroelasticity to improve the design and accuracy in electroelastic applications such as grippers, sensors and seals.

In order to overcome the large computational costs ([20,21]) in shape optimisation and the fact that the set of internal variables is unknown at the deformed state, we propose, in this contribution, a new method for solving inverse form finding problems in isotropic elastoplasticity based on the inverse mechanical formulation originally proposed in [1].

The present work is organised as follows: In order to introduce the utilised notations, the kinematics of geometrically nonlinear continuum mechanics are presented at first. Furthermore a macroscopical phenomenological isotropic elastoplastic model based on the additive decomposition of the total strains in the logarithmic strain space is introduced. A direct and an inverse mechanical formulations for determining the deformed and the undeformed configurations of a workpiece are respectively presented. A recursive formulation for solving the inverse form finding problem in isotropic elastoplasticity is developed using both previously presented formulations. To illustrate the proposed recursive formulation three numerical examples are presented. The influence of the starting values for the set of internal variables at the beginning of the computation is finally discussed.

Methods

Kinematics of geometrically nonlinear continuum mechanics

In this section we introduce the notations, similar to [13], by briefly recalling the basic kinematic quantities of geometrically nonlinear continuum mechanics. Let $\mathcal{B}_0$ denote the material configuration or undeformed shape of a continuum body parameterised by material coordinates $\boldsymbol{X}$ at time $t = 0$ and $\mathcal{B}_t$ the corresponding spatial configuration or deformed shape parameterised by spatial coordinates $\boldsymbol{x}$ at time t, as depicted in Figure 1. The boundary of $\mathcal{B}_0$ and $\mathcal{B}_t$ is assumed to be decomposed into disjoint parts, so that

$$\partial\mathcal{B}_0 = \partial\mathcal{B}_0^{\overline{T}} \cup \partial\mathcal{B}_0^{\overline{\varphi}} \qquad \text{with} \qquad \partial\mathcal{B}_0^{\overline{T}} \cap \partial\mathcal{B}_0^{\overline{\varphi}} = \emptyset \tag{1}$$

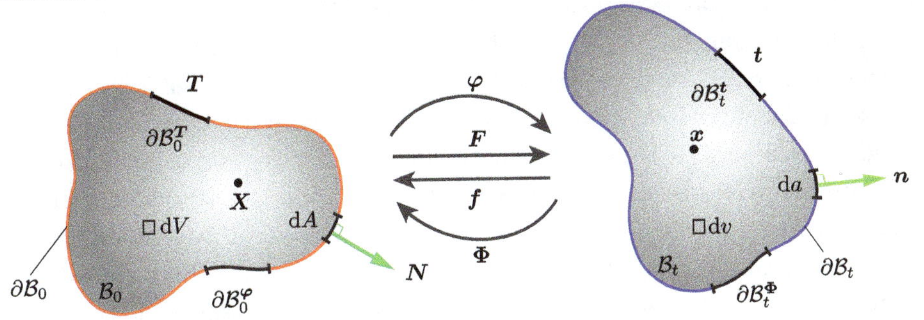

Figure 1 Material (red) and spatial (blue) configurations with boundary conditions.

and

$$\partial\mathcal{B}_t = \partial\mathcal{B}_t^{\bar{t}} \cup \partial\mathcal{B}_t^{\bar{\Phi}} \qquad \text{with} \qquad \partial\mathcal{B}_t^{\bar{t}} \cap \partial\mathcal{B}_t^{\bar{\Phi}} = \emptyset, \tag{2}$$

where $\partial\mathcal{B}_0^{\bar{T}}$ and $\partial\mathcal{B}_t^{\bar{t}}$ are the Neumann type boundary conditions and $\partial\mathcal{B}_0^{\bar{\varphi}}$ and $\partial\mathcal{B}_t^{\bar{\Phi}}$ are the Dirichlet type boundary conditions. In the usual direct mechanical formulation, the material configuration is given and the objective is to determine the direct deformation map $\boldsymbol{\varphi}$ as

$$\boldsymbol{x} = \boldsymbol{\varphi}(\boldsymbol{X}) : \mathcal{B}_0 \longrightarrow \mathcal{B}_t. \tag{3}$$

The corresponding direct deformation gradient together with its Jacobian determinant are defined as

$$\boldsymbol{F} = \nabla_X \boldsymbol{\varphi}, \qquad J = \det \boldsymbol{F} > 0. \tag{4}$$

∇_X denotes the gradient operator with respect to the material coordinates X. On the contrary, in the inverse mechanical formulation, the spatial configuration is given and the inverse deformation map $\boldsymbol{\Phi}$ has to be determined as

$$\boldsymbol{X} = \boldsymbol{\Phi}(\boldsymbol{x}) : \mathcal{B}_t \longrightarrow \mathcal{B}_0. \tag{5}$$

The corresponding inverse deformation gradient together with its Jacobian determinant are given by

$$\boldsymbol{f} = \nabla_x \boldsymbol{\Phi}, \qquad j = \det \boldsymbol{f} > 0. \tag{6}$$

Again, ∇_x denotes the gradient operator with respect to the spatial coordinates $\boldsymbol{x}$. It follows immediately from the above definitions, that the inverse deformation map denotes a nonlinear map inverse to the deformation map of the direct problem

$$\boldsymbol{\Phi} = \boldsymbol{\varphi}^{-1}. \tag{7}$$

Thus, the inverse and direct deformation gradients together with their Jacobian determinants are simply related through an algebraic inversion

$$\boldsymbol{f} = \boldsymbol{F}^{-1} \qquad \text{and} \qquad j = J^{-1}. \tag{8}$$

Nonlinear isotropic elastoplastic material model in the logarithmic strain space

Several discussions occurred in the last decades in the mechanical community on the use or not of the additive decomposition of the total strain proposed by [27] for large strains as in the small strain theory in comparison with the use of the multiplicative decomposition of the deformation gradient proposed by [28], see for example [29] or [30]. In his book Ibrahimbegovic [31] wrote on page 338 "abandon any formulation of large strain plasticity using an additive decomposition of the the total strain". For these reasons when dealing with elastoplasticity at large strains the multiplicative decomposition of the deformation gradient is widely used, see for example [31-33] or [34]. Nevertheless [35] and [36] proposed in their papers an alternative formulation for elastoplasticity at large strains which is a modular macroscopic phenomenological approach formulated in Lagrangian logarithmic strain. They also showed that in metal plasticity (small elastic but large plastic deformations), as in our case, the obtained results with the multiplicative formulation and the additive decomposition in the logarithmic strain space are close to each other. Apel, 2004 [36] and [37] compared as well the performances of the two approaches for sheet drawing processes and concluded that for the range of metal plasticity at moderate elastic strains the results are closed to each other, while the additive formulation provides simpler and more efficient settings. In the subsequent, we present the material model in the logarithmic strain space as proposed in [35] and [36], that we used for a matter of convenience and for a better utilisation of our recursive formulation for solving inverse form finding problems. Three steps are required for the use of the modular approach in large strain: a "geometric preprocessor", the constitutive model and a "geometric postprocessor". In the logarithmic strain space, the total strain is first written as a function of the right Cauchy–Green tensor $\boldsymbol{C}$

$$\boldsymbol{E} = \frac{1}{2}\ln(\boldsymbol{C}) = \frac{1}{2}\ln(\boldsymbol{F}^T \cdot \boldsymbol{F}) = \frac{1}{2}\sum_{i=1}^{3}\ln(\lambda_i)\boldsymbol{M}_i, \tag{9}$$

where $\lambda_{i=1,2,3}$ are the eigenvalues of $\boldsymbol{C}$, i.e. squares of the principal stretches and $\boldsymbol{M}_{i=1,2,3}$ the associated eigenvalue bases (see for example [38]). The total strains are then decomposed into an elastic and a plastic part using an additive Lagrangian formulation

$$\boldsymbol{E} = \boldsymbol{E}^e + \boldsymbol{E}^p. \tag{10}$$

It can be seen that the structure adopted from the geometrically linear theory is conserved. The first and second derivatives of the logarithmic strain with respect to the right Cauchy–Green strain [38] are defined by

$$\mathbb{P} = 2\frac{\partial \boldsymbol{E}}{\partial \boldsymbol{C}} \quad \text{and} \quad \mathbb{L} = 2\frac{\partial \mathbb{P}}{\partial \boldsymbol{C}}. \tag{11}$$

In a second step the total and logarithmic strains enter in the additive format a constitutive model, that defines the stresses and consistent tangents work-conjugate to the logarithmic strain measure. Considering the first and second law of thermodynamics, the reduced Clausius–Duhem inequality is written as

$$\mathcal{D} = \boldsymbol{T} : \dot{\boldsymbol{E}} - \dot{\psi} \geq 0, \tag{12}$$

where $\dot{(-)}$ denotes the material time derivative. The total free energy density ψ is decomposed into an elastic and a plastic part

$$\begin{aligned} \psi(\boldsymbol{E}, \boldsymbol{E}^p, \alpha) &= \psi^e(\boldsymbol{E} - \boldsymbol{E}^p) + \psi^p(\alpha) \\ \Rightarrow \psi(\boldsymbol{E}^e, \alpha) &= \psi^e(\boldsymbol{E}^e) + \psi^p(\alpha), \end{aligned} \tag{13}$$

where $(\boldsymbol{E}^p, \alpha)$ is the set of internal variables, α denotes a scalar variable that models isotropic hardening. The elastic part of the free energy density depends only on the elastic part of the total strains and on the material parameters λ and μ

$$\psi^e(\boldsymbol{E}^e) = \frac{1}{2}\lambda\,\mathrm{tr}(\boldsymbol{E}^e)^2 + \mu\,\mathrm{tr}((\boldsymbol{E}^e)^2). \tag{14}$$

λ and μ are the Lamé parameters and $\mathrm{tr}(\cdot)$ is the trace of the corresponding tensor. The plastic part of the free energy density which models nonlinear isotropic hardening reads

$$\psi^p(\alpha) = \frac{1}{2}\,h\,\alpha^2 + [\sigma_\infty - \sigma_0]\left[\alpha + \frac{e^{-w\alpha}}{w}\right], \tag{15}$$

where h, σ_0, σ_∞ and w are also material parameters, i.e., the isotropic hardening parameter, the initial yield stress, the infinite yield stress and the saturation parameter, which defines the nonlinearity of the hardening, respectively. $\boldsymbol{T}$ in Equation 12 is defined as the Lagrangian stress tensor work-conjugate to the logarithmic strain measure $\boldsymbol{E}$

$$\boldsymbol{T} = \frac{\partial \psi^e}{\partial \boldsymbol{E}^e} = \lambda\,\mathrm{tr}(\boldsymbol{E}^e)\,\boldsymbol{I} + 2\mu\,\boldsymbol{E}^e. \tag{16}$$

With the definition of the free energy density in Equation 13, the Clausius-Duhem inequality can be reduced to

$$\mathcal{D} = \boldsymbol{T} : \dot{\boldsymbol{E}}^p - \frac{\partial \psi}{\partial \alpha}\,\dot{\alpha} \geq 0. \tag{17}$$

The yield surface is defined by

$$\mathcal{Y} = \{\boldsymbol{T} \mid \Phi(\boldsymbol{T}, \frac{\partial \psi}{\partial \alpha}) - \sqrt{\frac{2}{3}}\sigma_0 = 0\}, \tag{18}$$

where Φ is the yield function defined as

$$\Phi(\boldsymbol{T}, \frac{\partial \psi}{\partial \alpha}) = ||\mathrm{dev}(\boldsymbol{T})|| - \sqrt{\frac{2}{3}}\frac{\partial \psi}{\partial \alpha}. \tag{19}$$

The evolution laws for the internal variables, with an associative plasticity model, are determined with the principle of maximum plastic dissipation. The following plastic flow rule and hardening law are characterised by

$$\dot{\boldsymbol{E}}^p = \dot{\gamma}\,\frac{\mathrm{dev}(\boldsymbol{T})}{||\mathrm{dev}(\boldsymbol{T})||} \tag{20}$$

and

$$\dot{\alpha} = \dot{\gamma}\sqrt{\frac{2}{3}}, \tag{21}$$

where $\dot{\gamma}$ is the plastic multiplier determined by the Kuhn–Tucker-type loading-unloading conditions and $\mathrm{dev}(\boldsymbol{T})$ denotes the deviatoric part of the tensor $\boldsymbol{T}$. In the subsequent numerical examples, the isotropic elastoplastic constitutive initial value problem is solved by a return mapping algorithm (or plastic corrector step) following the one presented in [33] for J2 plasticity. In a third step the components of the logarithmic space are mapped back to the nominal stresses and moduli by a "geometric postprocessor". The second Piola–Kirchhoff stress $\boldsymbol{S}$ is then expressed by

$$\boldsymbol{S} = \boldsymbol{T} : \mathbb{P}. \tag{22}$$

The associated elastoplastic modulus $\mathbb{C}^{ep}$ that is defined by setting the rate of the Piola–Kirchhoff tensor as a function of the Lagrangian rate $\dot{\boldsymbol{C}}/2$ is given by

$$\mathbb{C}^{ep} = \mathbb{P}^T : \mathbb{E}^{ep} : \mathbb{P} + \boldsymbol{T} : \mathbb{L}. \tag{23}$$

The transposition symbol refers to an exchange of the first and last pairs of index. $\mathbb{E}^{ep}$ is the fourth-order elastoplastic tangent modulus (see for example [32]).

Direct mechanical problem for determining the deformed shape from equilibrium

Before introducing the inverse boundary value problem in the subsequent section we present briefly the direct mechanical problem, where the undeformed configuration of the workpiece, the applied forces and boundary conditions are given, whereas the deformed configuration of the workpiece is sought. As in [13,15] the direct mechanical problem for determining the deformed shape from equilibrium is defined by

$$\begin{aligned} \mathrm{Div}(\boldsymbol{P}) &= \boldsymbol{0} \quad \text{in} \quad \mathcal{B}_0, \\ \boldsymbol{P}\cdot\boldsymbol{N} &= \overline{\boldsymbol{T}} \quad \text{on} \quad \partial\mathcal{B}_0^{\overline{\boldsymbol{T}}}, \\ \boldsymbol{\varphi} &= \overline{\boldsymbol{\varphi}} \quad \text{on} \quad \partial\mathcal{B}_0^{\overline{\boldsymbol{\varphi}}}, \end{aligned} \tag{24}$$

where $\overline{\boldsymbol{T}}$ is a given traction per unit area in the material configuration (Neumann boundary condition) and $\overline{\boldsymbol{\varphi}}$ is a given boundary deformation (Dirichlet boundary condition), which are illustrated in Figure 1. $\boldsymbol{P}$ is the Piola stress tensor. The weak form of the direct boundary value problem reads

$$G(\boldsymbol{\varphi},\boldsymbol{\eta};\boldsymbol{X}) = \int_{\partial\mathcal{B}_0^{\overline{\boldsymbol{T}}}} \boldsymbol{\eta}\cdot\overline{\boldsymbol{T}}\mathrm{d}A - \int_{\mathcal{B}_0} \mathrm{Grad}\,\boldsymbol{\eta} : \boldsymbol{P}\mathrm{d}V = 0, \tag{25}$$

where $\boldsymbol{\eta}$ is an arbitrary weighting function with the property $\boldsymbol{\eta}\in\mathcal{V}=\{\boldsymbol{\eta}\mid\boldsymbol{\eta}=\boldsymbol{0} \text{ on } \partial\mathcal{B}_0^{\overline{\boldsymbol{\varphi}}}\}$. The determination of the deformed configuration $\mathcal{B}_t$ is performed by the Finite Element Method (FEM). $\mathcal{B}_0$ is discretised into n_{el} elements. The weak form of the direct boundary value problem becomes thereby a nonlinear system of equations, which is solved by the Newton–Raphson method. A linearisation of the weak form is thus performed and gives the needed tangent stiffness matrix of the direct problem. Here we recall the tangent stiffness matrix for the direct mechanical formulation

$$\boldsymbol{k}^{(ij)} := \mathop{\mathbf{A}}_{e=1}^{n_{el}} \int_{\mathcal{B}_0^e} \nabla_X N^{(i)} \overset{2}{\cdot} \frac{\partial \boldsymbol{P}}{\partial \boldsymbol{F}} \cdot \nabla_X N^{(j)} \mathrm{d}V \tag{26}$$

with

$$\frac{\partial \boldsymbol{P}}{\partial \boldsymbol{F}} = [\boldsymbol{F}\overline{\otimes}\boldsymbol{I}] : \mathbb{C}^{ep} : [\boldsymbol{F}^t\overline{\otimes}\boldsymbol{I}] + \boldsymbol{I}\overline{\otimes}\boldsymbol{S}, \tag{27}$$

where (i,j) are the node numbers, $\overset{2}{\cdot}$ denotes the contraction with the second index of the corresponding tangent operator and $\overline{\otimes}$ denotes a non-standard dyadic product with $[\boldsymbol{A}\overline{\otimes}\boldsymbol{B}]_{ijkl} = A_{ik}B_{jl}$. Due to the computation of the direct mechanical formulation the path-dependency, which has to be considered in elastoplasticity, is ensured.

Inverse mechanical problem for determining the undeformed shape from equilibrium

The inverse form finding problem consists in the determination of the undeformed configuration of a functional component, when only the deformed configuration of a workpiece, the applied forces and the boundary conditions are previously known. Written as an optimisation problem (for more details see the subsequent "Form finding optimisation scheme" Section) the goal is to find the undeformed configuration of a workpiece by minimising the difference between the target deformed configuration and the computed deformed configuration with the direct mechanical problem as presented above. The undeformed configuration of the workpiece represents the vector of variables.

The minimisation problem is also subjected to constraints which fulfill the kinematics, the stresses, the boundary value problem, the Karush–Kuhn–Tucker conditions, the consistency condition and the evolution law. The inverse form finding problem can also be formulated from the equilibrium by defining the subsequent inverse boundary value problem, as in [13,15],

$$\begin{aligned} \operatorname{div}(\boldsymbol{\sigma}) &= \mathbf{0} && \text{in} \quad \mathcal{B}_t, \\ \boldsymbol{\sigma} \cdot \boldsymbol{n} &= \bar{\boldsymbol{t}} && \text{on} \quad \partial \mathcal{B}_t^{\bar{t}}, \\ \boldsymbol{\Phi} &= \overline{\boldsymbol{\Phi}} && \text{on} \quad \partial \mathcal{B}_t^{\overline{\Phi}}, \end{aligned} \tag{28}$$

where $\bar{\boldsymbol{t}}$ is a given traction per unit area in the spatial configuration (Neumann boundary condition) and $\overline{\boldsymbol{\Phi}}$ is a given boundary deformation (Dirichlet boundary condition), which are illustrated in Figure 1. The symmetric Cauchy stress $\boldsymbol{\sigma}$ in the inverse boundary value problem is obtained from the Piola–Kirchhoff stress $\boldsymbol{S}$ by a push-forward according to

$$\boldsymbol{\sigma} = j \boldsymbol{F} \cdot \boldsymbol{S} \cdot \boldsymbol{F}^T. \tag{29}$$

The weak form of the inverse boundary value problem reads

$$g(\boldsymbol{\Phi}, \boldsymbol{\eta}; \boldsymbol{x}) = \int_{\partial \mathcal{B}_t^{\bar{t}}} \boldsymbol{\eta} \cdot \bar{\boldsymbol{t}} \mathrm{d}a - \int_{\mathcal{B}_t} \operatorname{grad} \boldsymbol{\eta} : \boldsymbol{\sigma} \mathrm{d}v = 0, \tag{30}$$

where $\boldsymbol{\eta}$ is an arbitrary weighting function with the property $\boldsymbol{\eta} \in \mathcal{V} = \{\boldsymbol{\eta} \mid \boldsymbol{\eta} = \mathbf{0}$ on $\partial \mathcal{B}_t^{\overline{\Phi}}\}$. The particular feature of this formulation is that all integrals extend over the spatial configuration, which is given, and all quantities are parameterised in the given spatial coordinates $\boldsymbol{x}$. The determination of the undeformed configuration $\mathcal{B}_0$ is performed by the Finite Element Method (FEM) as for the direct boundary value problem. $\mathcal{B}_t$ is discretised into n_{el} elements as $\mathcal{B}_0$. The weak form of the inverse boundary value problem becomes also a nonlinear system of equations, which is also solved by the Newton–Raphson method. A linearisation of the weak form is thus performed and gives the needed tangent stiffness matrix of the inverse problem

$$\boldsymbol{K}^{(ij)} := \mathop{\mathbf{A}}_{e=1}^{n_{el}} \int_{\mathcal{B}_t^e} \nabla_{\boldsymbol{x}} N^{(i)} \overset{2}{\cdot} \frac{\partial \boldsymbol{\sigma}}{\partial \boldsymbol{f}} \cdot \nabla_{\boldsymbol{x}} N^{(j)} \mathrm{d}v, \tag{31}$$

with

$$\frac{\partial \boldsymbol{\sigma}}{\partial \boldsymbol{f}} = \boldsymbol{\sigma} \otimes \boldsymbol{F}^t - \boldsymbol{F} \overline{\otimes} \boldsymbol{\sigma} + j \boldsymbol{F} \cdot [\frac{1}{2} \mathbb{C}^{ep} : \frac{\partial \boldsymbol{C}}{\partial \boldsymbol{f}}] \cdot \boldsymbol{F}^t - \boldsymbol{\sigma} \underline{\otimes} \boldsymbol{F} \tag{32}$$

where (i, j) are the node numbers, $\overset{2}{\cdot}$ denotes the contraction with the second index of the corresponding tangent operator and $\underline{\otimes}$ denotes a non-standard dyadic product with $[\boldsymbol{A} \underline{\otimes} \boldsymbol{B}]_{ijkl} = A_{il} B_{jk}$. For more details see [13] or [15]. Furthermore [15,16] demonstrated that this inverse mechanical formulation might be used in elastoplasticity, when the set of internal variables at the deformed state is given and remains constant during the iterations. Since however in sheet metal forming processes the set of internal variables is not known at the deformed state, this formulation can not be used because the undeformed configuration will thus not be unique.

Form finding optimisation scheme

- Find $\boldsymbol{X}$ such that

$$f(\boldsymbol{X}) = \frac{1}{2} ||\boldsymbol{\varphi}(\boldsymbol{X}) - \boldsymbol{x}^{\text{target}}||^2 \rightarrow \min_{\boldsymbol{X}}$$

- Subject to:

1. $\boldsymbol{F} = \nabla_X \varphi, \boldsymbol{C} = \boldsymbol{F}^T \cdot \boldsymbol{F}, \boldsymbol{E} = \frac{1}{2} \ln \boldsymbol{C}, \boldsymbol{E} = \boldsymbol{E}^e + \boldsymbol{E}^p$
2. $\boldsymbol{T} = \mathbb{E} : \boldsymbol{E}^e, \boldsymbol{S} = \boldsymbol{T} : \mathbb{P}, \boldsymbol{P} = \boldsymbol{F} \cdot \boldsymbol{S}$
3. $\mathrm{Div}(\boldsymbol{P}) = \boldsymbol{0}, \boldsymbol{P} \cdot \boldsymbol{N} = \overline{\boldsymbol{T}}, \varphi = \overline{\varphi}$

- and along trajectory $\boldsymbol{x} = \boldsymbol{x}(t)$ with $\boldsymbol{x}(0) = \boldsymbol{X}, \forall t$:

1. $\Phi \leq 0, \quad \dot{\gamma} \geq, \dot{\gamma}\Phi = 0$ with $\Phi = \Phi(\boldsymbol{T}, \alpha)$ and $\dot{\alpha} = \dot{\gamma}\sqrt{\frac{2}{3}}$
2. $\dot{\gamma}\dot{\Phi} = 0$
3. $\dot{\boldsymbol{E}}^p = \dot{\gamma}\frac{\partial \Phi}{\partial \boldsymbol{T}}$

Recursive formulation for solving inverse form finding problems

In order to avoid this problem, we developed a recursive algorithm in which we used both direct and inverse mechanical formulations. At the beginning, the set of internal variables is initialised to a homogeneous field equal to zero, i.e., $(\boldsymbol{E}^p, \alpha) = (0, 0)$. The inverse mechanical formulation in elastoplasticity, as presented above, is performed with the target deformed configuration of the functional component $\boldsymbol{x}^{\text{target}}$ as well as the applied forces and boundary conditions. The total force is applied in one load step [16]. Thus, an undeformed configuration $\boldsymbol{X}^{\text{current}}$ is obtained. This undeformed configuration is then used as starting value in the direct mechanical formulation, as presented above. Total force is decomposed into several load steps in order to capture the path-dependency. The obtained internal variables are used as starting value in the next iteration and so on and so forth. Since the direct mechanical formulation in elastoplasticity gives an unique deformed configuration $\boldsymbol{x}^{\text{current}}$ the corresponding heterogeneous set of internal variables $(\boldsymbol{E}^p_{\text{current}}, \alpha_{\text{current}})$ is thus unique. The obtained deformed configuration $\boldsymbol{x}^{\text{current}}$ is then compared with the target deformed configuration $\boldsymbol{x}^{\text{target}}$ by calculating

$$\Delta = ||\boldsymbol{x}^{\text{target}} - \boldsymbol{x}^{\text{current}}(\boldsymbol{X}^{\text{current}})||^2. \tag{33}$$

If $\Delta < \varepsilon$ is verified with $\varepsilon = 10^{-8}$, for example, the target undeformed configuration of the functional component $\boldsymbol{X}^{\text{target}}$ is obtained. If the convergence tolerance yet is not reached, the target deformed configuration $\boldsymbol{x}^{\text{target}}$, the applied forces, the boundary conditions and now also the heterogeneous field of the internal variables, obtained from the direct mechanical problem, i.e., $(\boldsymbol{E}^p, \alpha) = (\boldsymbol{E}^p_{\text{current}}, \alpha_{\text{current}})$, are used as starting values in the next elastoplastic inverse mechanical formulation. This recursive procedure is continued until convergence is reached. Figure 2 resumes schematically the recursive formulation. Note that Equation 33 does not differ from the objective function used in [15,17] for solving inverse form finding problems based on shape optimisation.

Remark:

- If the set of internal variables is again set to a homogeneous field equal to zero and not updated to the heterogeneous field $(\boldsymbol{E}^p_{\text{current}}, \alpha_{\text{current}})$ in the inverse computation the wanted undeformed configuration will not be reached.

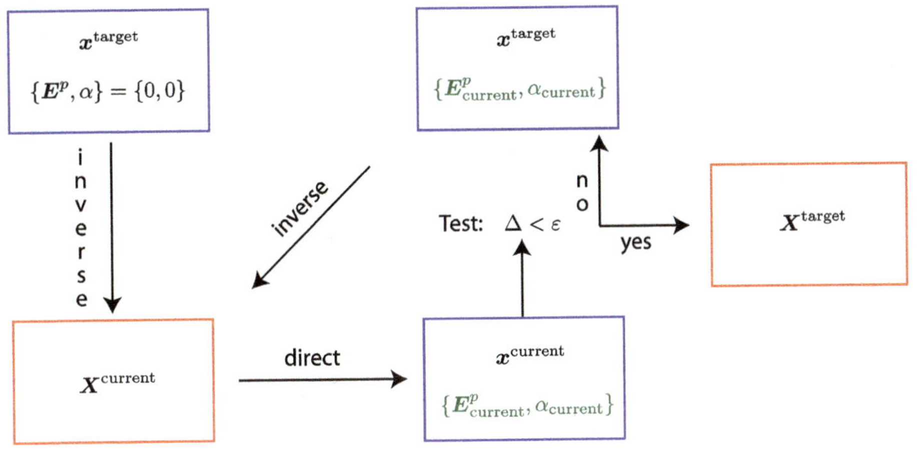

Figure 2 Schematic view of the recursive formulation.

Experiments and results

In this section, the previously presented method for solving inverse form finding problems in isotropic elastoplasticity is evaluated by three numerical examples. Since elastoplasticity is a path-dependent problem, the applied forces in the direct computation are decomposed in several load steps. After each load step the heterogeneous set of internal variables obtained at the equilibrium is used for the initialisation of the Newton–Raphson method in order to reach the equilibrium at the next load step. The inverse computation is performed with only one load step because the plastic strains are given and frozen, so that the problem remains elastic (Equation 10). The obtained final undeformed configurations are plotted. The undeformed configurations are subsequently taken as an input for the direct mechanical formulation. The evolution of the obtained deformed configuration after the last iteration, on which the equivalent plastic strain is plotted, is shown. The equivalent plastic strain is obtained according to:

$$E^p_{eq} = \sqrt{\frac{2}{3}[\,(E^p_{11})^2 + (E^p_{22})^2 + (E^p_{33})^2 + 2(E^p_{12})^2 + 2(E^p_{23})^2 + 2(E^p_{31})^2]}. \tag{34}$$

The internal variables are initialised to zero at the beginning of the computation. The convergence tolerance was set to $\varepsilon = 10^{-8}$. Each numerical example was computed on an Intel Core2 Duo (2533 MHz).

Numerical example 1: bar

The first example deals with a benchmark problem consisting of the traction of a bar. When using the newly presented method for solving inverse form finding problems in elastoplasticity, the straight form of the bar is considered as the deformed configuration. The deformed configuration of the bar is illustrated in Figure 3. The forces are applied on the top of the bar in vertical direction (red arrows). The bottom of the bar is fixed in the three directions (blue squares). The bar has a 10 mm square base and is 20 mm high. The discretisation is obtained with MSC.Patran2010.2, where hexahedral elements are used. The number of nodes is equal to 45 and the number of elements is 16. The applied force is set to 45000 units of force and decomposed in 20 load steps in the direct computation. The isotropic elastoplastic material parameters used in the simulation are summarised in

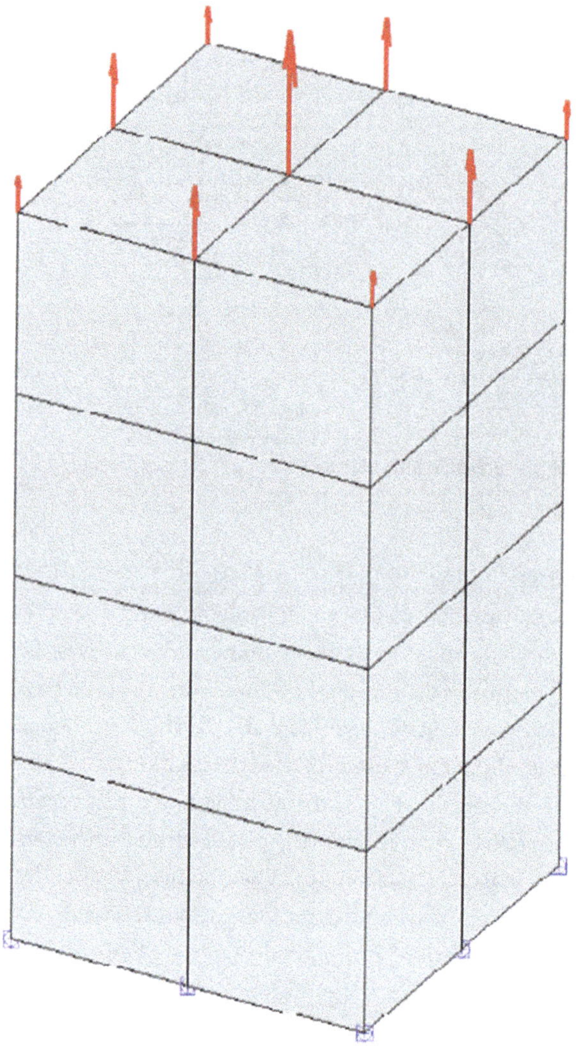

Figure 3 Bar: deformed configuration with applied forces (red) and boundary conditions (blue).

Table 1. The computation of the recursive method took 6 minutes 3 seconds. Six iterations were needed to reach ε. The Δ values introduced in Equation 33 computed after each iteration are shown in Table 2 and plotted on Figure 4 (blue curve). It can be observed that the rate of convergence is almost linear. The finally obtained undeformed configuration of the bar (after iteration six) is illustrated in Figure 5(A) with loads and boundary conditions. The direct mechanical computation with this undeformed configuration as an input is shown in Figure 5(B) with the equivalent plastic strain (-).

Table 1 Material parameters

Material parameters	
E	211000 MPa
ν	0.3
h	100 MPa
σ_0	415 MPa
σ_∞	750 MPa
w	15

Table 2 Bar: Convergence

	Bar: Calculation of Δ		
Iteration	$\{E^p, \alpha\} = \{0, 0\}$	$\{E^p, \alpha\} = \{0.1, 0\}$	$\{E^p, \alpha\} = \{0, 0.4\}$
1	$8.8532\ 10^{-3}$	53.7853	37.5161
2	$7.1592\ 10^{-5}$	2.3427	163.3108
3	$4.7781\ 10^{-6}$	6.3799	$9.0531\ 10^{-2}$
4	$3.303\ 10^{-7}$	10.0643	$1.4682\ 10^{-2}$
5	$1.4706\ 10^{-8}$	49.9705	$1.4356\ 10^{-5}$
6	$4.7759\ 10^{-10}$	9.6875	$1.2357\ 10^{-5}$
7		29.0923	$1.6002\ 10^{-8}$
8		7.8441	$1.2104\ 10^{-8}$
9		20.5272	$1.2903\ 10^{-11}$

Numerical example 2: Cook's membrane

The second example deals with the Cook's membrane in 3D. When using the newly presented method for solving inverse form finding problems in elastoplasticity, the straight form of the Cook's membrane is considered as the deformed configuration. The deformed configuration of the membrane is illustrated in Figure 6. The forces are applied on the right hand side acting from the bottom to the top (red arrows). The left hand side of the membrane is fixed in the three directions (blue squares). The dimensions of the membrane are the same as in [19]. The membrane is discretised by hexahedral elements with MSC.Patran2010.2. The number of nodes is equal to 780 and the number of elements is 528. The applied force is set to 1000 units of force and decomposed in 20 load steps in the direct computation. The isotropic elastoplastic material parameters used in the simulation are summarised in Table 1. The computation of the recursive method took 5 hours 58 minutes 53 seconds. Nine iterations were needed to reach ε. The Δ values introduced in Equation 33 computed after each iteration are summarised in Table 3 and plotted on Figure 7 (blue curve). It can be observed that the rate of convergence is almost linear. Compared to the algorithm presented in [19] for solving inverse form finding problems in elastoplasticity the recursive formulation proposed in this contribution is more efficient regarding the computational costs. The finally obtained undeformed configuration of the Cook's membrane (after iteration nine) is illustrated in Figure 8(A) with loads and boundary conditions. The direct mechanical computation with this undeformed configuration as an input is shown in Figure 8(B) with the equivalent plastic strain (-).

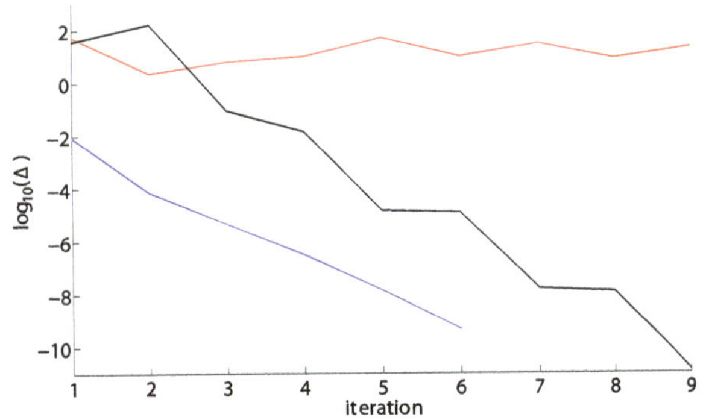

Figure 4 Convergence of the bar when $\{E^p, \alpha\} = \{0, 0\}$ (blue), $\{E^p, \alpha\} = \{0.1, 0\}$ (red) and $\{E^p, \alpha\} = \{0, 0.4\}$ (black).

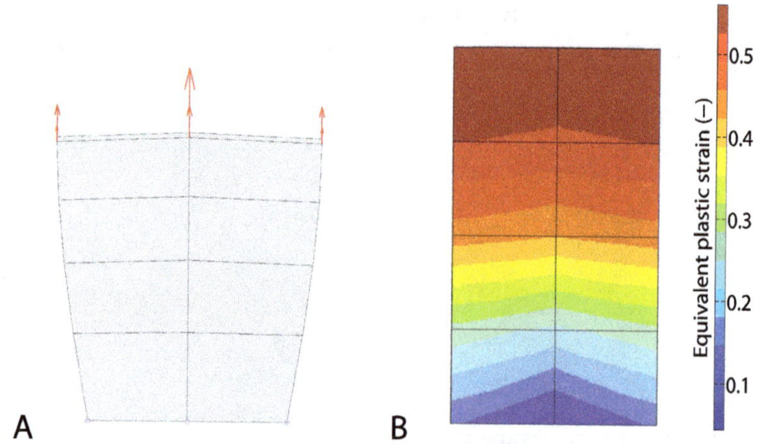

Figure 5 Bar obtained after iteration six: (A) undeformed configuration with applied forces (red) and boundary conditions (blue), (B) deformed configuration obtained with equivalent plastic strain (-).

Numerical example 3: circular, flat plate

The third example deals with a circular, flat plate in 3D. When using the newly presented method for solving inverse form finding problems in elastoplasticity, the circular, flat form of the plate is considered as the deformed configuration. The deformed configuration of the circular, flat plate is illustrated in Figure 9. The applied forces are plotted in red. The inner hole is fixed in three directions (blue squares). The outer hole is set to 95 mm, whereas the inner hole is equal to 75 mm. The circular, flat plate has a thickness of 3 mm.

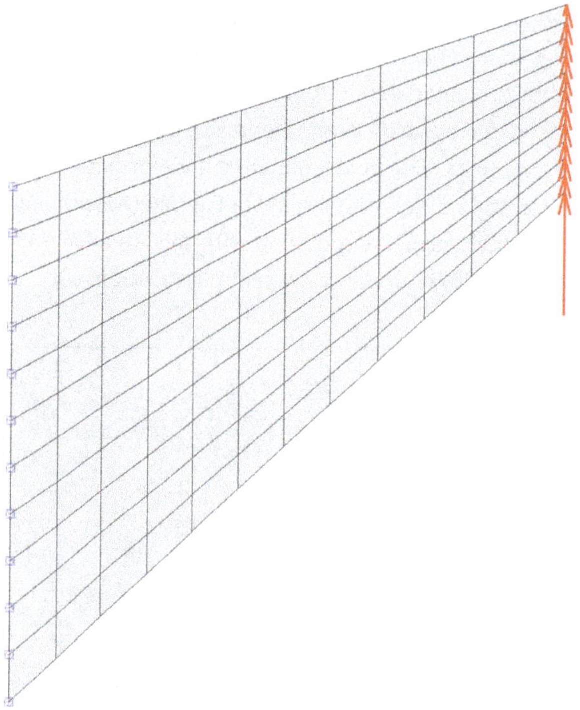

Figure 6 Cook's membrane: deformed configuration with applied forces (red) and boundary conditions (blue).

Table 3 Cook's membrane: Convergence

	Cook's membrane: Calculation of Δ		
Iteration	$\{E^p,\alpha\} = \{0,0\}$	$\{E^p,\alpha\} = \{0,0.1\}$	$\{E^p,\alpha\} = \{0.02,0.02\}$
1	84.5661	3124.0123	2811.6881
2	$5.8472\ 10^{-2}$	9.5801	257.2319
3	$2.6557\ 10^{-2}$	2.8556	49.982
4	$2.9812\ 10^{-3}$	$4.5033\ 10^{-1}$	61.4643
5	$2.4846\ 10^{-4}$	$3.6374\ 10^{-2}$	102.2576
6	$1.6952\ 10^{-5}$	$2.3578\ 10^{-3}$	229.8063
7	$9.859\ 10^{-7}$	$1.4145\ 10^{-4}$	477.2713
8	$5.7253\ 10^{-8}$	$7.924\ 10^{-6}$	1092.9303
9	$3.1401\ 10^{-9}$	$4.4152\ 10^{-7}$	2632.0609
10		$2.4095\ 10^{-8}$	
11		$1.3194\ 10^{-9}$	

The plate is discretised by hexahedral elements with MSC.Patran2010.2. The number of nodes is equal to 1440 and the number of elements is 600. The applied force is set to 3500 units of force and decomposed in 20 load steps in the direct computation. The isotropic elastoplastic material parameters used in the simulation are summarised in Table 1. The computation of the recursive method took 17 minutes 30 seconds. Five iterations were needed to reach ε. The Δ values introduced in Equation 33 computed after each iteration are summarised in Table 4 and plotted in Figure 10 (blue curve). It can be observed that the rate of convergence is almost linear. The finally obtained undeformed configuration of the plate is illustrated in Figure 11(A) with loads and boundary conditions is plotted. A zoom of the top of plate is plotted in Figure 11(B) to better see the deformation. The direct mechanical computation with this final undeformed configuration as an input is shown in Figure 11(C) with the equivalent plastic strain (-).

Discussion

In this section the influence in the choice of the starting value (initialisation of the recursive formulation) for the set of internal variables on the bar, the Cook's membrane and the circular plate is discussed. We used for a more convenient implementation constant single numerical values for the starting set of internal variables instead of a non homogeneous field.

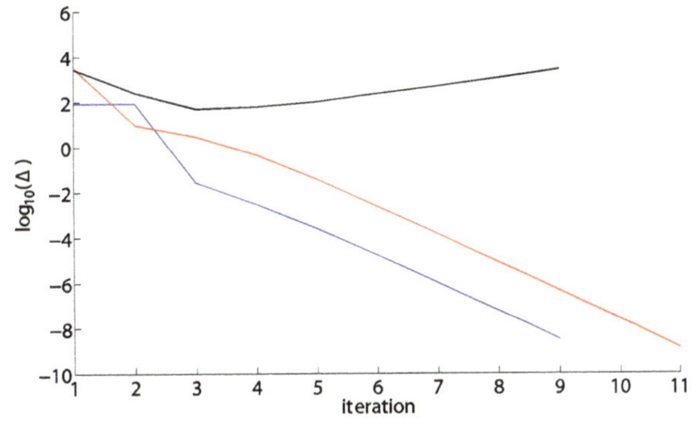

Figure 7 Convergence of the Cook's membrane when $\{E^p,\alpha\} = \{0,0\}$ (blue), $\{E^p,\alpha\} = \{0,0.1\}$ (red) and $\{E^p,\alpha\} = \{0.02,0.02\}$ (black).

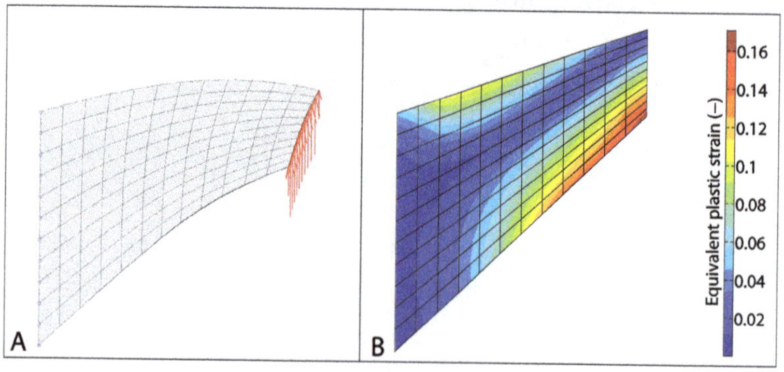

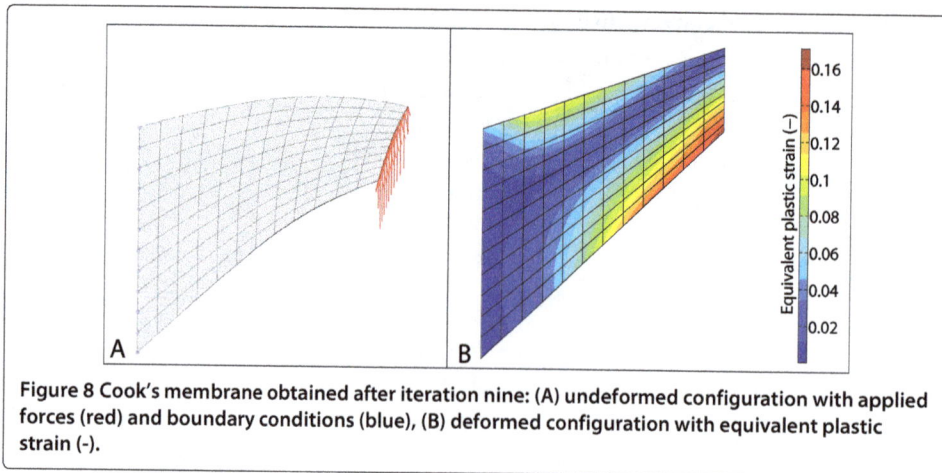

Figure 8 Cook's membrane obtained after iteration nine: (A) undeformed configuration with applied forces (red) and boundary conditions (blue), (B) deformed configuration with equivalent plastic strain (-).

Influence of the starting values of the internal variables on the bar

The presented method was computed as in the previous section but this time the internal variables were first initialised to $\{E^p, \alpha\} = \{0, 0.4\}$ and subsequently to $\{E^p, \alpha\} = \{0.1, 0\}$. It was found that the computation took 10 minutes 40 seconds when $\{E^p, \alpha\} = \{0, 0.4\}$. The convergence tolerance ε was obtained after nine iterations (Table 2). The values Δ after each iterations are plotted in Figure 4 (black curve).

For this case the computation took three additional iterations and additional 4 minutes 37 seconds to converge, than when the starting set of internal variables was initialised to zero. Furthermore, if the undeformed position of the nodes obtained with both starting values of the internal variables is compared, it was found that

$$\Delta = ||\boldsymbol{X}_{\{E^p,\alpha\}=\{0,0\}} - \boldsymbol{X}_{\{E^p,\alpha\}=\{0,0.4\}}||^2 = 2.4810^{-10}, \tag{35}$$

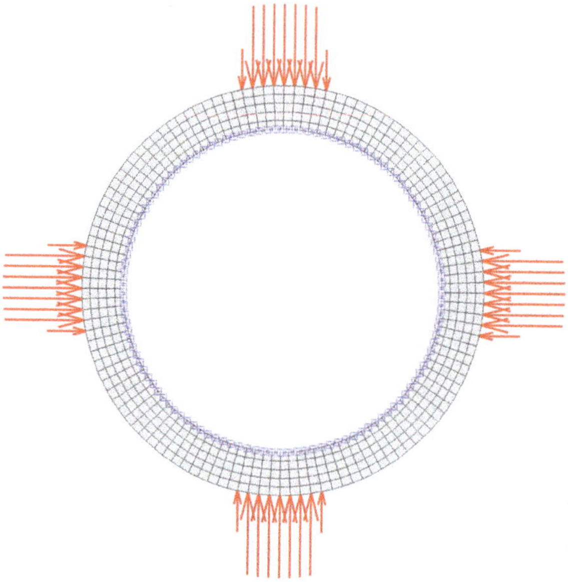

Figure 9 Circular, flat plate: deformed configuration with applied forces (red) and boundary conditions (blue).

Table 4 Circular plate: Convergence

	Circular plate: Calculation of Δ		
Iteration	$\{E^p, \alpha\} = \{0, 0\}$	$\{E^p, \alpha\} = \{0, 0.1\}$	$\{E^p, \alpha\} = \{-0.1, 0\}$
1	$1.897\ 10^{-2}$	13.3091	88.913
2	$2.0153\ 10^{-4}$	$5.9613\ 10^{-3}$	11.4455
3	$2.1253\ 10^{-6}$	$1.8077\ 10^{-4}$	22.8091
4	$4.1797\ 10^{-8}$	$1.3011\ 10^{-6}$	50.1605
5	$4.4199\ 10^{-10}$	$4.3398\ 10^{-8}$	129.4614
6		$9.6423\ 10^{-10}$	

i.e., the difference is negligible. In the case where $\{E^p, \alpha\} = \{0.1, 0\}$ the computation was stopped after nine iterations because of the divergence of the Δ values. The Δ values after each iterations are plotted in Figure 4 (red curve). We conclude that the initial values of the internal variables chosen for the computation of the recursive formulation have an influence on the computational costs and on the number of iterations needed to reach the convergence but not on the final result, i.e., on the sought undeformed configuration of the bar, if convergence can be achieved.

Influence of the starting values of the internal variables on the Cook's membrane

The presented method was computed as in the previous section but this time the internal variables were first initialised to $\{E^p, \alpha\} = \{0, 0.1\}$ and subsequently to $\{E^p, \alpha\} = \{0.02, 0.02\}$. It was found that the computation took 5 hours 23 minutes 34 seconds for the first case. The convergence tolerance ε was obtained after 11 iterations (Table 3). The values Δ after each iterations are plotted in Figure 7 (red curve). It can be observed that the rate of convergence is almost linear.

For this case the computation took two additional iterations and additional 35 minutes 19 seconds to converge, than when the starting set of internal variables was initialised to zero. Furthermore, if the undeformed position of the nodes obtained with both starting values of the internal variables is compared, it was found that

$$\Delta = ||X_{\{E^p,\alpha\}=\{0,0\}} - X_{\{E^p,\alpha\}=\{0,0.1\}}||^2 = 1.8610^{-7}, \tag{36}$$

i.e., the difference is negligible. In the case where $\{E^p, \alpha\} = \{0.02, 0.02\}$ the computation was stopped after nine iterations because of the divergence of the Δ values. The Δ values

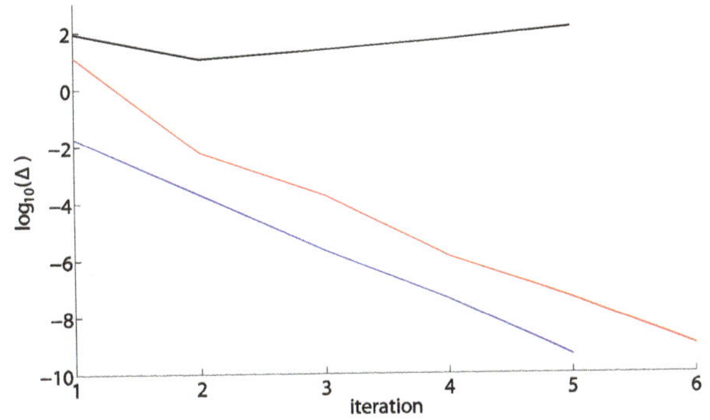

Figure 10 Convergence of the plate when $\{E^p, \alpha\} = \{0, 0\}$ (blue), $\{E^p, \alpha\} = \{0, 0.1\}$ (red) and $\{E^p, \alpha\} = \{-0.1, 0\}$ (black).

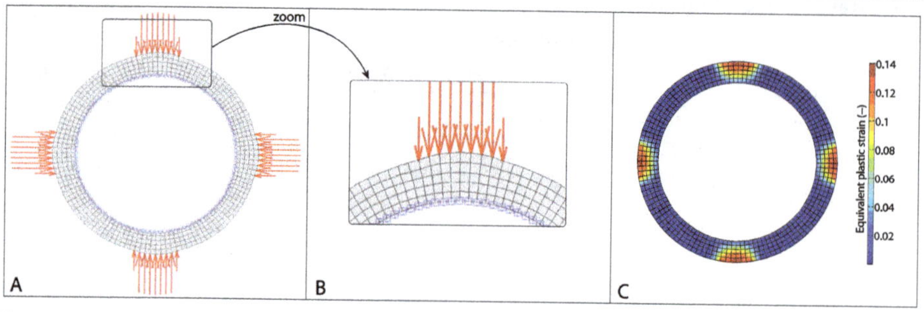

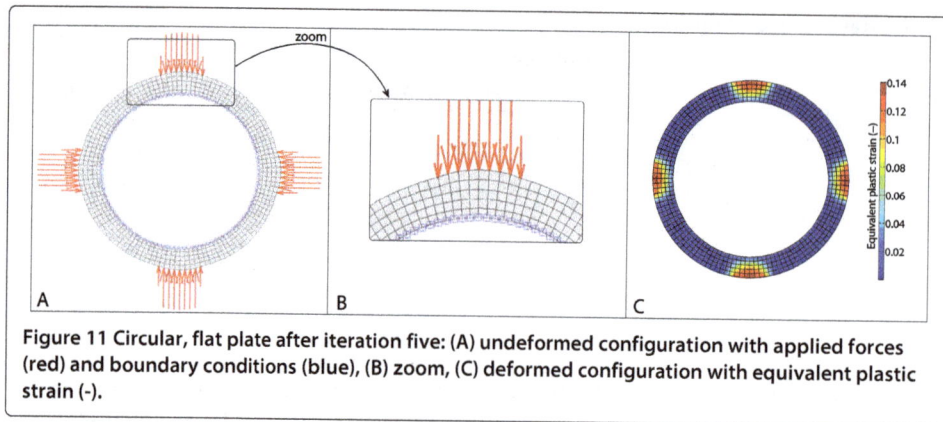

Figure 11 Circular, flat plate after iteration five: (A) undeformed configuration with applied forces (red) and boundary conditions (blue), (B) zoom, (C) deformed configuration with equivalent plastic strain (-).

after each iterations are plotted in Figure 7 (black curve). We conclude that the initial values of the internal variables chosen for the computation of the recursive formulation have an influence on the computational costs and on the number of iterations needed to reach the convergence but not on the final result, i.e., on the sought undeformed configuration of the Cook's membrane, if convergence can be achieved.

Influence of the starting values of the internal variables on the circular, flat plate

The presented method was computed as in the previous section but this time the internal variables were first initialised to $\{E^p, \alpha\} = \{0, 0.1\}$ and subsequently to $\{E^p, \alpha\} = \{-0.1, 0\}$. It was found that the computation took 21 minutes 57 seconds for the first case. The convergence tolerance ε was obtained after six iterations (Table 4). The values Δ after each iterations are plotted in Figure 10 (red curve). It can be observed that the rate of convergence is almost linear.

For this case the number of iterations is equal to the case, where the set of internal variables was initialised to zero, but the computation took 4 minutes 27 seconds longer. Furthermore, if the undeformed position of the nodes obtained with both starting values of the internal variables is compared, it was found that

$$\Delta = ||\boldsymbol{X}_{\{E^p,\alpha\}=\{0,0\}} - \boldsymbol{X}_{\{E^p,\alpha\}=\{0,0.1\}}||^2 = 3.1510^{-10}, \tag{37}$$

i.e., the difference is negligible. In the case where $\{E^p, \alpha\} = \{-0.1, 0\}$ the computation was stopped after five iterations because of the divergence of the Δ values . The Δ values after each iterations are plotted in Figure 10 (black curve). We conclude that the initial values of the internal variables chosen for the first computation of the recursive formulation have an influence on the computational costs, but not on the final result, i.e., on the sought undeformed configuration of the circular, flat plate, if convergence can be achieved.

Conclusion

In this contribution a new method for solving inverse form finding problems for isotropic elastoplastic materials is presented. To that end, a recursive formulation is deployed to find the desired undeformed configuration of the functional component. The inverse mechanical formulation in elastoplasticity is first performed on the target deformed configuration of the workpiece with the set of internal variables initialised to a homogeneous field equal to zero. Subsequently, a direct mechanical formulation on the computed undeformed configuration is used, which ensures the path-dependency in elastoplasticity. The obtained deformed configuration is furthermore compared with the target deformed configuration of the component. If the difference is negligible, the wanted undeformed configuration of the functional component is obtained. Otherwise the computation of the elastoplastic inverse mechanical formulation is started again with the target deformed configuration and the current heterogeneous state of internal variables obtained at the end of the computed direct formulation. This process is continued until convergence is reached. Three numerical examples, a bar, the Cook's membrane and a circular, flat plate in 3D illustrated this recursive formulation for finding the corresponding undeformed configurations in isotropic elastoplasticity. The convergence was reached after six, nine and five iterations, respectively, when initialising the set of internal variables to zero at the beginning of the computation. The influence of the starting values for the set of internal variables at the beginning of the computation was afterwards discussed. It was found that when the initial set of internal variables was initialised to zero at the beginning of the computation the convergence was reached after less iterations and less computational time than with other values. The rates of convergence were almost linear. The computation of the three numerical examples with the recursive formulation did not converge for one value of the set of internal variables and had to be stopped. Comparing the results of the numerical examples, it was demonstrated that different starting values for the set of internal variables have no influence on the obtained undeformed configuration. We conclude that the choice of the initial set of internal variables has an influence on the convergence evolution but not on the result, if convergence can be achieved. Therefore an initial homogeneous set of internal variables equal to zero, which is a natural choice in programming since the set of internal variables is unknown at the beginning of the computation, should be considered for such problems. An extension of the presented new method for solving inverse form finding problems to anisotropic elastoplasticity will be of great interest for metal forming processes.

Competing interests

The authors declare that they have no competing interests.

Authors' contributions

SG: conception and design of the study, analysis and interpretation of data, drafted the manuscript. PL: revised of the manuscript. PS: study supervision, revision of the manuscript. All authors read and approved the final manuscript.

Acknowledgements

This work is supported by the German Research Foundation (DFG) under the Transregional Collaborative Research Center SFB/TR73: "Manufacturing of Complex Functional Components with Variants by Using a New Sheet Metal Forming Process - Sheet-Bulk Metal Forming".

References

1. Govindjee S, Mihalic P (1996) Computational methods for inverse finite elastostatics. Comput Methods Appl Mech Eng 136(1–2): 47–57
2. Govindjee S, Mihalic P (1998) Computational methods for inverse deformations in quasi-incompressible finite elasticity. Int J Numerical Methods Eng 43(5): 821–838
3. RT Shield R (1967) Inverse deformation results in finite elasticity. Zeitschrift für angewandte Mathematik und Physik 18: 490–500
4. Govindjee S (1999) Finite deformation inverse design modeling with temperature changes, axis-symmetry and anisotropy. In Report Number UCB/SEMM-1999/01, University of California, Berkley
5. Koishi M, Govindjee S (2001) Inverse design methodology of a tire. Tire Sci Technol 29(3): 155–170
6. Yamada T (1998) Finite element procedure of initial shape deformation for hyperelasticity. Struct Eng Mech 6: 173–183
7. Fachinotti V, Cardona A, Jetteur P (2008) Finite element modelling of inverse design problems in large deformations anisotropic hyperelasticity. Int J Numerical Methods Eng 74(6): 894–910

8. Fachinotti V, Cardona A, Jetteur P (2006) A finite element model for inverse design problems in large deformations anisotropic hyperelasticity. In: Cardona A, Nigro N, Sonzogni V, Storti M (eds) Mecánica Computacional, November 2006, Santa Fe, pp 1269–1284
9. Albanesi A, Fachinotti V, Cardona A (2010) Inverse finite element method for large-displacement beams. Int J Numerical Methods Eng 84: 1166–1182
10. Albanesi A, Fachinotti V, Pucheta M (2010) A review on design methods for compliant mechanisms. In: Dvorkin E, Goldschmit M, Storti M (eds) Mecánica Computacional, 15–18 November 2010, Buenos Aires, pp 59–72
11. Lu J, Zhou X, Raghavan M (2007) Computational method of inverse elastostatics for anisotropic hyperelastic solids. Int J Numerical Methods Eng 69: 1239–1261
12. Zhou X, Lu J (2008) Inverse formulation for geometrically exact stress resultant shells. Int J Numerical Methods Eng 74: 1278–1302
13. S, Scherer M, Steinmann P (2010) On inverse form finding for anisotropic hyperelasticity in logarithmic strain space. Int J Struct Changes Solids - Mech Appl 2(2): 1–16
14. Germain S, Scherer M, Steinmann P (2010) On inverse form finding for anisotropic materials. In: Wieners C (ed) Proceedings in Applied Mathematics and Mechanics: 81st Annual Meeting of the International Association of Applied Mathematics and Mechanics (GAMM), 22–26 March 2010, Karlsruhe, pp 159–160
15. Germain S (2013) On inverse form finding for anisotropic materials in the logarithmic strain space. PhD thesis, Chair of Applied Mechanics, University of Erlangen-Nuremberg
16. Germain S, Steinmann P (2011) On inverse form finding for anisotropic elastoplastic materials. AIP Conf Proc 1353: 1169–1174
17. Germain S, Steinmann P (2011) Shape optimization for anisotropic elastoplasticity in logarithmic strain space. In: Oñate E, Owen D, Peric D, Suárez B (eds) Computational Plasticity XI - Fundamentals and Applications: 7-11 September 2011, Barcelona, pp 1479–1490
18. Germain S, Steinmann P (2012) Towards inverse form finding methods for a deep drawing steel DC04. Key Eng Mater 504–506: 619–624
19. Germain S, Steinmann P (2012) On a recursive algorithm for avoiding mesh distortion in inverse form finding. J Serbian Soc Comput Mech 6: 216–234
20. Germain S, Steinmann P (2011) A comparison between inverse form finding and shape optimization methods for anisotropic hyperelasticity in logarithmic strain space. In: Brenn G, Holzapfel GA, Schanz M, Steinbach O (eds) Proceedings in Applied Mathematics and Mechanics: 82nd Annual Meeting of the International Association of Applied Mathematics and Mechanics (GAMM), 18-21 April 2011, Graz, pp 367–368
21. Germain S, Steinmann P (2012) On two different inverse form finding methods for hyperelastic and elastoplastic materials. In: Alber HD, Kraynyukova N, Tropea C (eds) Proceedings in Applied Mathematics and Mechanics: 83rd Annual Meeting of the International Association of Applied Mathematics and Mechanics (GAMM), 26-30 March 2012, Darmstadt, pp 263–264
22. Ibrahimbegovic A, Knopf-Lenoir C, Kucerova A, Villon P (2004) Optimal design and optimal control of elastic structures undergoing finite rotations. Int J Numerical Methods Eng 61(14): 2428–2460
23. Ibrahimbegovic A, Knopf-Lenoir C (2003) Shape optimization of elastic structural systems undergoing large rotations: simultaneous solution procedure. Comput Model Eng Sci 4: 337–344
24. Sousa LC, Castro CF, António CAC, Santos AD (2002) Inverse methods in design of industrial forging processes. Journal 128: 266–273
25. Ponthot JP, Kleinermann JP (2006) A cascade optimization methodology for automatic parameter identification and shape/process optimization in metal forming simulation. Comput Methods Appl Mech Eng 195: 5472–5508
26. Ask A, Denzer R, Menzel A, Ristinmaa M (2013) Inverse-motion-based form finding for quasi-incompressible finite electroelasticity. Int J Numerical Methods Eng 94(6): 554–572
27. Green AE, Naghdi PM (1965) A general theory of an elastic-plastic continuum. Arch Ration Mech Anal 18: 251–281
28. Lee EH (1969) Elastic-plastic deformation at finite strains. J Appl Mech 36: 1–14
29. Naghdi PM (1990) A critical review of the state of finite plasticity. J Appl Math Phys 41: 315–394
30. Casey J, Naghdi PM (1980) A remark on the use of the decomposition F=FeFp in plasticity. J Appl Mech 47: 672–675
31. Ibrahimbegovic A (2009) Nonlinear solid mechanics: Theoretical formulations and finite element solution methods. Springer, Dordrecht Heidelberg London New York
32. de Souza Neto EA, Perić D, Owen DRJ (2008) Computational Methods for Plasticity - Theory and Applications. John Wiley & Sons Ltd, Chichester, UK
33. Simo JC, Hughes TJR (1998) Computational Inelasticity. Springer-Verlag New York, Inc., New York
34. Lubliner J (2008) Plasticity Theory. Dover Publications, Inc., Mineola, NY, USA
35. Miehe C, Apel N, Lambrecht M (2002) Anisotropic additive plasticity in the logarithmic strain space: modular kinematic formulation and implementation based on incremental minimization principles for standard materials. Comput Methods Appl Mech Eng 191(47–48): 5383–5425
36. Apel N (2004) Approaches to the description of anisotropic material behaviour at finite elastic and plastic deformations, theory and numerics. PhD thesis, Stuttgart University. Institute of Applied Mechanics (Chair I)
37. Miehe C, Apel N (2004) Anisotropicelastic-plastic analysis of shells at large strains. A comparison of multiplicative and additive approaches to enhances finite element design and constitutive modelling. Int J Numerical Methods Eng 61: 2067–2113
38. Miehe C, Lambrecht M (2001) Algorithms for computation of stresses and elasticity moduli in terms of Seth–Hill's family of generalized strain tensors. Commun Numerical Methods Eng 17: 337–353

A frontal approach to hex-dominant mesh generation

Tristan Carrier Baudouin[1*], Jean-François Remacle[1], Emilie Marchandise[1], François Henrotte[1] and Christophe Geuzaine[2]

*Correspondence: tristan.carrier@uclouvain.be
[1] Université catholique de Louvain, Institute of Mechanics, Materials and Civil Engineering, Bâtiment Euler, Avenue Georges Lemaître 4, Louvain-la-Neuve 1348, Belgium
Full list of author information is available at the end of the article

Abstract

Background: Indirect quad mesh generation methods rely on an initial triangular mesh. So called triangle-merge techniques are then used to recombine the triangles of the initial mesh into quadrilaterals. This way, high-quality full-quad meshes suitable for finite element calculations can be generated for arbitrary two-dimensional geometries.

Methods: In this paper, a similar indirect approach is applied to the three-dimensional case, i.e., a method to recombine tetrahedra into hexahedra. Contrary to the 2D case, a 100% recombination rate is seldom attained in 3D. Instead, part of the remaining tetrahedra are combined into prisms and pyramids, eventually yielding a mixed mesh. We show that the percentage of recombined hexahedra strongly depends on the location of the vertices in the initial 3D mesh. If the vertices are placed at random, less than 50% of the tetrahedra will be combined into hexahedra. In order to reach larger ratios, the vertices of the initial mesh need to be anticipatively organized into a lattice-like structure. This can be achieved with a frontal algorithm, which is applicable to both the two- and three-dimensional cases. The quality of the vertex alignment inside the volumes relies on the quality of the alignment on the surfaces. Once the vertex placement process is completed, the region is tetrahedralized with a Delaunay kernel. A maximum number of tetrahedra are then merged into hexahedra using the algorithm of Yamakawa-Shimada.

Results: Non-uniform mixed meshes obtained following our approach show a volumic percentage of hexahedra that usually exceeds 80%.

Conclusions: The execution times are reasonable. However, non-conformal quadrilateral faces adjacent to triangular faces are present in the final meshes.

Keywords: Advancing front methods; Tetrahedra recombination; Mixed hexahedral meshes

Background

Whether hex-meshing or tet-meshing is better for finite element computations is a long-standing controversy. This paper does not aim at deciding on that issue. Yet, it is a fact that a large number of finite element users would highly appreciate having automatic hex-meshing procedures for general 3D domains. A number of arguments can indeed be stated in favor of hex-meshing. For the same number of vertices, hex meshes have fewer elements, which speeds up the matrix/residual assembly. In solid mechanics, hexahedra exhibit higher accuracy than tetrahedra [1], which are plagued by locking problems [2].

In fluid dynamics, boundary layers made of hexahedra are effective for capturing large gradients and resolving viscous flows near the boundary, and semi-structured boundary-layer meshes attract significant interest (see, e.g. [3-5]).

Hexahedral mesh generation is still an ongoing research [6], and a major conclusion so far is that the generation of full-hex conforming meshes on arbitrary domains is beyond our reach nowadays. Relaxing a bit the requirements, hex-dominant meshes [7], in contrast with full-hex meshes, are allowed to aggregate a mixture of hexahedra, prisms, pyramids and tetrahedra. The goal of hex-dominant meshing is to generate meshes where hexahedral elements dominate, both in number and volume. This paper presents such an algorithm to automatically generate non-uniform isotropic hex-dominant meshes in arbitrary geometries. However, quadrilateral faces adjacent to triangular faces are usually found in the resulting meshes. Such non-conformities represent an additional complication for finite element methods. Various attempts at palliating the impact of these non-conformities have been discussed in the literature [8].

The proposed approach relies on an indirect strategy. The tetrahedra of an initial mesh are combined into hexahedra using Yamakawa-Shimada's algorithm [9], which works basically as follows: (*i*) All tetrahedra of an initial mesh are considered one after the other. (*ii*) The neighbors of each tetrahedron are visited in order to identify potential hexahedra. Candidate hexahedra are stored in an array and sorted with respect to their geometrical quality. (*iii*) The algorithm then iterates through this array, starting from the highest quality hexahedron, in order to effectively generate the hexahedral elements. Hexahedra that are composed of available tetrahedra (not marked for deletion) and that preserve hexahedral conformity are successively added to the mesh.

However, meshes obtained by applying a recombination algorithm to an arbitrary tetrahedral mesh fail to be hex-dominant. As an illustration, a mesh (depicted on Figure 1) was created using Yamakawa-Shimada's recombination algorithm from a tetrahedral mesh generated with the Delaunay refinement algorithm [10] of Gmsh [11]. The number of hexahedra in this mesh represents only 12.34% of the total number of elements.This low percentage of hexahedra results from the fact that the mesh vertices were not placed so as to favor recombination. In order to obtain higher ratios of good quality hexahedra, vertices need to be anticipatively aligned into a lattice-like structure that respects the user prescribed mesh size and the preferred directions of the mesh. This is what we are going to do with a specific frontal algorithm.

In a perfect hexahedral mesh, each interior vertex is linked with six other vertices: left-right, above-below, front-back (four vertices in case of a perfect quadrangular mesh). The main idea of our vertex placement algorithm is based on that observation. Knowing the prescribed local mesh size and the local preferred mesh directions, each interior vertex attempts to spawn six new vertices. A prospective vertex, however, is effectively created only if it lays inside the domain and if it is not too close to an existing vertex. This algorithm is applied to the boundaries of the geometry, prior to the volumes. When done, the vertices are tetrahedralized and Yamakawa-Shimada's algorithm can be applied.

Our approach has some similarities with the advancing front method. The vertices are created layer by layer toward the center of the geometry. However, contrary to the advancing front method, our algorithm does not construct a mesh topology along the way. All tetrahedra are built at the end. Figure 2 illustrates the various steps of our approach. The basic input is a CAD geometry file readable by the Gmsh free software.

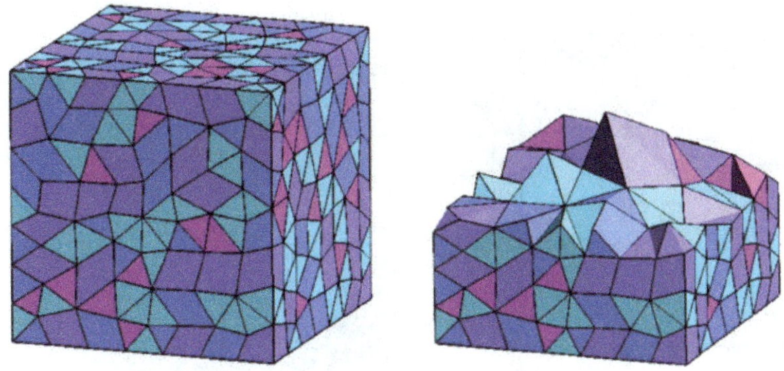

Figure 1 A mixed mesh obtained from an arbitrary tetrahedral mesh. Hexahedra are displayed in purple, prisms in dark blue, pyramids in pink and tetrahedra in light blue. **(2)** is a cutaway view of **(1)**.

The *Methods* section first examines more closely a number of data structures, tools and concepts that contribute to make the algorithm more efficient. It then describes the two- and three-dimensional versions of the frontal algorithm. Finally, it discusses Yamakawa-Shimada's algorithm and the issue of finite element conformity. The *Results and discussion* section presents a number of application examples, including mesh statistics and execution times.

Previous work

A large variety of procedures for hexahedral meshing have been proposed over the years. This section reviews a certain number of them, focusing on the automatic ones.

Several techniques extract hexahedral meshes from octree data structures [12]. These methods can generate non-uniform full-hex meshes in arbitrary geometries. The hexahedra orientation is defined by the octree root, which is a cube englobing the geometry. Boundary hexahedra perpendicular to the boundary surface can be achieved by projecting buffer layers [13,14]. The main limitations of octree-based approaches is that the generated hexahedra cannot be oriented flexibly, and that the quality of the hexahedra near boundaries is degraded [15].

Certain parametrization methods have been able to mesh complex three-dimensional domains [16,17]. These methods start by computing a three-dimensional direction field. The singularities of the direction field are then identified. Later on, the domain is cut in order to place all singularities on the boundary and to reduce the genus to 0 [18]. A parametrization minimizing the difference between the hexahedra orientation and the three-dimensional direction field is selected [19]. The supplementary cutting required to place all singularities on the boundary is necessary because it leads to a better parametrization [16]. Full hexahedral meshes of very good quality can be obtained for complicated domains. However, a variable mesh size cannot be prescribed [20]. Parametrization methods are very diverse. For example, a few algorithms capable of deforming a three-dimensional domain into a polycube model have been developed [21]. The polycube model is meshed and re-deformed back into its original shape.

Graph theory can be applied to the problem of creating hexahedra by recombining tetrahedra [22]. The starting point of the method is a tetrahedral mesh, which can

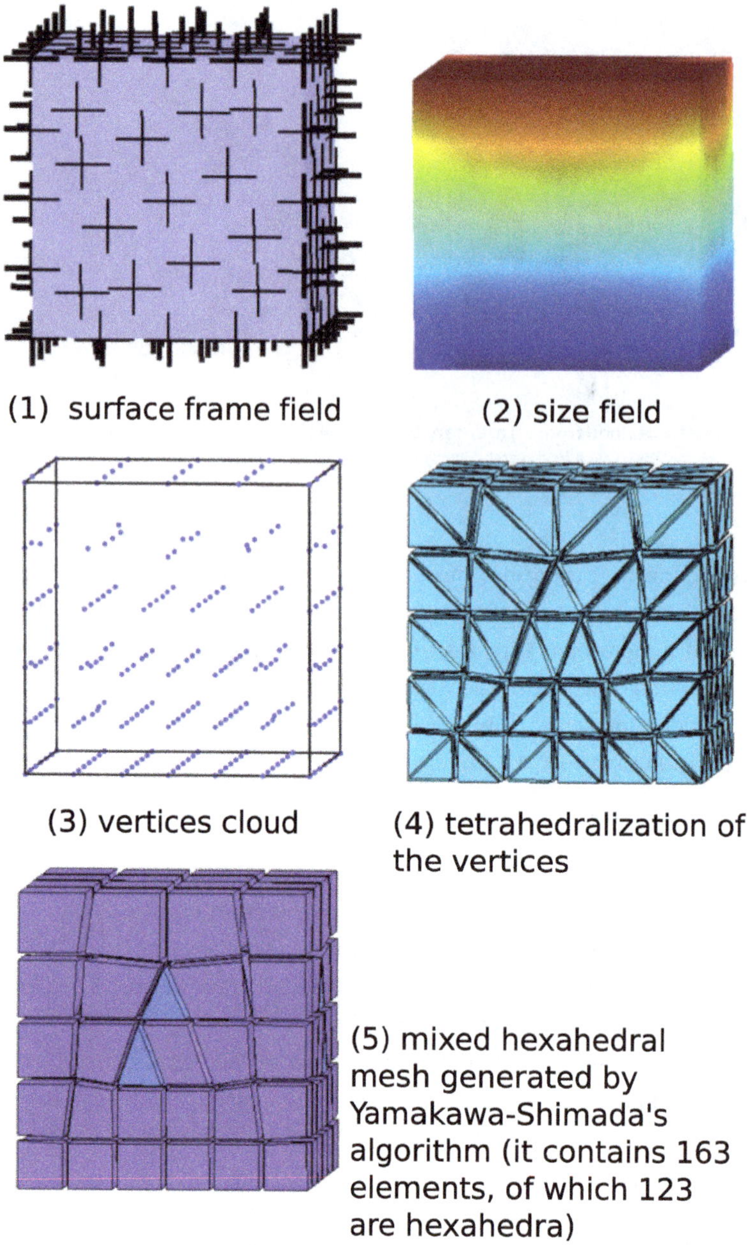

Figure 2 An overview of the steps necessary to generate a hex-dominant mesh.

be viewed as a graph. The method consists in searching through the mesh to identify subgraphs yielding hexahedra. There are six particular subgraphs to look for. Potential hexahedra can be constructed immediately or after classification along various criterions. This approach can be applied to prisms and pyramids as well. The recombination algorithm used throughout this article has been devised by Yamakawa and Shimada [9]. These authors have also designed an iterative procedure to align vertices in three dimensions. The first step of the procedure consists of filling the domain with crystal cells. Each crystal cell has the shape of a cube: there is one atom at the center and eight atoms in the corners. Crystal cells exert force on each other via their atoms. A set of equations of motion govern the cells positions. Throughout the process, cells can be added or removed depending on the local density. Once cells velocities have sufficiently decrease, the center of each crystal cell becomes a mesh vertex.

To improve the spatial distribution of the set of vertices (which has a large impact on the effectiveness of such graph-based approaches), several techniques have been proposed, such as a modified version of Lloyd's algorithm. Lloyd's algorithm repeatedly moves vertices to the centroid of their Voronoi cell [23]. As the number of iterations increases, the Voronoi cells take the shape of perfect hexagons. In fact, it has been proven that Lloyd's algorithm minimizes an energy functional equal to the sum of the moments of inertia of the Voronoi cells [24]. Lévy-Liu's algorithm consists of minimizing higher order moments of inertia: the Voronoi cells then become squarely instead of hexagonal, which has the effect of aligning vertices in precise directions. Tetrahedral meshes smoothed by Lévy-Liu's algorithm make excellent candidates for recombination. When used in conjunction with the graph method discussed above, Lévy-Liu's algorithm can generate mixed meshes with high hexahedra percentages [24]. However, the method presented in this article is frontal, not iterative. The vertices are created layer by layer.

Methods

We have pointed out above the importance of having the mesh vertices pre-aligned to ensure a good recombination rate. It is the purpose of the *vertex placement algorithm* to achieve this. This algorithm, however, relies on a number of data structures and geometrical concepts that are first introduced and developed below.

First, the vertex placement algorithm needs to know, at each point of the domain, the prescribed local mesh size and the local preferred mesh directions. In practice, those geometrical quantities are conjointly obtained by evaluating a specific field structure called *cross field*. The generation of direction fields was extensively studied in [25].

Secondly, the notion of distance itself represents another degree of freedom of the method. We shall show that it is particularly appropriate when dealing with hex-meshing, to compute distances with the *infinity norm*, instead of the standard Euclidean norm.

Finally, the algorithm is characterized by a large number of spatial searches, in order to check whether or not a prospective vertices is too close to any already existing vertex. To optimize the efficiency of this operation, an *R-tree* data structure is employed [26,27].

Cross fields

At each point of a region $\Omega \subset \mathcal{R}^3$, the *frame field* $(\mathbf{d}_1, \mathbf{d}_2, \mathbf{d}_3)$ represents the three local orthogonal preferred directions of the hexahedral mesh. Frame fields are usually required to satisfy many constraints [16,28]. On the geometrical edges of Ω, one of the three directions should be tangent to the edge itself [9]. On the surfaces of Ω, one of the three directions should be perpendicular to the surface [9,16]. A last requirement is that the frame field should be as smooth as possible.

On the other hand, at each point of Ω, the *size field* represents the prescribed local mesh size value. Mesh sizes h_1, h_2, h_3 are defined for every point of the volume in each of the directions $\mathbf{d}_1, \mathbf{d}_2, \mathbf{d}_3$. In this paper, the mesh size field at a point $\mathbf{x}$ is isotropic, i.e. $h(\mathbf{x}) = h_1(\mathbf{x}) = h_2(\mathbf{x}) = h_3(\mathbf{x})$. The extension to anisotropic meshing will be done in a forthcoming work.

The user fixes the mesh size at the geometrical vertices of the model. One-dimensional size fields are then computed along the geometrical edges. Because the surfaces are

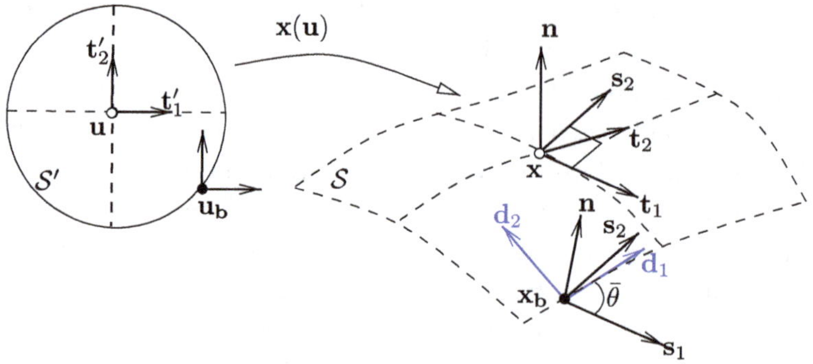

Figure 3 Surface parametrization and construction of the frame field (in blue).

bounded by geometrical edges, Dirichlet conditions can be imposed on the surfaces boundaries. A Laplace equation is used to obtain the size field over the surfaces. The size field over the volume is calculated in a similar manner. Continuous finite elements of the first order are employed in each case. The final size field is therefore a three-dimensional piecewise continuous field. The Laplace equation was chosen because it leads to smooth and gradual solutions.

The *cross field* $(h_1\mathbf{d}_1, h_2\mathbf{d}_2, h_3\mathbf{d}_3)$, now, combines both information into a single field. At each vertex of the mesh, the cross field evaluates into a symmetric real 3 by 3 tensor whose columns are the three orthogonal vectors parallel to the local preferred directions of the hexahedral mesh. Moreover, the norm of the vectors represent the local mesh size; the three norms are identical in case of an isotropic mesh (which is the case considered in this paper), but they may differ in case of an anisotropic mesh.

The construction of a frame field on a region Ω belongs to the category of elliptic problems. Boundary conditions must be imposed on the boundary $\partial\Omega$. We thus proceed logically by explaining first how the frame field is constructed on surfaces, and deal afterwards with the prolongation into the volumes.

Let

$$\mathbf{u} = \{u, v\} \in \mathcal{S}' \subset \mathcal{R}^2 \mapsto \mathbf{x}(\mathbf{u}) = \{x, y, z\} \in \mathcal{S} \subset \mathcal{R}^3, \tag{1}$$

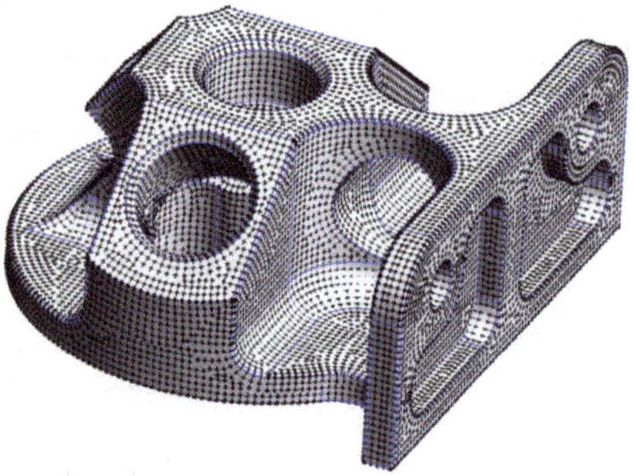

Figure 4 Cross fields on the surfaces of a mechanical part.

be a smooth parametrization of the surface $\mathcal{S}$ (see [29-31] for a review of parametrization techniques for surface remeshing). It should be noted that the parametrization does not need to be conformal, i.e the angles do not need to be conserved, for the algorithms presented in this paper. (This is a nice feature because guaranteed one-to-one conformal maps are more difficult to compute than bijective harmonic mapping.) For example, Figure 3 shows a harmonic parametrization of an arbitrary surface $\mathcal{S}$ onto a unit disk.

Consider the two tangent vectors

$$\mathbf{t}_1 = \frac{\partial \mathbf{x}}{\partial u} \text{ and } \mathbf{t}_2 = \frac{\partial \mathbf{x}}{\partial v},$$

which are the images in $\mathcal{S}$ of the basis vectors $\mathbf{t}'_1 = (1, 0)$ and $\mathbf{t}'_2 = (0, 1)$ of the parameter plane $\mathcal{S}'$. Because they are not parallel for any point of $\mathcal{S}'$, one can build the unit normal vector $\mathbf{n} = \mathbf{t}_1 \times \mathbf{t}_2 / \|\mathbf{t}_1 \times \mathbf{t}_2\|$. Each vector $\mathbf{t}$ tangent to $\mathcal{S}$ can be expressed as $\mathbf{t} = u\mathbf{t}_1 + v\mathbf{t}_2$ with (u, v) the covariant coordinates of $\mathbf{t}$. The tangent vector $\mathbf{t}$ is thus the image of a vector $\mathbf{t}' = (u, v)$ in the parameter plane. It is easy to compute covariant coordinates of any tangent vector $\mathbf{t}$ using the metric tensor of the parametrization. By definition, $\mathbf{t} = u\mathbf{t}_1 + v\mathbf{t}_2$. Then, $\mathbf{t} \cdot \mathbf{t}_1 = u\mathbf{t}_1 \cdot \mathbf{t}_1 + v\mathbf{t}_2 \cdot \mathbf{t}_1$ and $\mathbf{t} \cdot \mathbf{t}_2 = u\mathbf{t}_1 \cdot \mathbf{t}_2 + v\mathbf{t}_2 \cdot \mathbf{t}_2$, which reads in matrix form

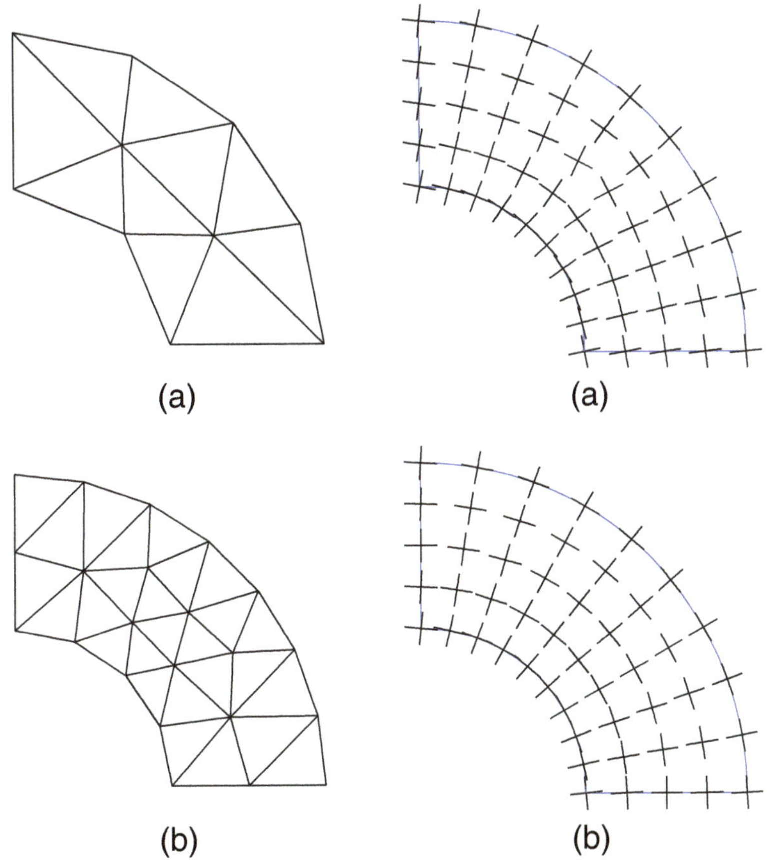

Figure 5 The mesh (a) was used to create the frame field (a) and the mesh (b) was used to create the frame field (b).

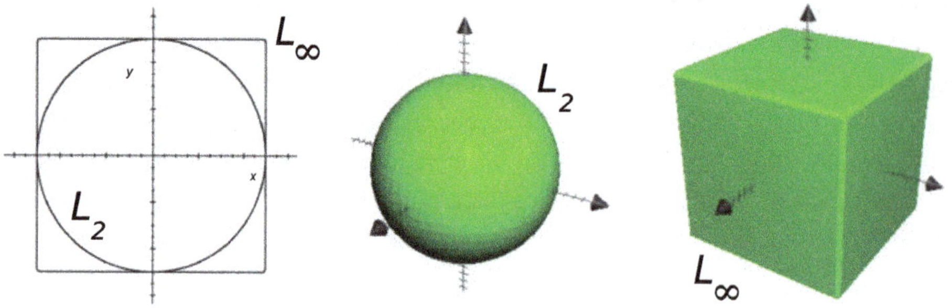

Figure 6 Unit circles and unit spheres in the L_2 and L_∞ norms.

$$\underbrace{\begin{bmatrix} \mathbf{t}_1 \cdot \mathbf{t}_1 & \mathbf{t}_1 \cdot \mathbf{t}_2 \\ \mathbf{t}_2 \cdot \mathbf{t}_1 & \mathbf{t}_2 \cdot \mathbf{t}_2 \end{bmatrix}}_{\mathcal{M}} \begin{bmatrix} u \\ v \end{bmatrix} = \begin{bmatrix} \mathbf{t} \cdot \mathbf{t}_1 \\ \mathbf{t} \cdot \mathbf{t}_2 \end{bmatrix} \tag{2}$$

where $\mathcal{M}$ is the metric tensor, invertible for any smooth parametrization.

For defining our frame field, a local orthonormal frame $(\mathbf{s}_1, \mathbf{s}_2, \mathbf{n})$ is first constructed at all points $\mathbf{x}$ of $\mathcal{S}$ with $\mathbf{s}_1 = \mathbf{t}_1/\|\mathbf{t}_1\|$, $\mathbf{s}_2 = \mathbf{n} \times \mathbf{t}_1/\|\mathbf{n} \times \mathbf{t}_1\|$. Next, the direction $\mathbf{d}_1$ of the frame field is computed at the points $\mathbf{x}_b$ of the boundaries of surface $\mathcal{S}$: $\mathbf{d}_1$ is the tangent vector to the boundary. The local orientation of the frame field $\bar{\theta}$ at the boundary can then be computed as the oriented angle between $\mathbf{s}_1$ and $\mathbf{d}_1$. Then, an elliptic boundary value problem is used to propagate the complex number $z(\mathbf{u}) = a(\mathbf{u}) + ib(\mathbf{u}) = e^{4i\theta(\mathbf{u})}$ in the parametric domain. More specifically, two Laplace equations with Dirichlet boundary conditions are solved in the parametric space $\mathcal{S}'$ in order to compute the real part $a(\mathbf{u}) = \cos 4\theta$ and the imaginary part $b(\mathbf{u}) = \sin 4\theta$ of z:

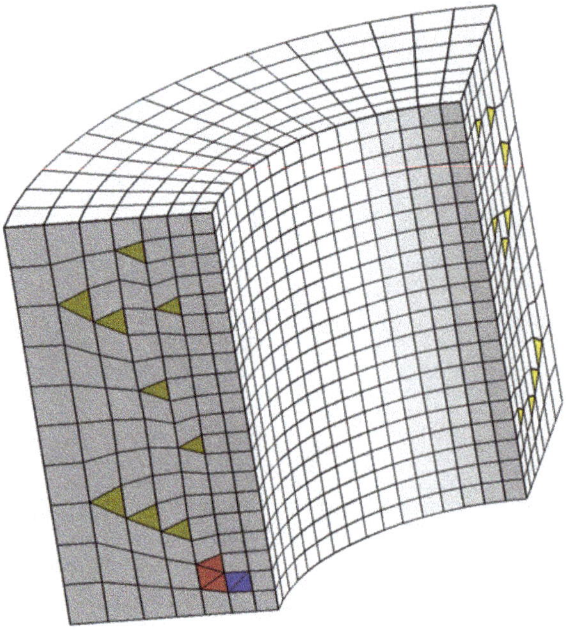

Figure 7 A mixed mesh created with the frontal algorithm using the L_∞ norm. The percentage of hexahedra by number is equal to 76.03% for L_∞ and 71.25% for L_2.

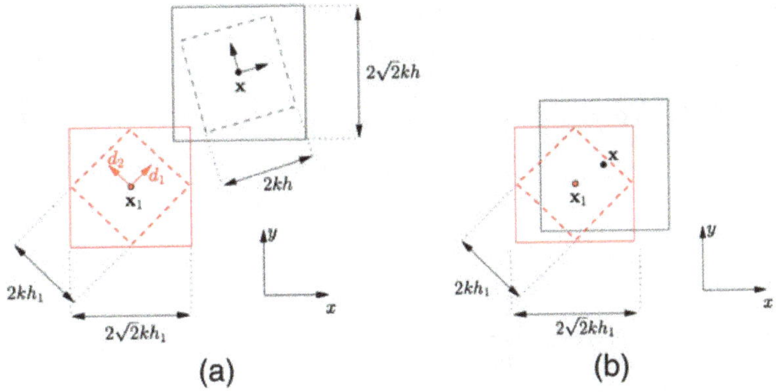

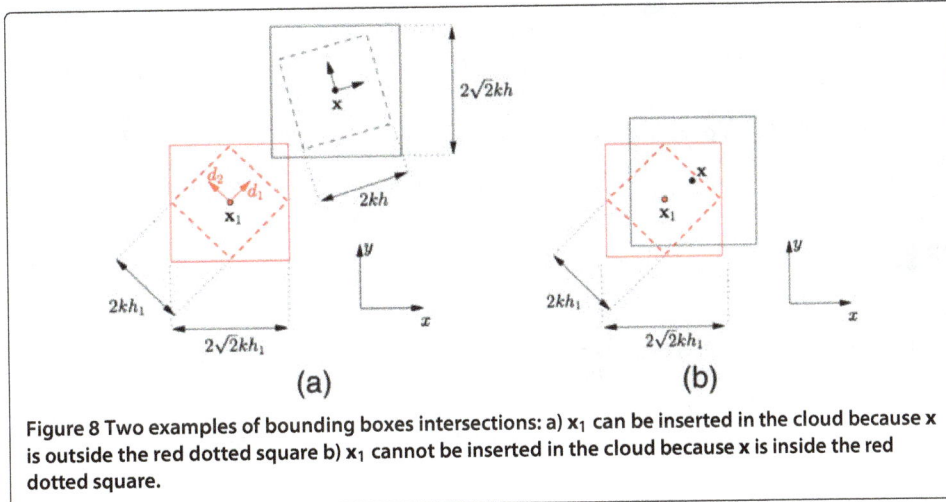

Figure 8 Two examples of bounding boxes intersections: a) $\mathbf{x}_1$ can be inserted in the cloud because $\mathbf{x}$ is outside the red dotted square b) $\mathbf{x}_1$ cannot be inserted in the cloud because $\mathbf{x}$ is inside the red dotted square.

$$\begin{aligned} \nabla^2 a = 0, \quad \nabla^2 b = 0 \ \text{ on } \quad & \mathcal{S}', \\ a = \bar{a}(\mathbf{u}), \quad b = \bar{b}(\mathbf{u}) \ \text{ on } \quad & \partial\mathcal{S}'. \end{aligned} \tag{3}$$

After solving those two PDEs, the frame field can be represented in the whole domain by the angle

$$\theta(\mathbf{u}) = \frac{1}{4}\text{atan2}(b(\mathbf{u}), a(\mathbf{u})).$$

The choice 4θ as the argument of z is motivated by symmetry arguments: frame fields are equivalent when they are rotated around $\mathbf{n}$ by any angle that is a multiple of $\pi/2$. Details of that procedure are given in [32]. Finally, the frame field $(\mathbf{d}_1, \mathbf{d}_2, \mathbf{d}_3)$ can be computed on the whole surface $\mathcal{S}$ as follows:

$$\mathbf{d}_1 = \mathbf{s}_1 \cos\theta + \mathbf{s}_2 \sin\theta, \quad \mathbf{d}_2 = -\mathbf{s}_1 \sin\theta + \mathbf{s}_2 \cos\theta, \quad \mathbf{d}_3 = \mathbf{d}_1 \times \mathbf{d}_2, \tag{4}$$

where θ is the solution of the elliptic boundary value problem (3).

As an example, Figure 4 presents the frame field computed on surfaces of a mechanical part. Figure 5 shows two triangular meshes of different coarseness and their resulting frame fields. Linear interpolation of the a and b components discussed earlier was used in

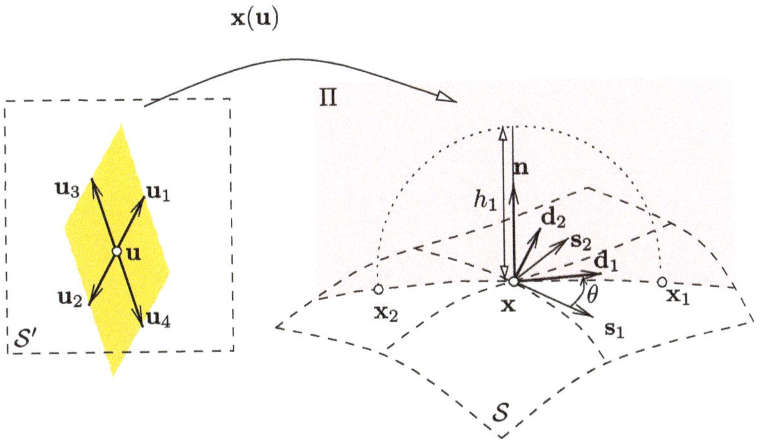

Figure 9 View of the different parameters of the point insertion scheme.

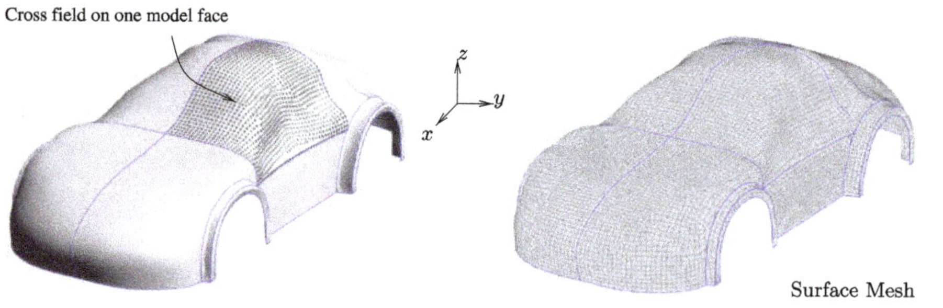

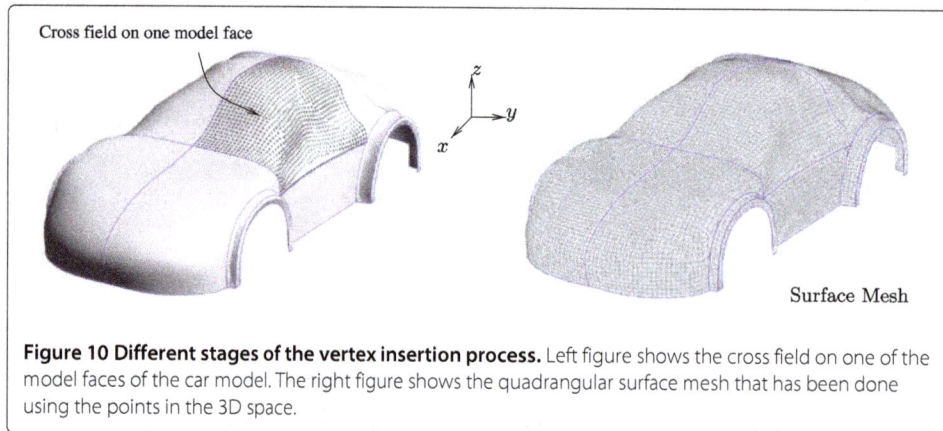

Figure 10 Different stages of the vertex insertion process. Left figure shows the cross field on one of the model faces of the car model. The right figure shows the quadrangular surface mesh that has been done using the points in the 3D space.

order to obtain the same number of frames regardless of the mesh density. As seen from the figure, the frame field (a) is not entirely radial and contains defects because the mesh (a) is too coarse.

The frame field at any point inside the volume is then chosen to be equal to the frame field at the closest surface vertex [24]. (ANN nearest neighbor library is employed for the queries [33].)

These frame fields are not going to be smooth whenever the distance function to the walls is not itself smooth. Recently, two methods capable of generating smooth frame fields have been developed [16,17]. Both of these methods employ LBFGS optimization to minimize energy functionals.

Measuring distances

For inserting a new mesh vertex in our frontal algorithm, the distance between a prospective vertex $\mathbf{x}_i$ and any already existing vertex $\mathbf{x}$ must be smaller than kh, where h is the local mesh size and k a free parameter of the algorithm ranging from 0 to 1. Parameter k absolutely needs to be inferior to one. If not, too many valid vertices will be missing from the cloud. In the implementation described in this work, k is equal to 0.7.

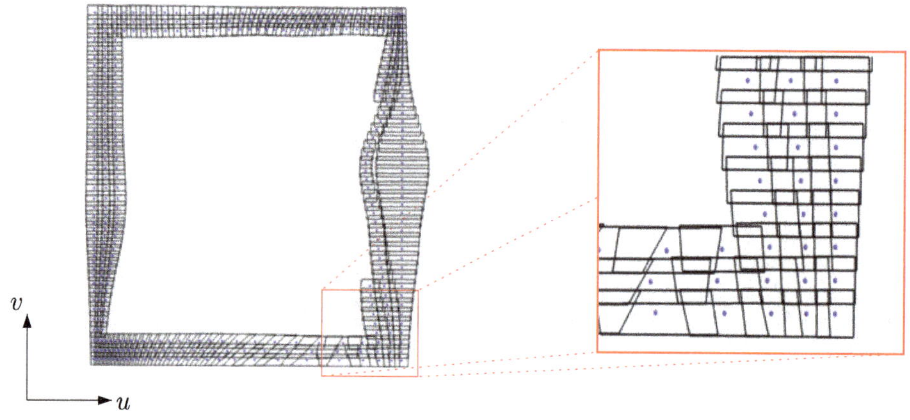

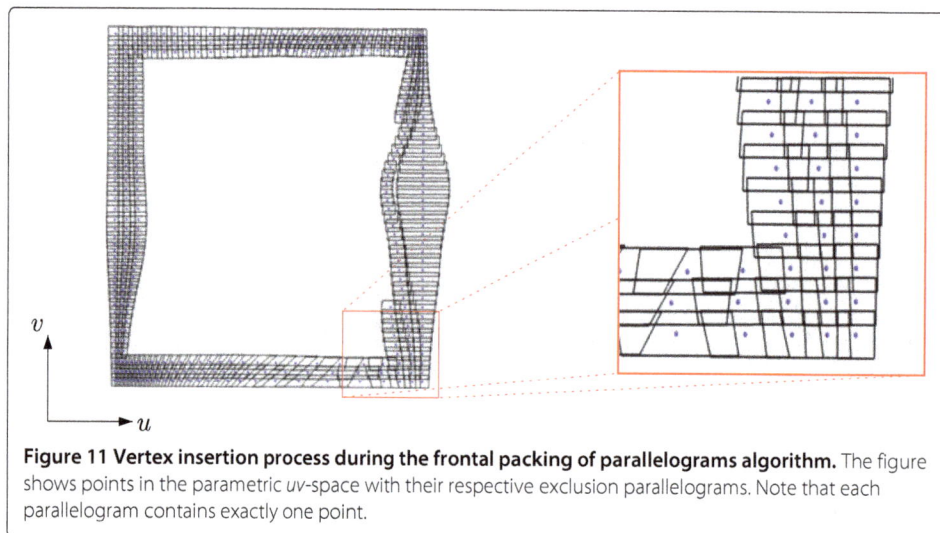

Figure 11 Vertex insertion process during the frontal packing of parallelograms algorithm. The figure shows points in the parametric *uv*-space with their respective exclusion parallelograms. Note that each parallelogram contains exactly one point.

The way distances between vertices are calculated is however a degree of freedom of the method. When dealing with hex-meshing, it turns out to be advantageous to compute distances in the *infinity norm*, instead of in the Euclidean norm:

$$||\mathbf{x}-\mathbf{y}||_2 = \sqrt{|x_1-y_1|^2+|x_2-y_2|^2+|x_3-y_3|^2} \quad (5)$$

$$||\mathbf{x}-\mathbf{y}||_\infty = \max(|x_1-y_1|,|x_2-y_2|,|x_3-y_3|). \quad (6)$$

In the infinity norm, the unit *sphere* is actually a cube, which reduces to a square in two dimensions (see Figure 6). The exclusion area around each prospective vertex is therefore a cube, resp. a square, which precisely matches the shape of the elements one wishes to build.

Contrary to the Euclidean norm, the infinity norm is not isotropic and, consequently, it has an orientation which is given by the frame field. In the parameter plane, due to the change of coordinates (1), the exclusion area is the parallelogram determined by

$$d_\infty^{orien}(\mathbf{x},\mathbf{y}) = ||M_\mathbf{x}(\mathbf{x}-\mathbf{y})||_\infty < kh. \quad (7)$$

where $M_\mathbf{x}$ is the Jacobian matrix of (1), evaluated at **x**.

The infinity distance is not a differentiable function [24]. However, this is not an issue, because the frontal algorithm does not require the computation of distance derivatives.

Using the infinity distance instead of the Euclidean distance can increase the hexahedra percentage. The quarter cylinder illustrated on Figure 7 provides an example where an improvement by 5% of the ratio of hexahedra is observed, by simply using the L_∞ norm instead of the L_2 norm in the R-tree spatial search algorithm described in the next section.

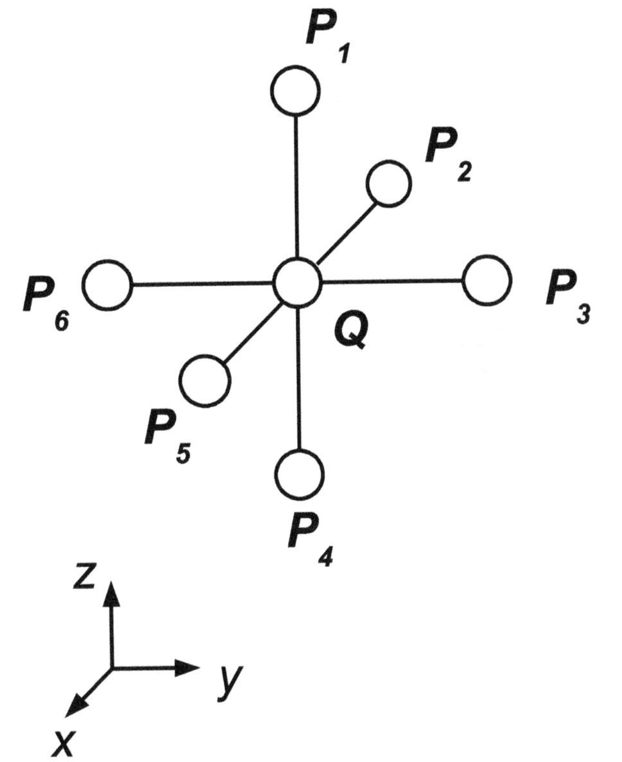

Figure 12 In three dimensions, vertex Q try to create six vertices: P_1, P_2, P_3, P_4, P_5 and P_6.

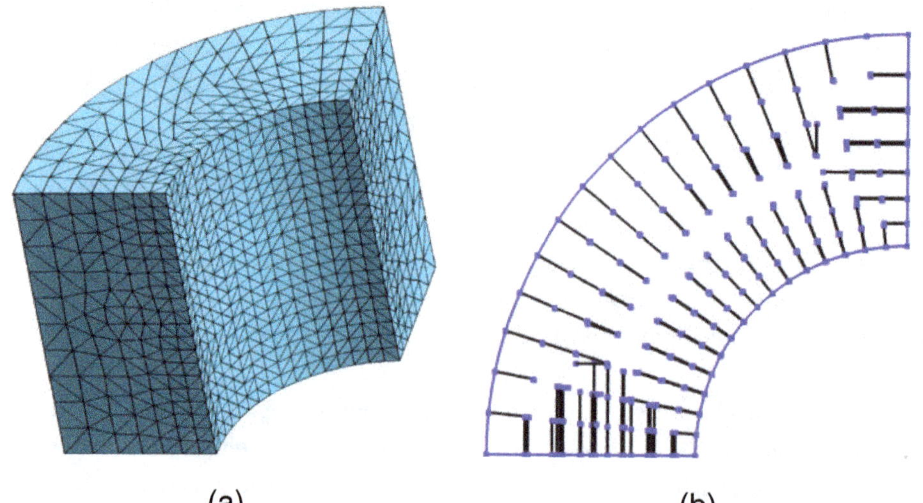

Figure 13 3D point insertion algorithm: (a) view of the surface mesh of a quarter cylinder, (b) top view of the quarter cylinder during the point insertion algorithm.

Using R-trees for spatial searches

As said before, a prospective vertex is effectively created only if there is enough unoccupied space around it. The size of this exclusion area or volume depends on the local mesh size. According to the dimension and the chosen norm, the shape of the exclusion region can be a parallelogram or an ellipse (in 2D), and a cube or a sphere (in 3D).

The computation of the distance between the prospective vertex and all the other vertices would have a quadratic complexity in time and would therefore be prohibitive in terms of computation time. The number of computations required to ensure the exclusions can however be considerably decreased if the exclusion cube of each vertex is enclosed in a bounding box whose edges are parallel to the coordinate axis. An R-tree

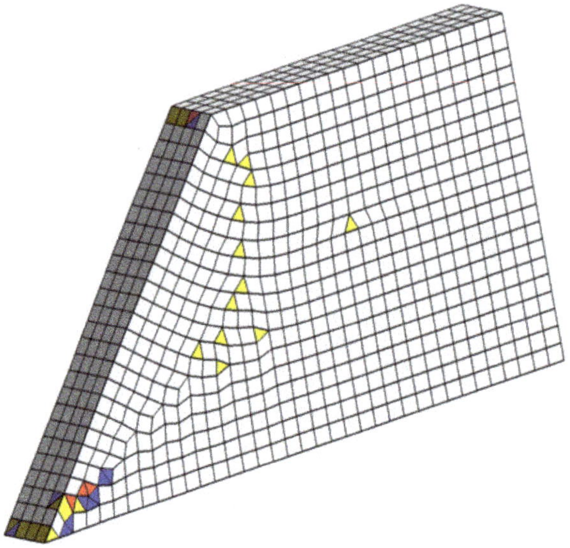

Figure 14 A mixed mesh containing hexahedra (white), prisms (yellow), pyramids (red) and tetrahedra (blue).

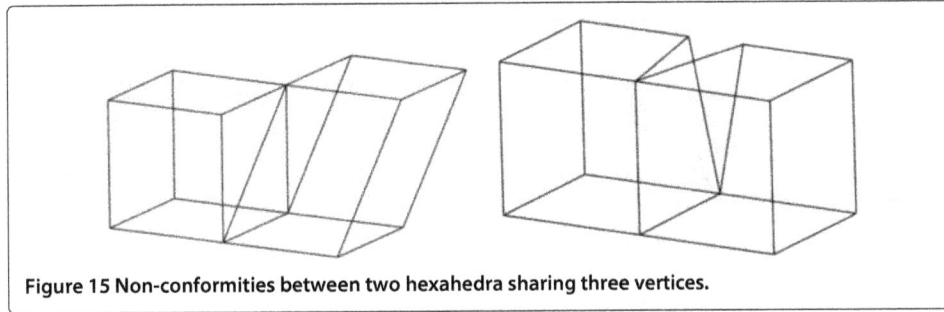

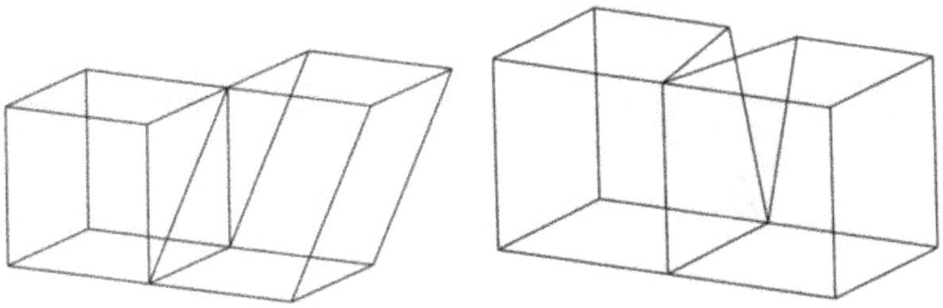
Figure 15 Non-conformities between two hexahedra sharing three vertices.

data structure [26,27] can efficiently determine bounding boxes intersections and, then, it is enough to compute the distance between pairs of vertices whose boxes intersect each other.

We now illustrate with a 2D example of a planar surface how to decide wether a prospective vertex can be inserted or not. For this example, we have chosen the infinity norm for computing distances. In Figure 8(a), $\mathbf{x}_1$ is the prospective vertex and $\mathbf{x}$ is an existing mesh vertex. The dotted square around $\mathbf{x}_1$ is the oriented exclusion area of vertex $\mathbf{x}_1$, that is computed from the surface cross field $(h_1\mathbf{d}_1, h_1\mathbf{d}_2)$ that has a uniform mesh size field h_1. The solid box surrounding the prospective vertex is the bounding box of the exclusion area that is parallel to the xy-coordinate axis. This bounding box should always include the oriented exclusion square of side $2kh$. This condition is satisfied in 2D if the box is of side $2\sqrt{2}kh$ and in 3D if the cube is of side $2\sqrt{3}kh$. Even if the boxes intersect each other in Figure 8(a), the distance between $\mathbf{x}_1$ and $\mathbf{x}$ is sufficiently large. Thus, $\mathbf{x}_1$ can be inserted in the cloud and added to the queue.

Figure 8(b) shows the same two vertices. Again, the boxes intersect each other. This time, however, $\mathbf{x}_1$ is too close to $\mathbf{x}$ and $\mathbf{x}_1$ cannot be added to the cloud or to the queue.

It should be noted that on Figures 8(a) and 8(b), $d_\infty^{orien}(\mathbf{x}_1, \mathbf{x})$ is not necessarily equal to $d_\infty^{orien}(\mathbf{x}, \mathbf{x}_1)$. The local mesh sizes at $\mathbf{x}_1$ and $\mathbf{x}$ can also be different as illustrated in Figure 8(a). However, if $\mathbf{x}$ is outside the dotted square of $\mathbf{x}_1$, it is considered sufficient.

For non-planar surfaces, the surfaces need to be parametrized. As the parametrization is not necessarily conformal, i.e. the angles between $\mathbf{d}_1$ and $\mathbf{d}_2$ are not conserved, the dotted squares (exclusion area) of Figure 8 become parallelograms in the parametric space. As far as the bounding boxes are concerned, they are computed in the same manner and are then parallel to the uv-coordinate axis of the parametric space.

Let's assume that on surfaces, each vertex attempts to create four vertices in the four cardinal directions. If the surface normal is not constant, these prospective vertices may

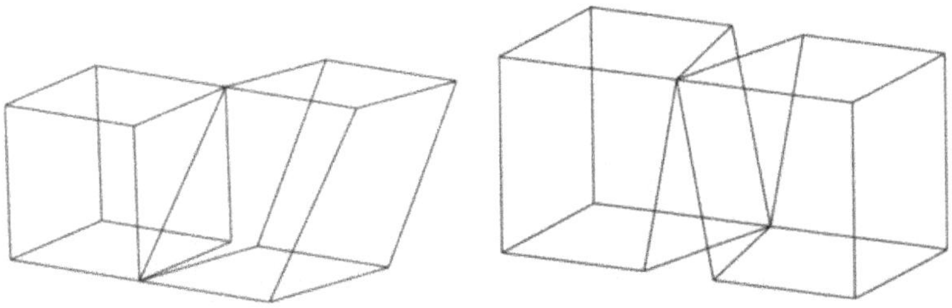
Figure 16 Non-conformities between two hexahedra sharing two vertices.

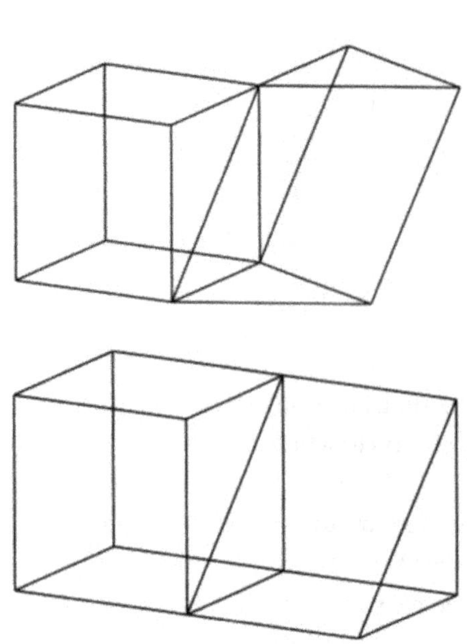

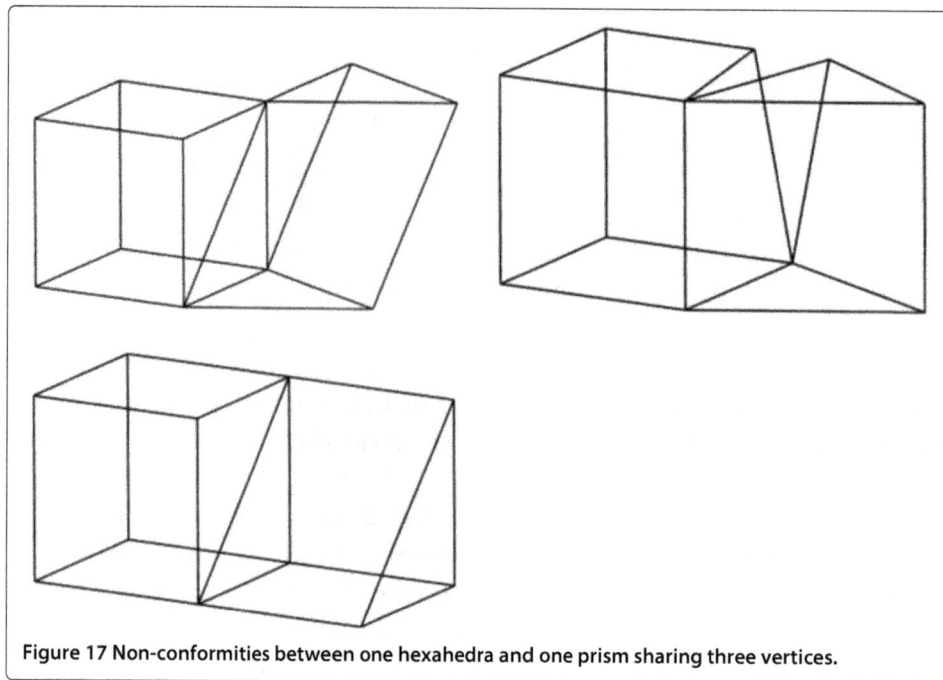

Figure 17 Non-conformities between one hexahedra and one prism sharing three vertices.

not rest on the surface. The next section describes a scheme capable of solving this issue by intersecting surfaces with circle arcs.

Surface meshing: the packing of parallelograms algorithm

The quadrilateral mesh algorithm presented here is a simpler variant of [32] that we call *packing of parallelograms*. Consider one vertex located at point $\mathbf{u} = (u, v)$ of the parameter plane $\mathcal{S}'$ which correspond to point $\mathbf{x}(u, v)$ in the 3D space (see Figure 9). The cross field at this point of the surface is $(h_1\mathbf{d}_1, h_2\mathbf{d}_2, h_n\mathbf{n})$, in terms of the three orthonormal preferred mesh directions, $\{\mathbf{d}_1, \mathbf{d}_2, \mathbf{n}\}$, and the three corresponding mesh sizes, $\{h_1, h_2, h_n\}$,.

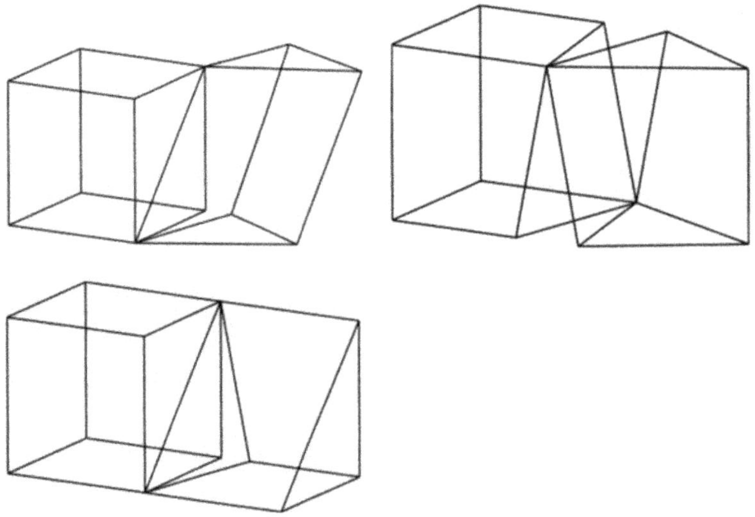

Figure 18 Non-conformities between one hexahedra and one prism sharing two vertices.

Figure 19 The arrow indicates a non-conformal face between a hexahedron and two pyramids. Tetrahedra are not shown to facilitate visibility.

In a perfect quad mesh, each vertex is connected to four neighboring vertices forming a cross parallel to the cross field. In our approach, four prospective points $\mathbf{x}_i$, $i = 1, \ldots, 4$ are constructed in the neighborhood of point $\mathbf{x}$ with the aim of generating the perfect situation.

Points $\mathbf{x}_1$ and $\mathbf{x}_2$ are constructed as the intersection of the surface $\mathcal{S}$ with a circle of radius h_1, centered on $\mathbf{x}$ and situated in the plane Π of normal $\mathbf{d}_2$ (see Figure 9). Points $\mathbf{x}_3$ and $\mathbf{x}_4$ are constructed as the intersection of the surface $\mathcal{S}$ with a circle of radius h_2, centered on $\mathbf{x}$ and situated in the plane of normal $\mathbf{d}_1$ (not in the figure for clarity).

Numerical difficulties associated with the surface-curve intersection are overcome by choosing a good initial guess for the intersection. If we approximate the surface by its tangent plane at $\mathbf{x}$, point $\mathbf{x}_1$ is situated at $\mathbf{x}_1 = \mathbf{x} + h_1\mathbf{d}_1$. A good initial guess in the parameter plane is $\mathbf{u}_1 = \mathbf{u} + d\mathbf{u}_1$ where $d\mathbf{u}_1 = (du_1, dv_1)$ is computed using (2) i.e.

$$\mathcal{M}\begin{pmatrix} du_1 \\ dv_1 \end{pmatrix} = \begin{pmatrix} h_1\mathbf{d}_1 \cdot \mathbf{t}_1 \\ h_1\mathbf{d}_1 \cdot \mathbf{t}_2 \end{pmatrix}.$$

This also gives $\mathbf{u}_2 = \mathbf{u} - d\mathbf{u}_1$, $d\mathbf{u}_3 = (du_3, dv_3)$

$$\mathcal{M}\begin{pmatrix} du_3 \\ dv_3 \end{pmatrix} = \begin{pmatrix} h_2\mathbf{d}_2 \cdot \mathbf{t}_1 \\ h_2\mathbf{d}_2 \cdot \mathbf{t}_2 \end{pmatrix}.$$

$\mathbf{u}_3 = \mathbf{u} + d\mathbf{u}_3$ and $\mathbf{u}_4 = \mathbf{u} - d\mathbf{u}_3$.

The algorithm works as follows. Each vertex of the boundary is inserted in a fifo queue. Then, the vertex $\mathbf{x}$ at the head of the queue is removed and its four prospective neighbors

Figure 20 A supplementary pyramid was created by Owen-Canann-Saigal's algorithm to fix the non-conformity.

Table 1 Mesh data

Figure	# vertices	H_{nbr}	H_{vol}	Q	CPU (s)	NC
21	126, 922	62.91%	85.70%	0.95	488 s.	7.19%
23	42, 263	75.42%	92.85%	0.96	164 s.	3.97%
26	11, 648	58.16%	82.68%	0.94	112 s.	7.70%
27	2737	60.23%	83.02%	0.94	45 s.	8.04%
28	6006	84.54%	94.10%	0.96	80 s.	2.25%
32	19, 284	86.19%	94.93%	0.98	51 s.	1.81%

Lévy-Liu's algorithm [24,36] was used for Figures 26, 27 and 28.

$\mathbf{x}_i$ are computed. A new vertex $\mathbf{x}_i$ is inserted at the tail of the queue if the following conditions are satisfied: (*i*) vertex $\mathbf{x}_i$ is inside the domain and (*ii*) vertex $\mathbf{x}_i$ is not too close to any of the vertices that have already been inserted.

As for the first condition, it is enough to check if the preimage $\mathbf{u}_i \in \mathcal{S}'$ of $\mathbf{x}_i$ is inside the bounds of the parameter domain. Concerning the second condition, the distances on the surface $\mathcal{S}$ should theoretically be measured in terms of geodesics, This is however clearly overkill from a mesh generation point of view. We define an exclusion zone for every vertex that has already been inserted (this includes boundary vertices). This exclusion zone is a parallelogram in the parameter plane (see the yellow parallelogram of Figure 9). This parallelogram is scaled down by a factor $k = 0.7$ in order to allow the insertion of (at least) points x_i. The different stages of the procedure for a non planar surface are presented on Figure 10 and Figure 11. Then, the surfaces are triangulated in the parameter plane

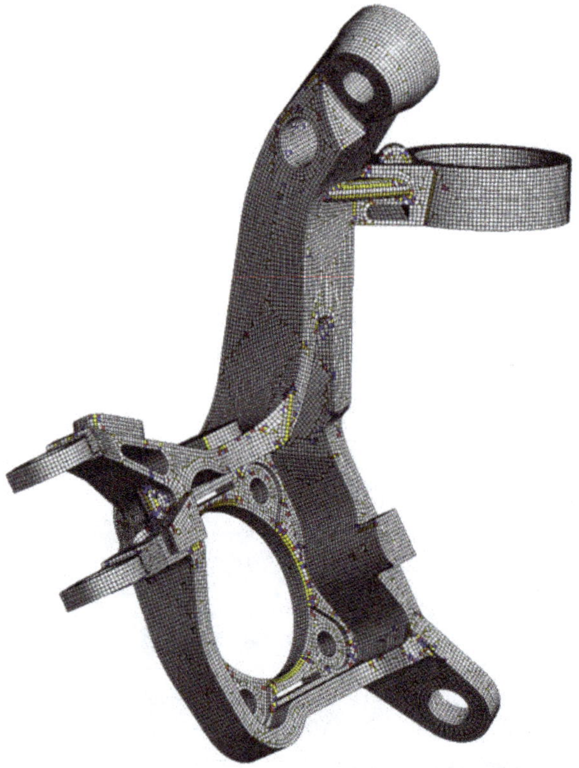

Figure 21 A mechanical piece.

using an anisotropic Delaunay kernel and the triangles are subsequently recombined into quadrilaterals using the Blossom-Quad algorithm [34].

As shown on Figure 11, exclusion areas can become anisotropic parallelograms in the parametric plane. However, they always correspond to squares in the three-dimensional space. The vertices are triangulated in the parametric plane. Anisotropic triangulation is therefore necessary in order to obtain the expected arrangement of right triangles.

Volume meshing: the 3D point insertion algorithm

Volume meshing proceeds in the same way as surface meshing. The procedure starts from a 2D triangular mesh that has been created using surfacic frame fields. A frontal algorithm is used to create well aligned vertices inside the volume, starting from surface points. The 3D point insertion algorithm works in the same manner as the one used for surfaces.

All boundary mesh vertices are initially pushed into a queue. The vertices are popped in order: each vertex $\mathbf{Q}$ popped out of the queue attempts to create six neighboring vertices in the six cardinal directions $\mathbf{P}_{1,2} = \mathbf{Q} \pm h\mathbf{d}_1$, $\mathbf{P}_{3,4} = \mathbf{Q} \pm h\mathbf{d}_2$, $\mathbf{P}_{5,6} = \mathbf{Q} \pm h\mathbf{d}_3$ at a distance h from itself (see Figure 12).

A prospective vertex is added to the vertices cloud and to the queue only if it satisfies the two following conditions:

1. It is inside the domain.
2. It is not too close to an existing mesh vertex, i.e. if the distance is smaller than kh.

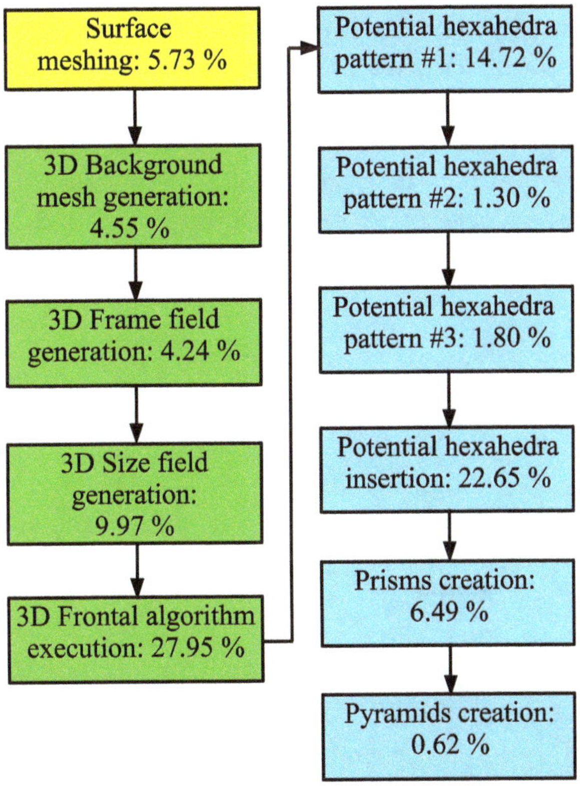

Figure 22 Global algorithm flowchart and execution time percentages.

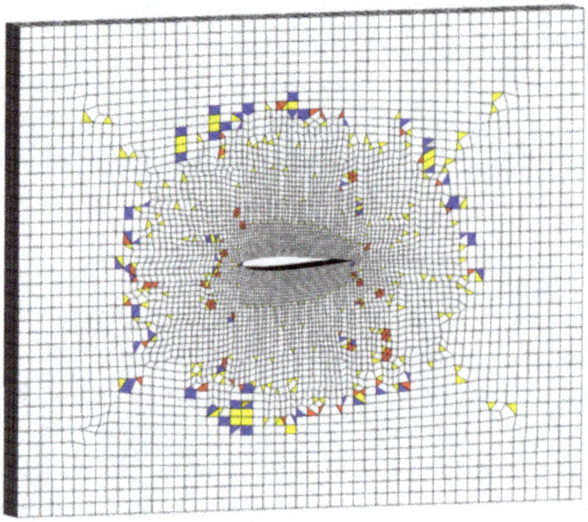

Figure 23 A NACA airfoil.

An octree data structure is again employed to efficiently determine if a vertex is inside the domain [11].

Eventually, no more prospective vertices can be added to the cloud without being too close to existing ones. The process then stops and the cloud is tetrahedralized with a Delaunay procedure [35].

The frontal algorithm was applied to the quarter cylinder starting from the surface mesh shown in Figure 13(a). In Figure 13(b), lines are traced between each vertex and its parent in order to observe the progression of the 3D point insertion algorithm.

The quality of the alignment inside the geometry is very dependent on the quality of the alignment on the boundaries. If the triangles on the boundaries are far from being right-angled, then the vertices inside the geometry will not be well aligned. Various algorithms are capable of generating sets of aligned vertices on surfaces, such as the Delquad

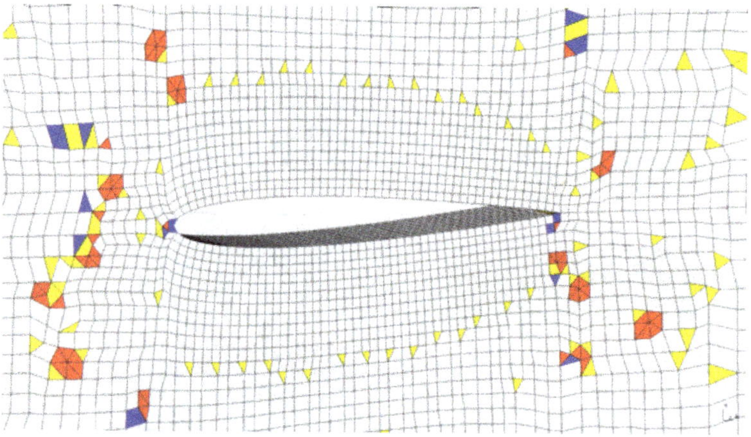

Figure 24 A zoom of the airfoil.

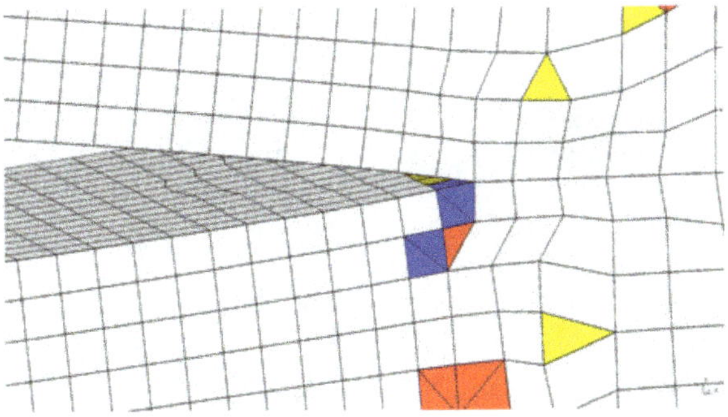

Figure 25 A zoom of the trailing edge.

algorithm [32] or Lévy-Liu's algorithm [24,36]. However, for the majority of the examples presented in this article, a two-dimensional version of the frontal algorithm was employed.

As explained earlier, each vertex attempts to create six other vertices at a distance $d = h$ from itself. For smoother size transitions, d can instead be an average between the local mesh size at the parent vertex and the local mesh size at the prospective vertex.

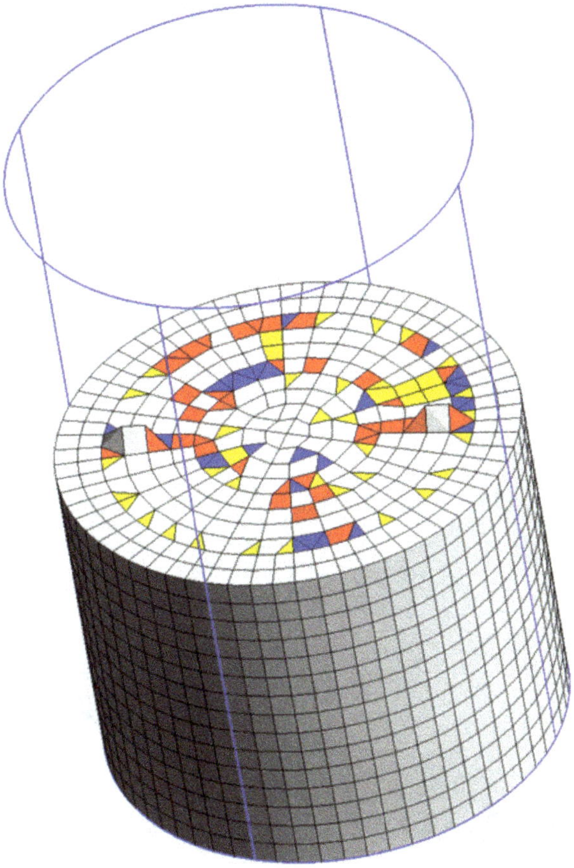

Figure 26 A cylinder. The blue lines illustrate the cylinder geometry.

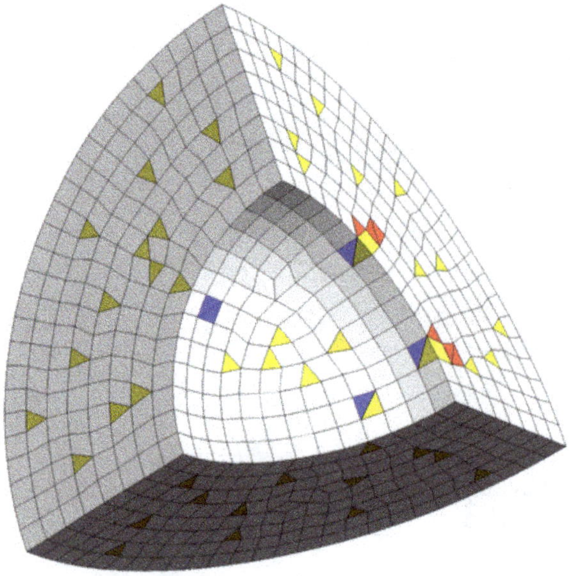

Figure 27 An hollow eighth of ball with a constant mesh size.

Volume meshing: Yamakawa-Shimada's algorithm and finite element conformity

This section briefly describes Yamakawa-Shimada's recombination algorithm. It then discusses the problem of finite element conformity in the case of mixed hex meshes.

Yamakawa-Shimada's algorithm begins by iterating through the tetrahedra of the initial mesh. For each tetrahedron, it attempts to find neighboring tetrahedra with which to construct a hexahedron. Five, six or seven tetrahedra are required to construct one hexahedron. Three patterns of assembly are considered. Two out of these three patterns are described in [9]. When a potential hexahedron is found, it is added to an array. However,

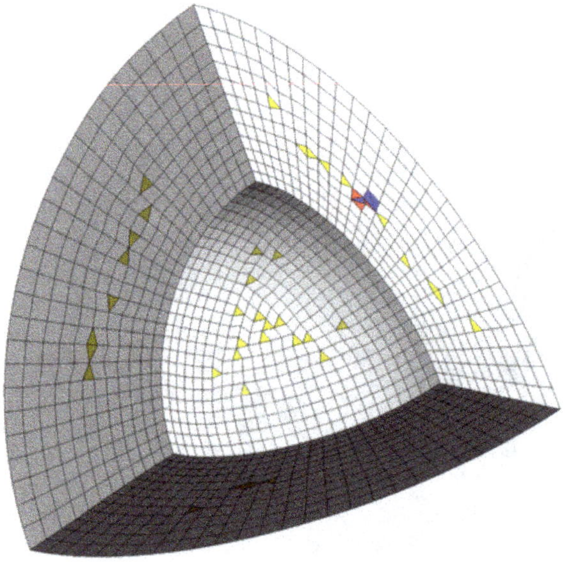

Figure 28 An hollow eighth of ball with a variable mesh size.

the hexahedron will not necessarily be part of the final mesh. Once all tetrahedra have been visited, the array is sorted by hex quality.

The quality Q is defined as follows:

$$Q = \min_{i=1..8} \left| \frac{\mathbf{v_{i1}} \cdot (\mathbf{v_{i2}} \times \mathbf{v_{i3}})}{||\mathbf{v_{i1}}||\,||\mathbf{v_{i2}}||\,||\mathbf{v_{i3}}||} \right|, \quad (8)$$

where i is the vertex number. For a hexahedron, i goes from 1 to 8; $\mathbf{v_{i1}}$, $\mathbf{v_{i2}}$ and $\mathbf{v_{i3}}$ are the three vectors parallel to the three edges connected to vertex i. Q is in fact the modulus of the minimum scaled Jacobian [9]. Evidently, Q is meaningless for invalid hexahedra. Invalid hexahedra are characterized by a null or negative Jacobian determinant, which renders the mesh improper for calculations.

Starting from the highest quality hexahedron, the algorithm then iterates through the array. Potential hexahedra composed of tetrahedra not yet marked for deletion are added to the mesh. The tetrahedra of the added hexahedron are then marked for deletion. It is to be noted that only a small fraction of potential hexahedra appear in the final mesh.

Prisms can later be added by following a similar procedure [9]. All prisms are composed of three tetrahedra. There is only one pattern of construction for prisms [9].

Figure 14 shows a mixed mesh created with Yamakawa-Shimada's algorithm.

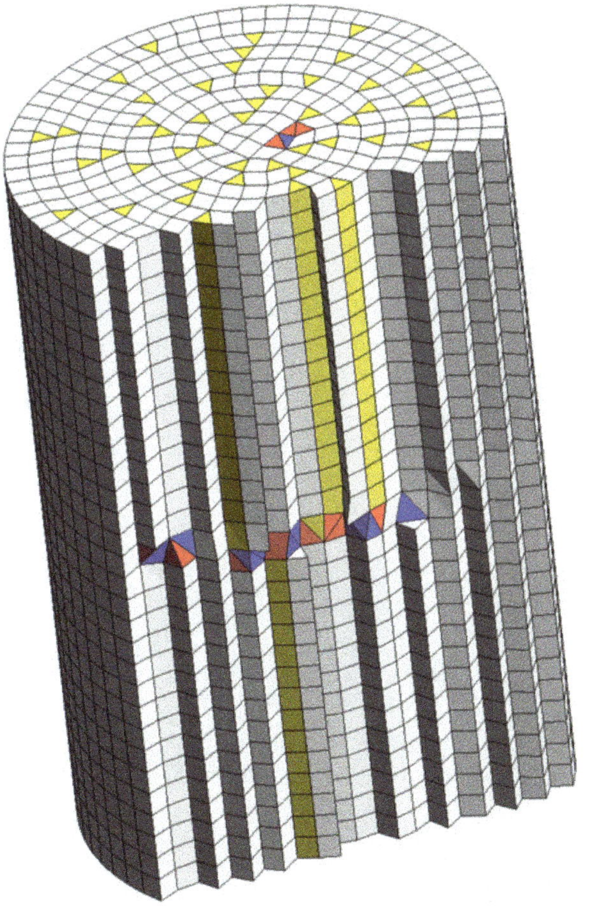

Figure 29 In this mesh, the front was initiated only from the lower and upper disks. The collision is visible in the middle of the cylinder.

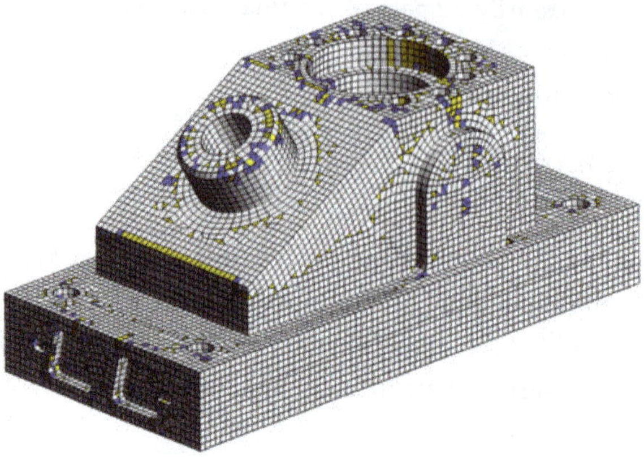

Figure 30 A mixed hexahedral mesh of the anc101 mechanical part.

Let's assume that finite elements of the first order are employed. The tetrahedral shape functions are therefore linear, while the hexahedral shape functions are trilinear [37]. On triangular faces, the interpolation is linear and takes into account three degrees of liberty. On quadrilateral faces, the interpolation is bilinear and takes into account four degrees of liberty [8,9]. If a nonplanar quadrilateral face is adjacent to a triangular face, there will be a gap or overlap [9]. The elements are not going to be a perfect partition of the domain anymore, which goes against the basic assumptions of the finite element method. Gaps or overlaps can also be created by several configurations of neighboring hexahedra or prisms. Figures 15 and 16 show four cases of non-conformities between hexahedra [9].

Figures 17 and 18 show six cases of non-conformities between one hexahedron and one prism. Non-conformities resulting from neighboring prisms can be deduced from these six cases.

Yamakawa-Shimada's algorithm should therefore avoid creating the configurations illustrated on Figures 15, 16, 17 and 18 while iterating through the sorted arrays of potential hexahedra and prisms. Non-conformities can be efficiently identified by employing hashing techniques.

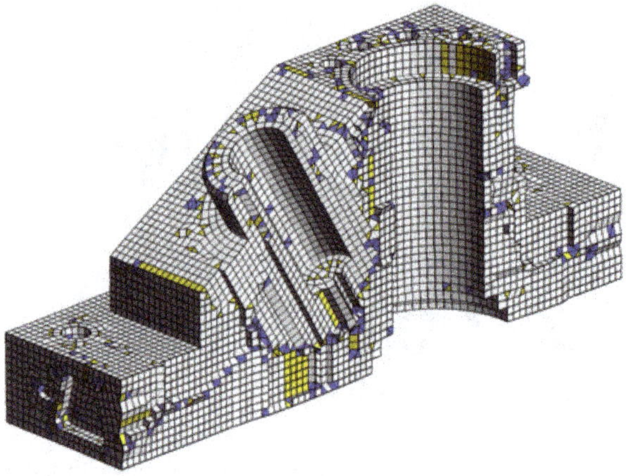

Figure 31 A cutaway view of the anc101 mesh.

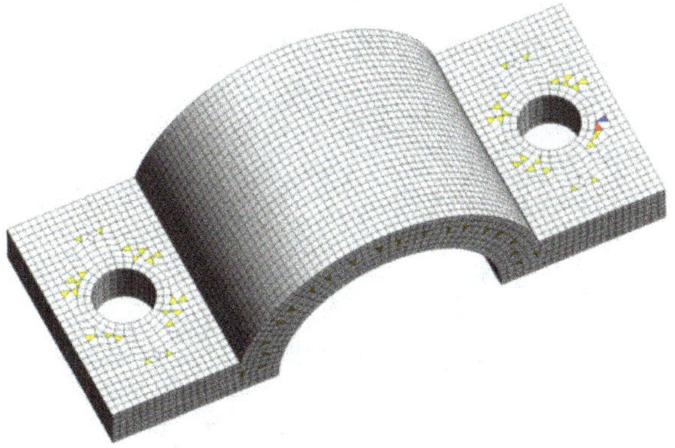

Figure 32 One of the mesh used for the convergence analysis. The mesh size is equal to $h = 1.0$.

After the creation of hexahedra and prisms, some tetrahedra are recombined into pyramids. Every pair of tetrahedra resting on a quadrilateral face is merged to form a pyramid. This step can fix many non-conformities. However, it does not resolve them all. As shown on Figure 19, many quadrilateral faces can still be adjacent to triangular faces belonging to either tetrahedra or pyramids.

These non-conformities can be fixed by Owen-Canann-Saigal's algorithm [38]. Owen-Canann-Saigal's algorithm first creates a flat pyramid on each non-conformal quadrilateral face. The apex of the pyramid is not initially present in the mesh, but it is added by the algorithm. Surrounding tetrahedra and pyramids need to be subdivided to accommodate this new vertex. The pyramid is then raised so it does not have a null volume. Figure 20 illustrates the pyramid constructed to correct the non-conformity on Figure 19.

Owen-Canann-Saigal's algorithm can render a mixed hexahedral mesh completely conformal. However, it has a drawback. It increases the number of tetrahedra and pyramids, which lowers the percentage of hexahedra by number. As a consequence, Owen-Canann-Saigal's algorithm was not used for the results presented below.

Some quadrilateral faces will be adjacent to one or two triangles. Finite element solvers capable of handling these type of non-conformities are required.

Results and discussion

This section presents several mixed hex meshes created with the frontal algorithm and Yamakawa-Shimada's algorithm. Three quantities are used to evaluate the quality of the meshes: the percentage of hexahedra by number H_{nbr}, the percentage of hexahedra by volume H_{vol} and the average hex quality Q defined in Eq. 8. In general, H_{vol} is higher than H_{nbr}. *CPU* designates the total execution time (in *s*) on a 2010 laptop computer. All the data is compiled in Table 1. The following convention is used throughout this section: the hexahedra are white, the prisms are yellow, the pyramids are red and the tetrahedra are

Table 2 Mesh convergence analysis (credits: Gaëtan Compère)

Mesh size	First frequency (Hz)	Second frequency (Hz)
2.0	770.11	1950.65
1.0	760.46	1928.52
0.5	757.55	1921.77
struct.	757.76	1921.93

Figure 33 A motorcycle hub.

blue. The variable *NC* represents the percentage of non-conformal interior quadrilateral faces.

Figure 21 shows a mesh containing 142,466 elements. The mesh size is constant throughout the domain. The CAD model is composed of 250 geometrical faces of various sizes. In order to avoid altering the geometrical edges, hexahedra or prisms whose facets lie on two different geometrical faces are not created.

Each module described in this article shares a certain percentage of the total execution time. These percentages are detailed in Figure 22. The mesh illustrated on Figure 21 was used for the analysis. The blue modules refer to tetrahedra recombination. They are

Figure 34 A marine propeller.

Figure 35 A filter mount.

the most time-consuming. The green and yellow modules pertain to volume and surface mesh vertices generation.

In Figure 23, a rectangular parallelepiped surrounding a NACA airfoil is meshed with 43,094 elements. Figures 24 and 25 are zoomed images of Figure 23.

Figure 26 is a cutaway view of the mixed mesh inside a cylinder. The mesh size is constant and the cardinal directions are radial.

Choosing a size field consistent with the geometry can improve the hexahedra percentage. For the spherical model shown on Figures 27 and 28, a mesh size proportional to the radius is more suitable than a constant one.

A mixed hexahedral mesh of a cylinder is displayed on Figure 29. However, this time the front was initiated only from the lower and upper disks, not from the curved face. In other words, the vertices on the curved face were not allowed to create prospective

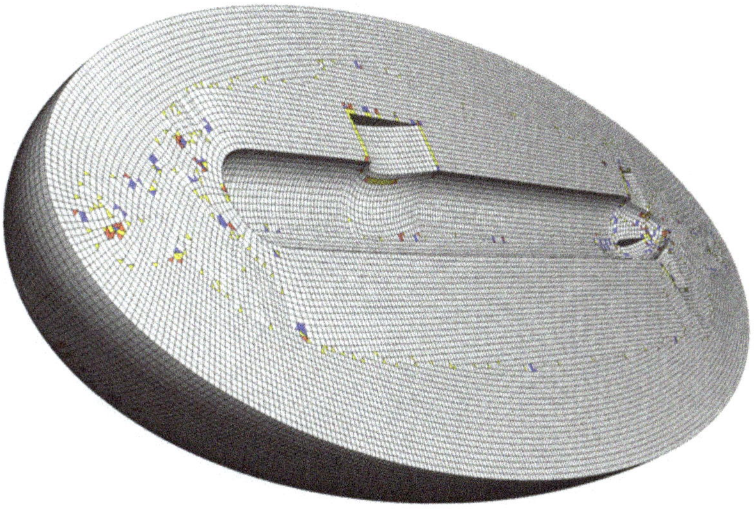

Figure 36 A submarine model.

Figure 37 A cutaway view of the motorcycle hub.

vertices. The mesh has the following statistics: H_{nbr} = 81.88%, H_{vol} = 91.44% and Q = 0.97. Because of the regularity of the curved face mesh, the hexahedra percentages are much higher than those of the previous cylinder.

Figures 30 and 31 show a mixed hexahedral mesh of the anc101 mechanical part. The mesh was generated in 194 seconds and contains 92,282 elements. It has a H_{nbr} of 59.39%. The anc101 part was designed by Computer Aided Manufacturing Inc. [39] and is commonly used in mesh generation literature, in particular in Lévy and Liu's article. For approximately the same mesh size, they obtain a H_{nbr} of 77.14% and an execution time of 12 minutes. Pyramid recombination was not employed in both case.

Figure 38 A cutaway view of the marine propeller.

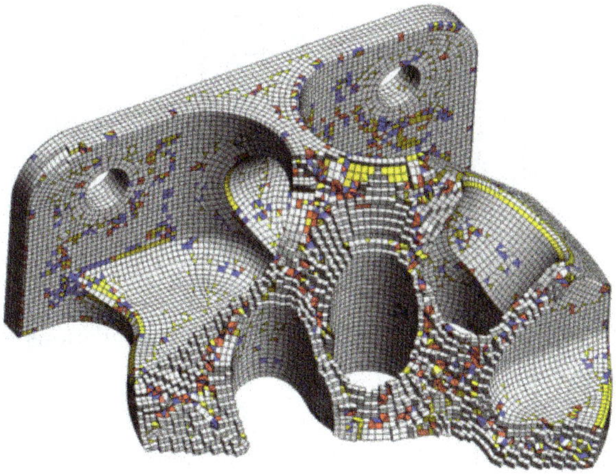

Figure 39 A cutaway view of the filter mount.

A mesh convergence analysis was performed on the mechanical piece displayed on Figure 32. The mechanical piece is made of steel and one of its extremities is fixed. The first and second frequencies are computed, as shown on Table 2. Mixed hexahedral meshes of various densities are employed. According to a similar finite element calculation performed on a structured mesh, the first frequency is equal to 757.76 Hz and the second one is equal to 1921.93 Hz. The results appear to converge.

Figures 33, 34, 35 and 36 present additional examples of mixed hexahedral meshes. Figures 37, 38, 39 and 40 show cutaway views of these meshes. Table 3 contains the corresponding mesh data. The CAD models of these meshes come from an online repository [40-42].

A frequency-domain computational acoustic simulation was performed on the submarine model SUB of Table 3, under plane wave incidence. Figure 41 shows a cutaway view

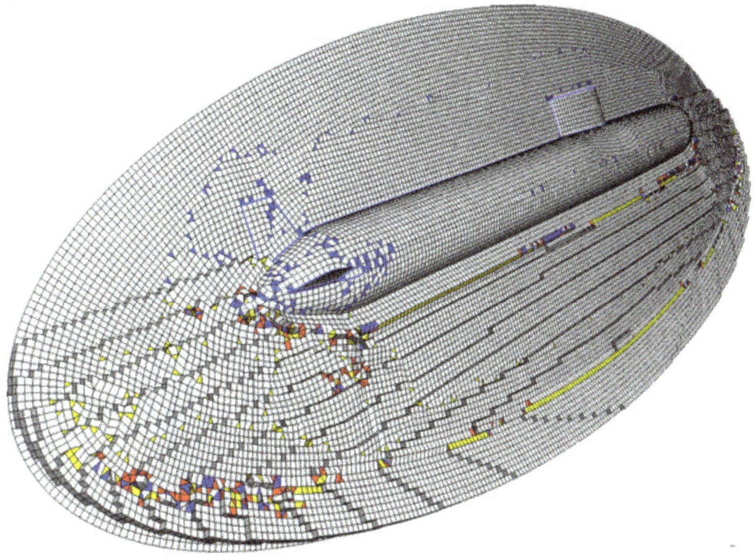

Figure 40 A cutaway view of the submarine model.

Table 3 Mesh data

Figure	# vertices	H_{vol}	CPU (s)
33	133,436	89.74%	247 s.
34	133,678	83.65%	268 s.
35	102,946	78.55%	225 s.
36	598,514	90.28%	1287 s.

The execution times were measured on a 2012 laptop computer.

of the iso-surfaces of the diffracted pressure field. Again, the simulations were shown to converge with mesh refinement.

Surfaces and volumes are meshed sequentially. The two surface meshes bounding a thin region may also not be identical. As a consequence, many non-hexahedral elements can be created. The algorithm is usually less effective for geometrical models featuring many thin regions.

Conclusion

A method capable of generating mixed hexahedral meshes has been presented. The first step consists of covering geometrical boundaries with aligned vertices using a frontal process. The interior is treated in a similar fashion. Vertices creation are guided by a direction field and a size field. The interior vertices are eventually tetrahedralized with a Delaunay procedure. All tetrahedra combinations yielding hexahedra are identified. They are sorted by quality and the highest quality hexahedra are created first. The same approach is applied to prisms. The final mesh contains hexahedra, prisms, pyramids and remaining tetrahedra.

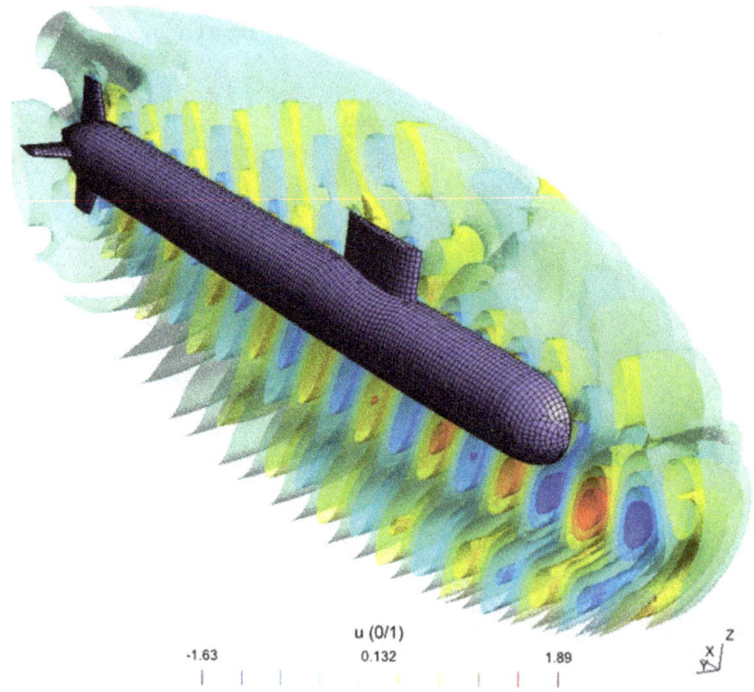

Figure 41 Diffracted acoustic waves around the submarine model.

The method has obvious drawbacks. First, there are no guarantees regarding the hexahedra percentage. It can be higher for certain geometries and lower for others. Secondly, the hexahedra are not anisotropic. For many geometries, well chosen anisotropy could increase the number of hexahedra. Finally, the resulting meshes are useful only to solvers capable of handling a certain number of non-conformal faces.

Competing interests

The authors declare that they have no competing interests.

Authors' contributions

TCB worked on the algorithms and drafted some parts of the manuscript. JFR, EM and FH worked on the algorithms, drafted some parts of the manuscript and carried out detailed revisions. CG performed acoustic finite element analyses with hex-dominant meshes and carried out detailed revisions of the manuscript. All authors read and approved the final manuscript.

Acknowledgements

This work has been partially supported by the Belgian Walloon Region under WIST grants ONELAB 1017086 and DOMHEX 1017074. The authors wish to thank Gaëtan Compère for the mesh convergence analysis and Jonathan Lambrechts for suggesting the use of R-trees. The authors also appreciate the reviewers' efforts and suggestions.

Author details

[1] Université catholique de Louvain, Institute of Mechanics, Materials and Civil Engineering, Bâtiment Euler, Avenue Georges Lemaître 4, Louvain-la-Neuve 1348, Belgium. [2] Université de Liège, Dept. of Electrical Engineering and Computer Science, Montefiore Institute, Bâtiment B28, Sart-Tilman, Liège 4000, Belgium.

References

1. Benzley SE, Perry E, Merkley K, Clark B, Sjaardema G (1995) A comparison of all hexagonal and all tetrahedral finite element meshes for elastic and elasto-plastic analysis. In: Tautges T (eds) Proceedings of the 4th International Meshing Roundtable. Sandia National Laboratories, Albuquerque
2. Puso MA, Solberg J (2006) A stabilized nodally integrated tetrahedral. Int J Numer Meth Eng 67:841–867
3. Ito Y, Nakahashi K (2004) Improvements in the reliability and quality of unstructured hybrid mesh generation. Int J Numer Meth Fl 45:79–108
4. Pirzadeh S (1996) Three-dimensional unstructured viscous grids by the advancing-layers method. AIAA J 34:43–49
5. Kallinderis Y, Ward S (1993) Prismatic grid generation for three-dimensional complex geometries. AIAA J 31(10):1850–1856
6. Shepherd JF, Johnson CR (2008) Hexahedral Mesh Generation Constraints. Eng Comput 24:195–213. https://dl.acm.org/citation.cfm?id=1394234.
7. Meyers RJ, Tautges TJ, Tuchinsky PM (1998) The 'Hex-Tet'hex-dominant meshing algorithm as implemented in CUBIT. In: Freitag L (ed) Proceedings of the 7th international meshing roundtable. Sandia National Laboratories, Dearborn, pp 151–158
8. Dewhirst DL, Grinsell PM, Tucker JR, Mahajan A (1993) Joining tetrahedra to hexahedra. In: Proceedings of MSC World Users' Conference. MSC Software, Arlington. http://web.mscsoftware.com/support/library/conf/wuc93/p04593.pdf.
9. Yamakawa S, Shimada K (2003) Fully-automated hex-dominant mesh generation with directionality control via packing rectangular solid cells. Int J Numer Meth Eng 57:2099–2129
10. Shewchuk JR (1998) Tetrahedral mesh generation by Delaunay refinement. In: Proceedings of the fourteenth annual symposium on Computational geometry. ACM, Minneapolis, pp 86–95
11. Geuzaine C, Remacle JF (2009) Gmsh: a three-dimensional finite element mesh generator with built-in pre- and post-processing facilities. Int J Numer Meth Eng 79:1309–1331
12. Schneiders R, Schindler R, Weiler F (1996) Octree-based Generation of Hexahedral Element Meshes. In: Proceedings of the 5th International Meshing Roundtable. Sandia National Laboratories, Pittsburgh
13. Ito Y, Shih AM, Soni BK (2008) Octree-based reasonable-quality hexahedral mesh generation using a new set of refinement templates. Int J Numer Meth Eng 77:1809–1833
14. Maréchal L (2009) Advances in Octree-Based All-Hexahedral Mesh Generation: Handling Sharp Features. In: Clark BW (ed) Proceedings of the 18th International Meshing Roundtable. Springer, Salt Lake City
15. Blacker TD (2000) Meeting the Challenge for Automated Conformal Hexahedral Meshing. In: Owen S (ed) Proceedings of the 9th International Meshing Roundtable. Sandia National Laboratories, New Orleans
16. Huang J, Tong Y, Wei H, Bao H (2011) Boundary aligned smooth 3D cross-frame field. In: Bala K (ed) Proceedings of ACM SIGGRAPH Asia. Association for Computing Machinery, Hong Kong, p 2011
17. Li Y, Liu Y, Xu W, Wang W, Guo B (2012) All-hex meshing using singularity-restricted field. In: Sloan P (ed) Proceedings of ACM SIGGRAPH Asia. Association for Computing Machinery, Singapore, p 2012
18. Huang J, Jiang T, Wang Y, Tong Y, Bao H (2012) Automatic Frame Field Guided Hexahedral Mesh Generation. Tech. rep., Zhejiang University. http://www.cad.zju.edu.cn/home/hj/12/hex/techreport/hex-techreport.pdf.
19. Nieser M, Reitebuch U, Polthier K (2011) CubeCover - parameterization of 3D volumes. Comput Graph Forum 30:1397–1406
20. Kowalski N, Ledoux F, Frey P (2012) A PDE Based Approach to, Multidomain Partitioning and Quadrilateral Meshing. In: Jiao X, Weill J (ed) Proceedings of the 21th International Meshing Roundtable. Springer, San Jose

21. Gregson J, Sheffer A, Zhang E (2011) All-Hex mesh generation via volumetric polyCube deformation. Comput Graph Forum 30:1407–1416
22. Meshkat S, Talmor D (2000) Generating a mixed mesh of hexahedra, pentahedra and tetrahedra from an underlying tetrahedral mesh. Int J Numer Meth Eng 49:17–30
23. Du Q, Faber V, Gunzburger M (1999) Centroidal Voronoi Tessellations: Applications and Algorithms. Siam Rev 41:637–676
24. Lévy B, Liu Y (2010) L_p Centroidal Voronoi Tessellation and its Applications. In: Hoppe H (ed) Proceedings of ACM SIGGRAPH 2010. Association for Computing Machinery, Los Angeles
25. Ray N, Vallet B, Li WC, Lévy B (2008) N-Symmetry direction field design. ACM T Graph 27:1–25
26. Guttman A (1984) R-Trees: A Dynamic Index Structure for Spatial Searching. In: ACM Special Interest Group on Management of Data. Association for Computing Machinery, Boston, pp 47–57
27. Douglas G, Green M, Guttman A, Stonebraker M (2004) R-trees: a dynamic index structure for spatial searching. http://www.superliminal.com/sources/RTreeTemplate.zip.
28. Vyas V, Shimada K (2009) Tensor-Guided Hex-Dominant Mesh Generation with Targeted All-Hex Regions. In: Clark BW (ed) Proceedings of the 18th International Meshing Roundtable. Springer, Salt Lake City
29. Remacle JF, Geuzaine C, Compère G, Marchandise E (2010) High quality surface meshing using harmonic maps. Int J Numer Meth Eng 83:403–425
30. Marchandise E, de Wiart CC, Vos WG, Geuzaine C, Remacle JF (2011) High-quality surface remeshing using harmonic maps-Part II: Surfaces with high genus and of large aspect ratio. Int J Numer Meth Eng 86:1303–1321
31. Marchandise E (2013) Remacle JF. Eng Comput. doi:10.1007/s00366-012-0309-3
32. Remacle JF, Henrotte F, Baudouin TC, Geuzaine C, Béchet E, Mouton T, Marchandise E (2011) A Frontal Delaunay quad mesh generator using the L^{∞} norm. In: Quadros W (ed) Proceedings of the 20th International Meshing Roundtable. Springer, Paris
33. Mount DM, Arya S (1997) ANN: A library for approximate nearest neighbor searching. In: CGC Workshop on Computational Geometry. The Center for Geometric Computing, Durham, pp 33–40
34. Remacle JF, Lambrechts J, Seny B, Marchandise E, Johnen A, Geuzaine C (2012) Blossom-Quad: a non-uniform quadrilateral mesh generator using a minimum cost perfect matching algorithm. Int J Numer Meth Eng 89:1102–1119
35. Si H (2006) TetGen: A Quality Tetrahedral Mesh Generator and Three-Dimensional Delaunay Triangulator. http://tetgen.berlios.de/files/tetgen-manual.pdf
36. Baudouin TC, Remacle JF, Marchandise E, Lambrechts J, Henrotte F (2012) Lloyd's energy minimization in the L_p norm for quadrilateral mesh generation. Eng Comput. DOI:10.1007/s00366-012-0290-x
37. Fortin A, Garon A (2014) Les Eléments finis : de la théorie à la pratique. http://giref.ulaval.ca/files/afortin/Publications/elements_finis.pdf.
38. Owen SJ, Canann SA, Saigal S (1997) Pyramid Elements for Maintaining Tetrahedra to Hexahedra Conformability. In: Canann S, Saigal S (ed) ASME Trends in Unstructured Mesh Generation. American Society of Mechanical Engineers, Evanston, pp 123–129
39. Agoston MK (2005) Computer Graphics and Geometric Modeling. Springer-Verlag, USA
40. Rieling R (2013) Yamaha XTZ-125. http://grabcad.com/library/yamaha-xtz-125.
41. 5 Bladed Propeller (2011). http://grabcad.com/library/5-bladed-propeller.
42. Hall C (2013) CV HP1. http://grabcad.com/library/cv-hp1.

6

Reduced yield stress for zirconium exposed to iodine: reactive force field simulation

Matthew L Rossi[1,4*], Christopher D Taylor[2,4] and Adri CT van Duin[3]

* Correspondence:
Matthew.Rossi@nrc.gov
[1]Nuclear Regulatory Commission, Rockville, MD 20852, USA
[4]Los Alamos National Laboratory, Los Alamos, NM 87545, USA
Full list of author information is available at the end of the article

Abstract

Iodine-induced stress-corrosion cracking (ISCC), a known failure mode for nuclear fuel cladding, occurs when iodine generated during the irradiation of a nuclear fuel pellet escapes the pellet through diffusion or thermal cracking and chemically interacts with the inner surface of the clad material, inducing a subsequent effect on the cladding's resistance to mechanical stress. To complement experimental investigations of ISCC, a reactive force field (ReaxFF) compatible with the Zr-I chemical and materials systems has been developed and applied to simulate the impact of iodine exposure on the mechanical strength of the material. We show that the material's resistance to stress (as captured by the yield stress of a high-energy grain boundary) is related to the surface coverage of iodine, with the implication that ISCC is the result of adsorption-enhanced decohesion.

Keywords: Zirconium; Iodine; Stress corrosion cracking; Failure; Molecular dynamics; Simulation; Reactive force field; Grain boundary

Correspondence/Findings

Iodine-induced stress-corrosion cracking (ISCC) is a complex phenomenon with significant interest to the nuclear industry. One of the major fission products of uranium is iodine, which is corrosive to zirconium-based cladding. The van-Arkel process [1], developed in 1925, catalogued the reaction between zirconium and iodine for use as a method of metal purification, utilizing the volatility of the tetra-coordinated iodide, ZrI_4 [2]. Despite this known reactivity between zirconium and iodine, zirconium was chosen as a cladding material due to its general corrosion resistance, good thermal conductivity, and low cross-section to thermal neutrons [3,4]. There have been a significant number of studies regarding the effects of iodine on Zr-based cladding materials and the pellet-cladding interactions (PCI) which occur on the inner surface of the cladding, such as those by Lyons et al. [5] and Atrens et al. [6]. In work by Lyons et al. [5], and others [7-11], fission products of interest to PCI corrosion have included cesium and cadmium in addition to iodine, all of which have a weakening effect on the mechanical strength of zirconium, as supported in theoretical work by Wimmer et al. [12]. It can be difficult to collect reactor-relevant experimental data due to the irradiation damage, exposure to high-energy atom bombardment, and reactor conditions the cladding is subjected to during its lifecycle. Much of the available experimental data regarding ISCC has been in controlled conditions, utilizing halide-solutions, such as the work of Francon [13], Goryachev [14], and Park [15]. However, it remains unknown how the stress-corrosion response initiates [14,16], since post-failure analysis precludes examination of this critical first step.

ISCC can involve multiple, non-linear, and possibly parallel, reaction steps with regards to the overall process. The complex nature of ISCC lies in the governing kinetics and potential pathways of the reaction. The cladding material itself is usually comprised of Zircaloy-4 or Zirlo. The general corrosion process of zirconium alloys has been documented in previous experimental work by Farina [17-19], Francon [13] and Kim [20], among others. Parallel reaction paths can arise due to the iodine decay chain, introducing mechanical effects as iodide phases decompose to chemically inert xenon. Considerations should also be given to the presence of the 8–10 μm thick passive oxide layer that protects the inner surface of the cladding [18]. This oxide layer is the first defense of the cladding material from ISCC, and the means via which it is compromised during an ISCC event is unclear [21]. It is possible that synergistic interactions with the electropositive fission products (Cs, Cd, Sr, among others) can chemically modify and/or mechanically weaken the oxide, undermining its passivity. The significance of grain-boundaries and triple-points was made apparent in recent investigations by Park [15]. A fundamental approach to modeling ISCC, therefore, should involve a deconstruction of these effects, to assess their individual contributions, prior to a final integration.

In our previous work [22-24], several mechanisms were investigated to help construct such a model. Beginning with the chemical interaction, the series of iodine intermediary states was examined, under the general scheme of iodine aggregation:

$$\mathrm{Zr} + 2\mathrm{I}_2 \leftrightarrow \mathrm{ZrI}_2 + \mathrm{I}_2 \leftrightarrow \mathrm{ZrI}_3 + \frac{1}{2}\mathrm{I}_2 \leftrightarrow \mathrm{ZrI}_4$$

In addition to molecular iodine aggregation, the properties of systems such as surface adsorbates, dimers of molecules, and mixed crystal systems were also modeled as being pertinent to the overall reaction scheme. The focus of the present work is now on the integration of the molecular and crystallographic studies into a molecular dynamics (MD) based model for simulating the interactions between zirconium and iodine. Although the interactions between the oxide and Cs and I, as primary fission products, are also of great interest, the challenges inherent to the development of interatomic potentials for ternary systems and beyond necessitate that we begin with the Zr-I binary. Specifically, we perform molecular dynamic simulations for the interactions of iodine with the 0001 (basal) plane of zirconium, which is known to provide the greatest resistance to ISCC.

While density-functional theory (DFT) was previously utilized to examine the molecular, solid state, and gaseous interactions pertinent to the Zr-I system, [22-24] the approach is computationally expensive and thus impractical for the study of the interaction of I with grain-boundaries that intersect the surface, and its effect on the stress–strain response. ReaxFF, a reactive force field developed by van Duin et al. [25], enables simulation of large systems (*i.e.* $>10^4$ atoms), based on a training set created by molecular and solid-state DFT calculations. ReaxFF MD simulations can predict chemical reactions (including changes in bonding) and diffusion pathways and materials properties while remaining computationally tractable. The specific details of computational models used have been published previously [22-24], and were used in the training set for the ReaxFF force field parameters [26]. The Large-scale Atomic/Molecular Massively Parallel Simulator, LAMMPS [27,28], was used to perform the ReaxFF-MD [29] simulations described herein.

Given that ISCC is a joint chemical–mechanical effect, it is critical to explore the role iodine plays in modifying the mechanical response of zirconium metal. Since ISCC primarily occurs in the intergranular mode, we consider the resistance of a tilt grain-boundary that intersects the (0001) surface plane to a systematically applied strain rate during molecular dynamics simulation. In order to select the grain boundary most susceptible to this kind of effect, we compare the unrelaxed grain-boundary energies for the slab bicrystals as a function of θ, the angle of rotation about the (0001) axis, Figure 1. Here we compare the grain-boundary energies by scaling the simulations to the equivalent energy per atom (hence units eV/atom):

$$\gamma \propto \frac{E_{gb} - E_{gb,\theta=0^{\circ}}}{N_{gb}}$$

Due to symmetry the energies are periodic every 60°, with reflective symmetry and a local minimum about 30°, at which angle every other plane parallel to the surface is coincident. Based on the relations in Figure 1, we selected a grain boundary with 15° tilt angle, as this creates a high-energy grain boundary that would be anticipated to be the most vulnerable to reaction with iodine [30]. Since the scope of this study is to demonstrate the utility of Reax-MD simulations to explore the synergies between chemical and mechanical effects, we defer systematic study of other grain boundary systems for future work.

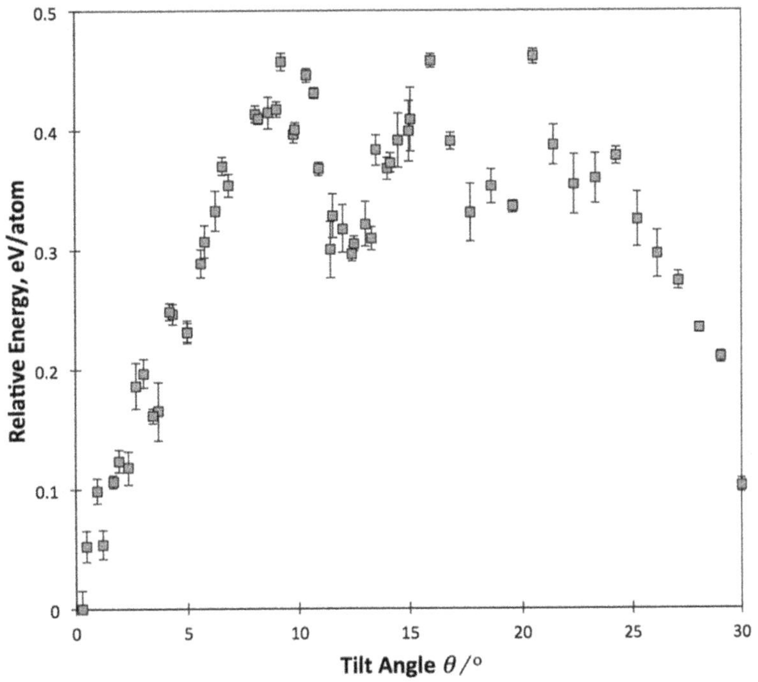

Figure 1 Relative grain boundary energy as a function of tilt angle about the (0001) axis calculated via ReaxFF. The error bars indicate the standard deviation.

Reax-MD simulations were then performed by introducing varying amounts of iodine into the vacuum space between the periodic slabs of the Zr(0001)-oriented bicrystals. Each simulation contained 31,500 Zr atoms divided into two single crystals 20 atomic layers thick, separated by the grain-boundary. An energy minimization was performed and then the volume and atom positions in the simulation supercell were equilibrated to 500°C and 0 MPa in the plane of the bicrystal. Due to the heavy mass of the atoms, a large step size could be used for the MD of 10 fs/timestep. Given the equilibrium cell volume and the number of iodine atoms introduced, the effective partial pressure of iodine was determined. Iodine was introduced in the atomic state, simulating radiolytic release. Iodine may also be present in other chemical states: molecular I_2, cesium iodide, or zirconium iodide vapors, but these were not treated within this study. Since it is known that molecular iodine adsorbs onto zirconium metal surfaces without a barrier, it is expected that atomic iodine should behave similarly to molecular iodine [31]. Following this step, MD simulations were performed under conditions of constant strain rate and at a constant temperature of 500°C. The temperature was selected to be representative of fission gas environments encountered during fuel pellet power cycles. The strain was applied in the direction perpendicular to the grain boundary at a rate of 10^8/s, or 10% over the 1 ns simulation. The stress in this direction was computed and normalized to the slab cross-sectional area to obtain the effective stress corresponding to the strain every 50 time steps. Subsequently, the stress–strain curve for the bicrystal was obtained for a range of iodine partial pressures ranging from 0–12 MPa. Pressures in this range are representative of the fission gas environment within the fuel-pellet gap. Stress–strain data is available directly as an excel file in the Additional file 1.

During our simulations, the maximum yield stress was obtained around 4% applied strain. The maximum yield stress was captured for each concentration of iodine and the results are plotted in Figure 2 (left axis). The results indicate a rapid fall-off of the grain-boundary resistance to applied stress up to 20% of the yield stress without iodine, when iodine pressures up to 0.5 MPa are applied. Following this point, the reduction

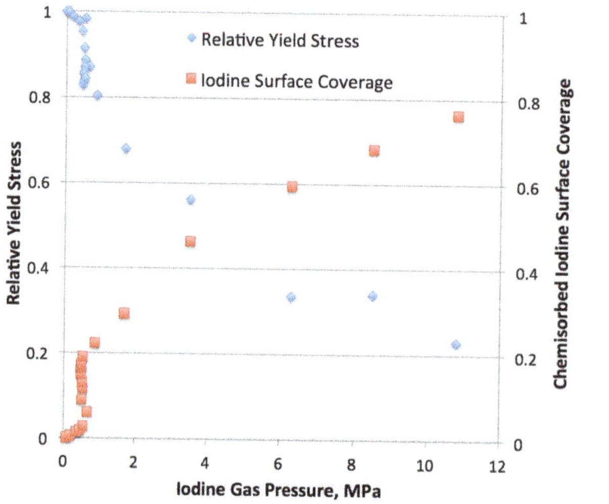

Figure 2 Relative yield stress compared to the iodine-free case (left axis) and the chemisorbed iodine surface coverage (right axis) plotted against the iodine gas pressure (iodine as I_2 molecules).

becomes more gradual, with up to 80% reduction in grain boundary yield stress at 11 MPa.

The MD trajectories were then examined to explore the mechanisms for this non-linear behavior. In the low-pressure region (<0.5 MPa) it was observed that the small number of iodine atoms introduced into the vapor space are almost entirely chemisorbed onto the surface, a process that interferes with the ability to accurately equilibrate the pressure in the gas phase. We use the term chemisorption because it is known that molecular iodine dissociates without a barrier onto the zirconium surface, and so the bonding relation between the undercoordinated Zr and I is stronger than the covalent I_2 bond [31]. As seen in Figure 2 (right axis) the surface chemisorption follows a similar non-linear response as a function of pressure, indicating that there is a direct connection between the surface chemisorption phenomena and the reduction in the applied stress. We, therefore, chose to discard pressure as a significant variable for pressures <0.5 MPa and evaluated the relative yield stress as a function of chemisorbed iodine surface coverage. When the data is presented in this way (Figure 3) the relation becomes strongly linear, indicating that the extent to which the monolayer iodide film is allowed to form has a significant impact on the stress required to initiate materials failure at the point of the grain-boundary.

Examination of the molecular dynamics trajectory about the yield point (Figure 4) indicates that yielding is initiated by the creation of excess surface area at the point where the grain boundary intersects the surface (i.e. crack initiation). In both of the cases shown in Figure 4, the formation of that yield point is accompanied by ingress of iodine

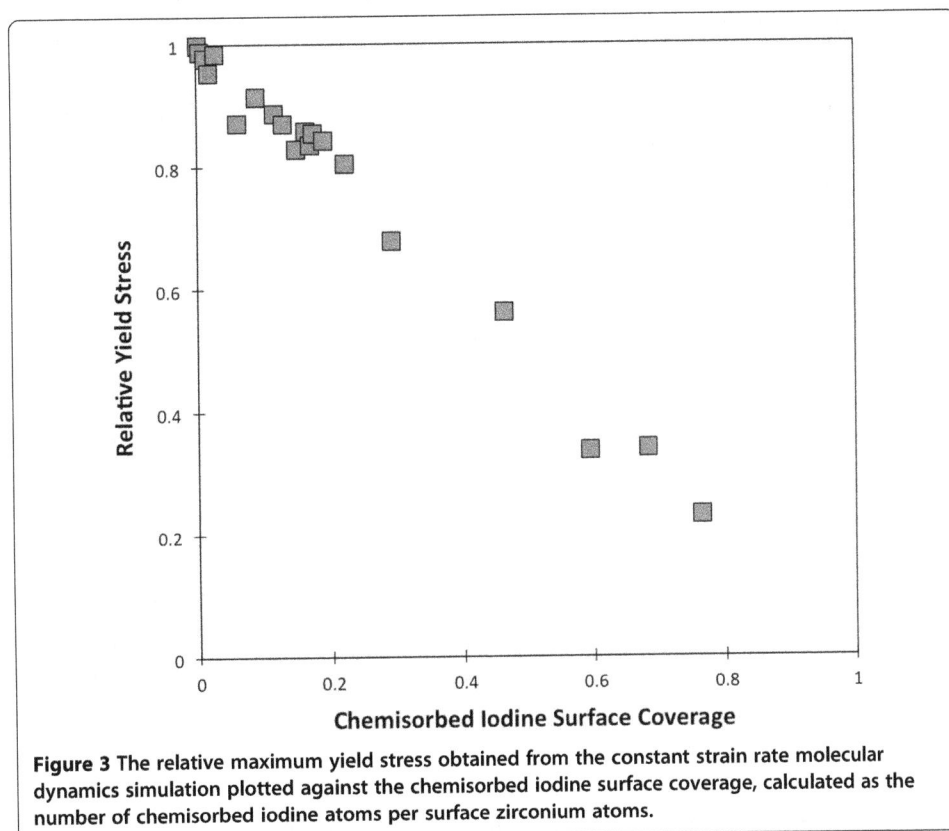

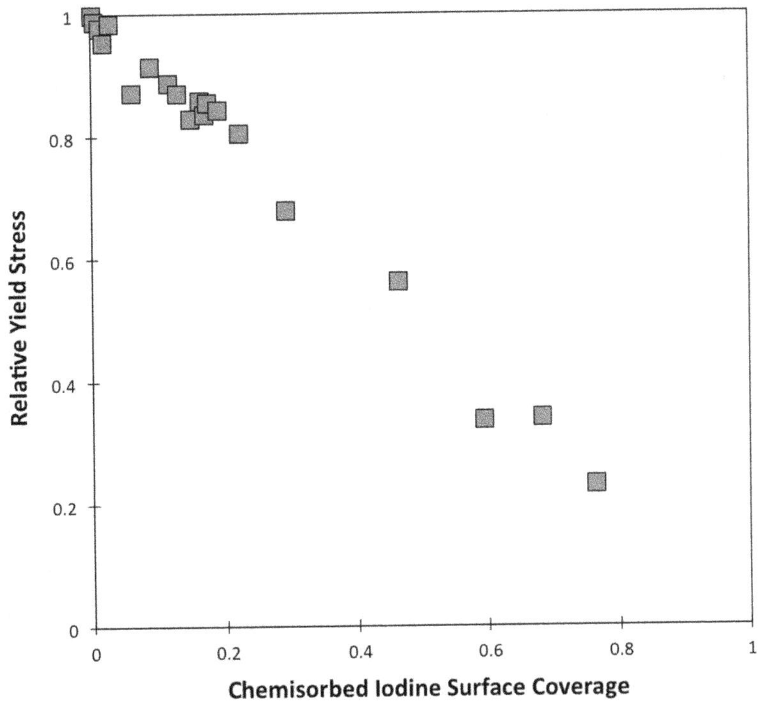

Figure 3 **The relative maximum yield stress obtained from the constant strain rate molecular dynamics simulation plotted against the chemisorbed iodine surface coverage, calculated as the number of chemisorbed iodine atoms per surface zirconium atoms.**

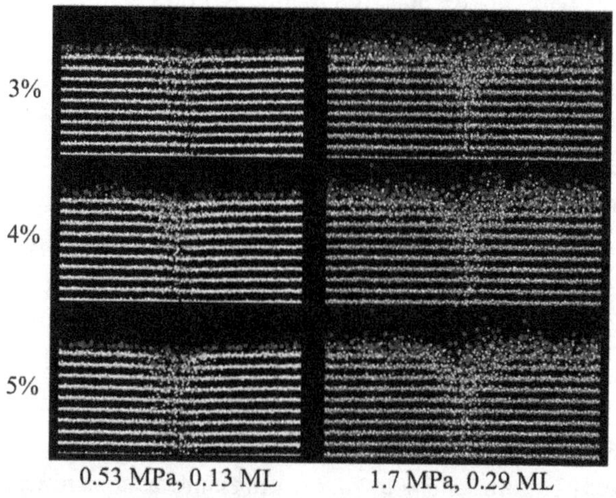

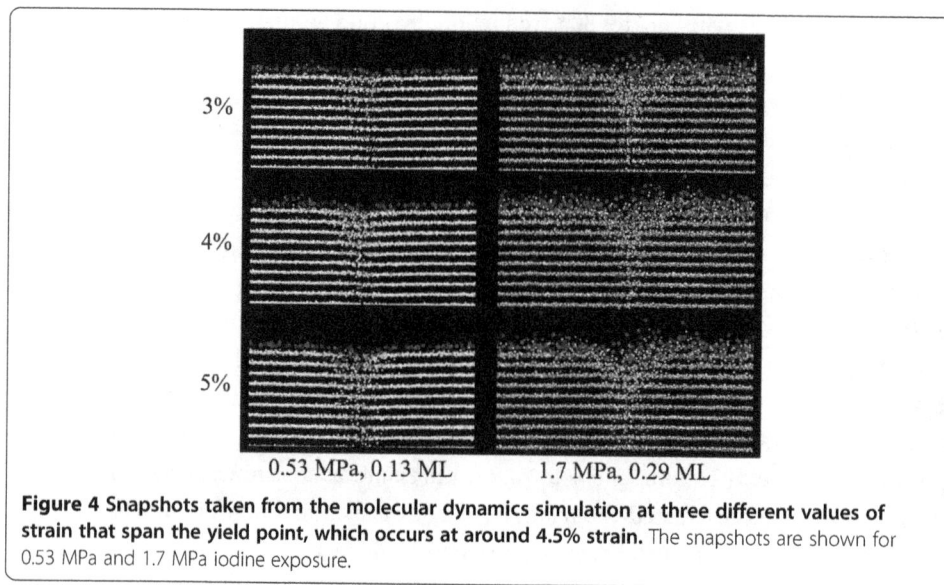

Figure 4 Snapshots taken from the molecular dynamics simulation at three different values of strain that span the yield point, which occurs at around 4.5% strain. The snapshots are shown for 0.53 MPa and 1.7 MPa iodine exposure.

into the crack region. The ability for iodine to migrate rapidly across the surface, [31] in addition to its propensity to form multicoordinate bonds to the freshly exposed Zr atoms [22] lowers the surface energy, and, consequently, the energy penalty associated with the yield phenomenon. In essence, this is similar to the enhanced decohesion mechanism of hydrogen-induced cracking [32].

A further distinction noted in Figure 4 relates to the change in iodine film structure between 0.53 MPa and 1.7 MPa: In the lower pressure case, the iodide film constitutes a partial monolayer, at the higher pressure, a phase transformation has occurred in the surface phase, resulting in the intermingling of surface Zr and I atoms, according to the agglomeration pathway outlined above [2]. The reorganization of the surface layers that accompanies the surface transformation provides an additional impetus for initiating the crack-opening at the grain-boundary/surface intersection, although the mechanistic aspects associated with this should be explored more thoroughly in future work. While very little direct incorporation of iodine into the grain boundary was observed, it has been speculated that grain boundaries facilitate diffusion of iodine into the material due to their excess volume. Relative to the chemisorption effect, however, this mechanism is slow and the material weakening observed herein is a consequence of the iodide film formation and subsequent disorder that it introduces into the surface layers of the material, which result in overall weakening and enhanced susceptibility to mechanical deformation.

In summary, the reactive force field simulations demonstrate that iodine chemisorption leads to a reduction in the resistance of a high-energy grain-boundary in zirconium metal to applied stress, and undergoes a chemisorption mediated reaction to form a zirconium iodide film. The extent to which the relative yield stress is lowered is linearly related to the surface coverage of adsorbed iodine. The methodology demonstrated in this work can now be extended to explore the mechanistic interactions between iodine, iodide-films and stress states at the intersection between the grain-boundary and the surface, as well as the volatilization of zirconium iodide under reactor conditions.

Additional file

Additional file 1: Stress–strain data and relevant computations based on the raw data and the stress–strain figures are presented in this additional excel file.

Abbreviations

DFT: Density functional theory; ISCC: Iodine-induced Stress Corrosion Cracking; LAMMPS: Large-scale Atomic/Molecular Massively Parallel Simulator; MD: Molecular Dynamics; PCI: Pellet-clad interaction; ReaxFF: Reactive Force Field.

Competing interests

The authors declare that they have no competing interests. This journal article was prepared, in part, by an employee of the U.S. Nuclear Regulatory Commission on his own time apart from his regular duties. NRC has neither approved nor disapproved its technical content.

Authors' contributions

MR developed the ReaxFF potential and performed DFT and LAMMPS simulations along with data analysis. CT also performed LAMMPS simulations and analysed the results. AvD provided assistance with development of the ReaxFF potential. All authors played contributing roles in the writing and approval of this final manuscript.

Acknowledgment

This research was performed at Los Alamos National Laboratory and was supported by the Consortium for Advanced Simulation of Light Water Reactors (www.casl.gov), an Energy Innovation Hub (http://www.energy.gov/hubs) for Modeling and Simulation of Nuclear Reactors under U.S. Department of Energy Contract No. DE-AC05-00OR22725. The Los Alamos National Security LLC for the National Nuclear Security Administration of the U.S. Department of Energy under contract DE-AC52-06NA25396. The authors would like to acknowledge Robert Montgomery (PNNL) for his valuable input.

Author details

[1]Nuclear Regulatory Commission, Rockville, MD 20852, USA. [2]Fontana Corrosion Center, The Ohio State University, Columbus, OH 43210, USA. [3]Mechanical and Nuclear Engineering, Pennsylvania State University, University Park, PA 16802, USA. [4]Los Alamos National Laboratory, Los Alamos, NM 87545, USA.

References

1. van Arkel AE, de Boer JH (1925) Darstellung von reinem Titanium-, Zirkonium-, Hafnium- und Thoriummetall. ZAAC 148(1):345–350
2. Cubicciotti D, Lau KH, Ferrante MJ (1978) Thermodynamics of vaporization and high temperature enthalpy of zirconium tetraiodide. J Electrochem Soc 125:972
3. (1994) Concise Encyclopedia Chemistry. Walter De Gruyter Inc, trans: Eagleson M
4. Fairchild HB (1949) The Properties of Zirconium and Its Possibilities for Thermal Reactors. Laboratory, Oak Ridge National Laboratory, Oak Ridge, TN
5. Lyons MF, Coplin DH, Jones GG (1963/1964) High Performance UO_2 Fuel Program. Quarterly Progress Reports. General Electric Company, GEAP-3771-15
6. Atrens A, Dannhäuser G, Bäro G (1984) Stress-corrosion-cracking of Zircaloy-4 cladding tubes. J Nuc Mater 126:91–102
7. Cox B, Surette BA, Wood JC (1981) In: McNitt RP, Sisson RD (eds) Environmental Degradation of Engineering Materials in Aggressive Environments. Pennsylvania State University, State College, PA, p 293
8. Cox B (1990) Pellet-Clad Interaction (PCI) Failures of Zirconium Alloy Fuel Cladding - A Review. J Nuc Mater 172:249–292
9. Götzmann O (1982) Thermochemical Evaluation of PCI Failures in LWR Fuel Pins. J Nuc Mater 107:185–195
10. Konashi K, Kamimura K, Yokouchi Y (1984) Estimation of Irradiation Induced Iodine Pressure in an LWR Fuel Rod. J Nuc Mater 125:244–247
11. Grimes RW, Ball RGJ, Catlow CRA (1992) Site preference and binding of iodine and caesium in uranium dioxide. J Phys Chem Solids 53(4):475–484
12. Wimmer E, Najafabadi R, Young GA Jr, Ballard JD, Angeliu TM, Vollmer J, Chambers JJ, Niimi H, Shaw JB, Freeman C, Christensen M, Wolf W, Saxe P (2010) *Ab initio* calculations for industrial materials engineering: successes and challenges. J Phys Condens Matter 22:384215
13. Francon V, Fregonese M, Abe H, Watanabe Y (2012) Iodine-Induced Stress Corrosion Cracking of Zircaloy-4: Identification of Critical Parameters Involved in Intergranular to Transgranular Crack Propagation. Solid State Phenomena 183:49–56
14. Goryachev SB, Gritsuk AR, Prasolov FF, Snegirev MG, Shestak VE (1992) Iodine induced SCC of Zr alloys at constant strain rate. J Nuc Mater 199:50–60
15. Park SY, Kim JH, Choi BK, Jeong YH (2007) Crack initiation and propagation behavior of zirconium cladding under an environment of iodine-induced stress corrosion. Metals and Mat Intl 13(2):155–163
16. Likhanskii VV, Matweev LV (2002) The development of the crack growth model in zirconium claddings in iodine environment. Nuc Eng Des 213:133–140
17. Farina SB, Duffo GS, Galvele JR (2002) Localized corrosion of zirconium and zircaloy-4 in iodine alcoholic solutions. LAAR 32:295–298

18. Farina SB, Duffo GS, Galvele JR (2002) Stress Corrosion Cracking of Zircaloy-4 in Halide Solutions. Effect of Temperature. Mat Res 5:107–112
19. Farina SB, Duffo GS (2004) Intergranular to transgranular transition in the stress corrosion cracking of Zircaloy-4. Corr Sci 46:2255–2264
20. Kim H-G, Jeong Y-H (2006) Effect of Annealing Conditions on the Microstructure and Corrosion Characteristics of Zr-xNb Alloys. Solid State Phenomena 118:77–82
21. Lewis BJ, Thompson WT, Kleczek MR, Shaheen K, Juhas M, Iglesias FC (2011) Modelling of iodine-induced stress corrosion cracking in CANDU fuel. Journal of Nuclear Materials 408(3):209–223, doi:10.1016/j.jnucmat.2010.10.063
22. Rossi ML, Taylor CD (2011) Atomistic Simulations of Formation of Elementary Zr-I Systems. O J Phys Chem 1:104–108
23. Rossi ML, Taylor CD (2013) Equations of State for Crystalline Zirconium Iodide: The Role of Dispersion. J Nuc Mater 433(1–3):30–36
24. Rossi ML, Taylor CD (2013) First-principles Insights into the Nature of Zirconium-Iodine Interactions and the Initiation of Iodine-induced Stress-Corrosion Cracking [in revision]. J Nuc Mater
25. van Duin ACT, Dasgupta S, Lorant F, Goddard WA 3rd (2001) ReaxFF: A reactive force field for hydrocarbons. J Phys Chem A 105:9396–9409
26. ReaxFF potentials available upon request. Contact acv13@psu.edu.
27. Large-scale Atomic/Molecular Massively Parallel Simulator. http://lammps.sandia.gov. Accessed August 6 2012
28. Plimpton SJ (1995) Fast Parallel Algorithms for Short-Range Molecular Dynamics. J Comp Phys 117:1–19
29. Chenoweth K, van Duin ACT, Goddard WA III (2008) ReaxFF Reactive Force Field for Molecular Dynamics Simulations of Hydrocarbon Oxidation. J Phys Chem A 112(5):1040–1053
30. Musienko A, Cailletaud G (2009) Simulation of inter- and transgranular crack propagation in polycrystalline aggregates due to stress corrosion cracking. Acta Mater 57:3840–3855
31. Legris A, Domain C (2005) Ab initio atomic-scale modeling of iodine effects on hcp zirconium. Phil Mag 85:589–595
32. Van der Ven A, Ceder G (2002) The thermodynamics of decohesion. Acta Mat 52:1223–1235

7

Treatment of nearly-singular problems with the X-FEM

Grégory Legrain* and Nicolas Moës

*Correspondence:
gregory.legrain@ec-nantes.fr
LUNAM Université, GeM, UMR CNRS 6183, Ecole Centrale de Nantes, 1 rue de la Noë, 44321 Nantes, France

Abstract

Background: In recent years, lot of research have been conducted on fictitious domain approaches in order to simplify the meshing process for computed aided analysis. The behaviour of such non-conforming methods is studied in the case of the approximation of nearly singular solutions. Such solutions appear when problems involve singularities whose center are located outside (but close) of the domain of interest. These solutions are common in industrial structures that usually involve rounded re-entrant corners.

Methods: The performance of the finite element method is evaluated in this context by means of a simple unidimensional example. Both numerical and theoretical estimates are considered in order to assess the behaviour of the numerical approximation. It is demonstrated that despite being regular, the convergence of the approximation can be bounded to an algebraic rate that depends on the solution. Reasons for such behaviour are presented, and two complementary strategies are proposed in order to recover optimal convergence rates. The first strategy is based on a proper enrichment of the approximation thanks to the X-FEM, while the second is based on a proper mesh design that follows a geometric progression. Finally, the proposed strategies are extended and validated in 2D.

Results: The performance of the two strategies is highlighted for both 1D and 2D examples. Both methods allow to recover proper convergence rates (optimal algebraic rate for *h*-convergence, exponential for *p*-convergence) in 1D and 2D.

Conclusions: The proposed strategies allow for a very accurate solution for such solutions. The enrichment strategy is valid for both *h* and *p* refinement, whereas the mesh-design strategy is only usable for *p* refinement. However, such enrichment functions can be tedious to derive.

Keywords: X-FEM; Non-conforming; p-fem; Singularity; Convergence

Background

Industrial structures usually involve re-entrant corners with possibly small fillets. Accurate stress analysis for such structures requires the proper treatment of these geometrical features. The size of these fillets is highly dependent on the manufacturing process, and depending on the quantity of interest and the size of the fillets with respect to global scale of the structure, it may be neglected in the definition of the mathematical model which is used for the computation. The problem is that such areas with high curvature necessitates the use of very small elements, unless blending mapping [1] or Nurbs-Enhanced FEM [2,3] are used. These very small elements have a high impact of the computational cost of the analysis, so that these geometrical features are usually discarded in the analysis and replaced by acute corners. In this case, the numerical solution is not consistent anymore with the mathematical model: the verification of the model is no longer possible, especially if the quantity of interest are related to stresses or strains in the fillet's area. The use of non-conforming approaches such as the X-FEM [4] or fictitious domain [5] can be considered in order to solve this mesh-density issue. Indeed, conforming meshing can be avoided, the price being paid in the integration process. In addition, one has to take care of the correct geometrical description: for example, the use of a level-set for representing the geometry won't allow to obtain an accurate geometry, unless a mesh with a density of the order of the curvature radius is used. Otherwise, one has to consider "sub-grid" level-sets as advocated in [6-8], a fine pixelized representation, as in [9], or the so called Nurbs-Enhanced X-FEM [10]. By means of these strategies, the size of the computational mesh does not have to be related to the size of the geometrical features. The solution being regular, optimal convergence rates are expected. However, albeit being regular, mechanical fields can be very rough in the fillet area. As highlighted in the following, this quasi-singularity prevents an optimal convergence of the solution when using non conforming "engineering" meshes i.e. meshes with a moderate number of elements. The objective of this contribution concerns the quantification of strategies for improving the convergence rate of low and high-order non-conforming finite element methods. Two paths can be followed: (i) using the partition of unity [11] and enrich the finite element approximation with adapted functions ; or (ii) using p-fem strategies that are based on non-conforming meshes with proper grading near the singularities [1]. These two strategies are first motivated in a one-dimensional settings, then validated and compared in a 2D setting.

This work is organized as follows: first, mechanical fields near fillets are presented, and their nearly singular behaviour highlighted. In a second part, the eXtended Finite Element Method is introduced together with some recent improvement in the field of high-order approximations. Then, a 1D model problem is introduced in order to highlight the influence of nearly singular fields on the convergence properties of the finite element method. A close study in the error contribution of the elements of the mesh enables us to propose strategies in order to improve the convergence. These strategies are extended in the 2D setting, and validated by means of various numerical examples. Finally, performances of these strategies are compared before concluding.

Near-fillets mechanical fields

As stated in the introduction, the example of a traction-free blunt re-entrant corner is considered as an example of nearly-singular problem. The geometry of interest is depicted in Figure 1(a). The opening angle of the corner is denoted as 2α, the fillet radius as ρ and the local coordinate system as (r, θ). For comparison purpose, we shall also consider the same problem, but with an acute angle in Figure 1(b). In order to understand the behaviour of the numerical schemes in the presence of blunt corners, the asymptotic mechanical fields near the apex of the corner are presented. The domain is assumed infinite and subjected to a remote uniform tension σ_0. The solution in the case of a sharp notch has been derived by [12]. The closed-form solution in case of a rounded notch has been presented recently by Lazzarin and Tovo [13] and Filippi et al. [14] among

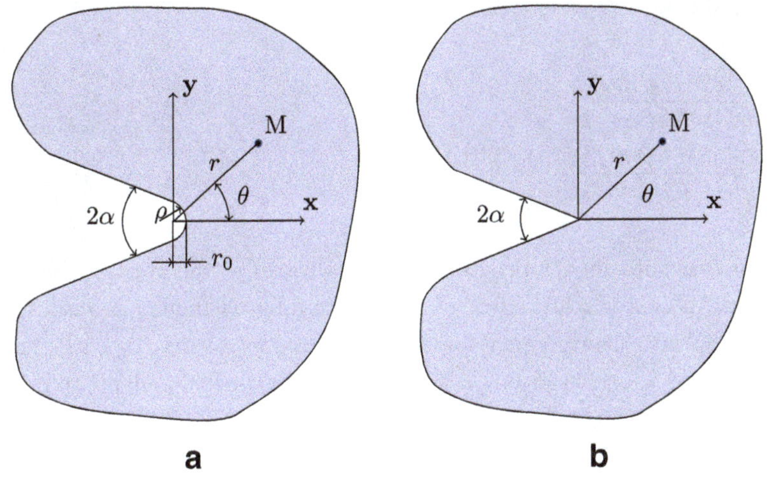

Figure 1 Semi-infinite notch: (a) rounded notch; (b) sharp notch.

others. The solution of the problem is obtained by making use of Kolosov-Muskhelishvili's potential. Both solutions are presented, so that their specific features can be highlighted.

Sharp corner

Following the work of Williams [12], the expression of the stress fields near a traction-free re-entrant corner is written as the sum of two modes that have a different singularity, λ_1 and λ_2. These eigenvalues verify the following equations:

$$\sin(\lambda_1\alpha) + \lambda_1 \sin(\alpha) = 0 \tag{1}$$

$$\sin(\lambda_2\alpha) - \lambda_2 \sin(\alpha) = 0 \tag{2}$$

λ_1 (resp. λ_2) is related to a symmetric (resp. anti-symmetric) mode. The expression of the asymptotic stress field is thus [15]:

$$\sigma_{ij}(r,\theta) = K_1^* r^{\lambda_1 - 1} f_{ij}^1(\alpha,\theta) + K_2^* r^{\lambda_2 - 1} f_{ij}^2(\alpha,\theta) \tag{3}$$

Factors K_i^* are called notch stress intensity factors (N-SIFs), and functions f_{ij}^k have the following expression:

$$\begin{bmatrix} f_{rr}^1 \\ f_{\theta\theta}^1 \\ f_{r\theta}^1 \end{bmatrix} = \frac{1}{\lambda_1 + 1 + \chi_{b1}(1-\lambda_1)} \begin{bmatrix} (3-\lambda_1)\cos((1-\lambda_1)\theta) - \chi_{b1}(1-\lambda_1)\cos((1+\lambda_1)\theta) \\ (\lambda_1+1)\cos((1-\lambda_1)\theta) + \chi_{b1}(1-\lambda_1)\cos((1+\lambda_1)\theta) \\ (1-\lambda_1)\sin((1-\lambda_1)\theta) + \chi_{b1}(1-\lambda_1)\sin((1+\lambda_1)\theta) \end{bmatrix} \tag{4}$$

and:

$$\begin{bmatrix} f_{rr}^2 \\ f_{\theta\theta}^2 \\ f_{r\theta}^2 \end{bmatrix} = \frac{1}{\lambda_2 + 1 + \chi_{b2}(1-\lambda_2)} \begin{bmatrix} (3-\lambda_2)\sin((1-\lambda_2)\theta) - \chi_{b1}(1-\lambda_2)\sin((1+\lambda_2)\theta) \\ (\lambda_2+1)\sin((1-\lambda_2)\theta) + \chi_{b1}(1-\lambda_2)\sin((1+\lambda_2)\theta) \\ (1-\lambda_2)\cos((1-\lambda_2)\theta) + \chi_{b1}(1-\lambda_2)\cos((1+\lambda_2)\theta) \end{bmatrix} \tag{5}$$

where (and using the notation $\alpha = q\pi$):

$$\chi_{b1} = -\frac{\sin((1-\lambda_1)q\pi/2)}{\sin((1+\lambda_1)q\pi/2)} \tag{6}$$

$$\chi_{b2} = -\frac{\sin((1-\lambda_2)q\pi/2)}{\sin((1+\lambda_2)q\pi/2)} \tag{7}$$

It can be seen that both modes produce singular stresses at the apex of the corner when $\lambda_i < 1$, which causes a loss of convergence for finite elements, as discussed in sections 'Discussion' and 'Convergence for nearly singular problems'. In particular, for $2\alpha = \pi/2$, one finds $\lambda_1 = 0.5448$ and $\lambda_2 = 0.9085$: Mode I is more singular than mode II.

Rounded corner

The determination of the asymptotic stresses near a rounded corner has been studied by Creager and Paris [16], Glinka [17], Lazzarin and Tovo [13], and improved recently by Filippi et al. [14]. The two last contributions are based on the use of Kolosov-Muskhelishvili's potentials together with an auxiliary system of curved coordinates that mimics the rounded corner's geometry. This conformal mapping approximation of the rounded notch has an hyperbolic shape which will have to be taken into account in the numerical examples (see Figure 2).

Following [14], the stress distribution near the rounded notch is seen to be:

$$\sigma_{ij}(r,\theta) = K_1^* r^{\lambda_1 - 1}\left[f_{ij}^1(\alpha,\theta) + g_{ij}^1(r,\alpha,\theta)\right] + K_2^* r^{\lambda_2 - 1}\left[f_{ij}^2(\alpha,\theta) + g_{ij}^2(r,\alpha,\theta)\right] \tag{8}$$

It is important to note that functions f_{ij}^k are the same as in the case of the sharp corner. Nevertheless, the resulting stress field is not singular as the origin of the frame is out of the domain (see Figure 1(a)). The expression of functions g_{ij}^k are the following:

$$\begin{bmatrix} g_{rr}^1 \\ g_{\theta\theta}^1 \\ g_{r\theta}^1 \end{bmatrix} = \frac{q\left(\frac{r}{r_0}\right)^{\mu_1 - \lambda_1}}{4(q-1)\left[\lambda_1 + 1 + \chi_{b1}(1-\lambda_1)\right]} \times \begin{bmatrix} \chi_{d1}\left[(3-\mu_1)\cos((1-\mu_1)\theta)\right] - \chi_{c1}\cos((1+\mu_1)\theta) \\ \chi_{d1}\left[(\mu_1+1)\cos((1-\mu_1)\theta)\right] + \chi_{c1}\cos((1+\mu_1)\theta) \\ \chi_{d1}\left[(1-\mu_1)\sin((1-\mu_1)\theta)\right] + \chi_{c1}\sin((1+\mu_1)\theta) \end{bmatrix} \tag{9}$$

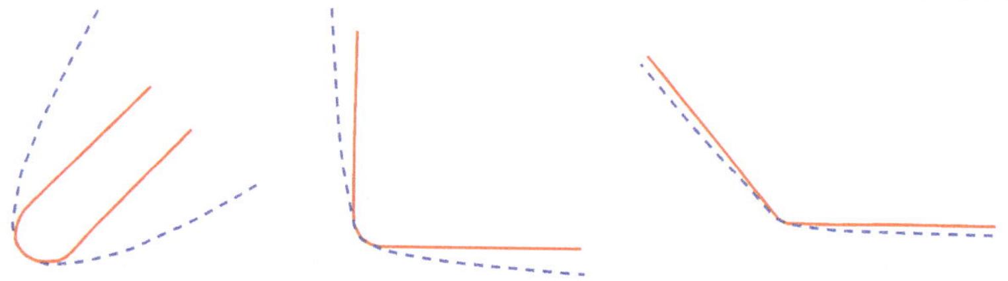

Figure 2 Analytical (dashed blue) and real (solid red) geometries for various opening angles. It can be noted that the larger the angle, the better the approximation of the real geometry.

and:

$$\begin{bmatrix} g_{rr}^2 \\ g_{\theta\theta}^2 \\ g_{r\theta}^2 \end{bmatrix} = \frac{q\left(\frac{r}{r_0}\right)^{\mu_2-\lambda_2}}{4(q-1)\left[\lambda_2+1+\chi_{b2}(1-\lambda_2)\right]}$$
$$\times \begin{bmatrix} \chi_{d2}\left[(3-\mu_2)\sin((1-\mu_2)\theta)\right] - \chi_{c2}\sin((1+\mu_2)\theta) \\ \chi_{d2}\left[(\mu_2+1)\sin((1-\mu_2)\theta)\right] + \chi_{c2}\sin((1+\mu_2)\theta) \\ \chi_{d2}\left[(1-\mu_2)\cos((1-\mu_2)\theta)\right] + \chi_{c2}\cos((1+\mu_2)\theta) \end{bmatrix} \tag{10}$$

Factors $\chi_{c1}, \chi_{c2}, \chi_{d1}$ and χ_{d2} are given in [14], and μ_1 and μ_2 are solutions of the following non-linear equations:

$$\left\{\frac{1-q(1+\mu_1)}{q}\left[3-\lambda_1-\chi_{b1}(1-\lambda_1)\right]-\epsilon_1\right\}(1+\mu_1)\cos\left[(1-\mu_1)q\frac{\pi}{2}\right]$$
$$+\left\{\left[(1-\mu_1)^2-\frac{1+\mu_1}{q}\right]\left[3-\lambda_1-\chi_{b1}(1-\lambda_1)\right]-(3-\mu_1)\epsilon_1\right\}\cos\left[(1+\mu_1)q\frac{\pi}{2}\right]=0 \tag{11}$$

and:

$$\left\{\left[\frac{q(1+\mu_2)-2}{q}\right]\left[\lambda_2-1-\chi_{b2}(1+\lambda_2)\right]-\epsilon_2\right\}(1-\mu_2)\cos\left[q\frac{\pi}{2}(1-\mu_2)\right]$$
$$+\left\{(\mu_2-1)\left[\frac{q(\mu_2-3)-2}{q}\right]\left[\lambda_2-1-\chi_{b2}(1+\lambda_2)\right]+(1-\mu_2)\epsilon_2\right\}\cos\left[q\frac{\pi}{2}(1+\mu_2)\right] \tag{12}$$

where ϵ_1 and ϵ_2 are also given in [14]. All the coefficients presented before can be evaluated from the knowledge of the geometry of the rounded corner (i.e. its opening angle 2α and its radius of curvature ρ). Finally, it is possible to obtain the displacement field associated with this asymptotic stress field, by means of the constitutive law and proper integration.

The displacement field has the following form:

$$\begin{aligned} u_r(r,\theta) &= u_r^{s,1}(r,\theta) + u_r^{s,2}(r,\theta) + u_r^{b,1}(r,\theta) + u_r^{b,2}(r,\theta) \\ u_\theta(r,\theta) &= u_\theta^{s,1}(r,\theta) + u_\theta^{s,2}(r,\theta) + u_\theta^{b,1}(r,\theta) + u_\theta^{b,2}(r,\theta) \end{aligned} \tag{13}$$

The explicit expression of functions $u_r^{s,i}$ and $u_r^{b,i}$ are given in Appendix A, and a sketch of the first mode displacement and strain are presented in Figures 3 and 4.

Discussion

In order to discuss the expected behaviour of the asymptotic solution derived in the previous section, let us consider the terminology coined in [1]. This classification, based on

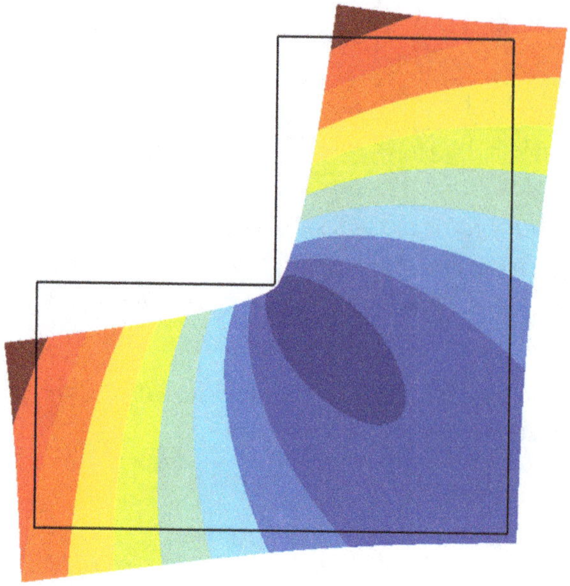

Figure 3 Displacement field associated to a blunt corner of radius 0.00625 subjected to a mode I loading.

the features of the analytical solution of a problem, enables to predict the behaviour of the finite element method. Solutions can be separated in three classes:

Category A	If the solution is analytic everywhere in the domain (including boundaries);
Category B	If the solution is analytic everywhere, except at a finite number of singular points (and edges in 3D);
Category C	If the solution does not belong to the previous categories (material interfaces for example);

Practical problems usually belong to category B. Note however that the solution is not necessarily singular near singular points: it depends on the eigenvalues of the expansion of the solution. If the eigenvalues are strictly smaller than one, then the solution is

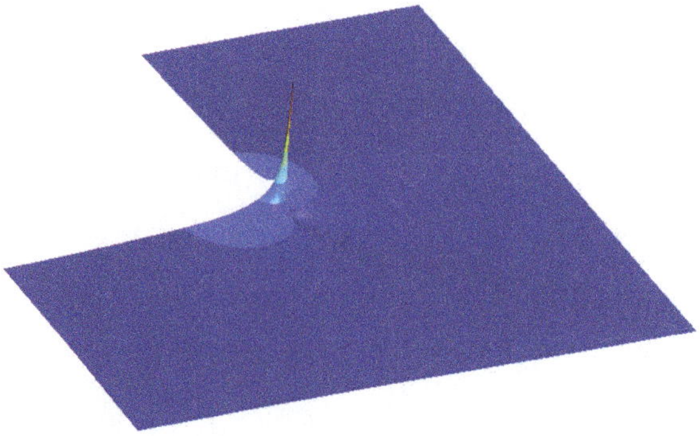

Figure 4 Strain field (Von-Mises norm) associated to a blunt corner of radius 0.00625 subjected to a mode I loading.

singular and the problem is said *strongly* in category B. Otherwise, it is qualified as *weakly* in category B.

As stated in section 'Sharp corner', the stress field associated with the sharp corner eqn. (3) is singular and problems involving these geometrical features belong strongly to category B. In this case, the convergence of the finite element method is bounded by the order of the singularity of the solution i.e. $\min(\lambda_1, \lambda_2)$. Note however that if $2\alpha \geq \pi$, then $\lambda_i \geq 1$, and the problem becomes regular (weakly in category B). The convergence is thus bounded by the polynomial order of the approximation (for h finite elements). On the contrary, the stress field related to the rounded corner eqn. (8) is regular, although it can be very rough if the radius of curvature ρ is small. The problem is then always weakly in category B, but if ρ is small then it tends to be strongly in category B. As the solution is regular, one would expect h convergence rates associated with the order p of the polynomial approximation (i.e. in $\mathcal{O}(h^p)$ in the energy norm). This is the case asymptotically, but not necessary for "engineering meshes" (meshes with a moderate number of elements), as illustrated in section '1D model problem'.

The eXtended finite element method

In the following, the eXtended Finite Element Method (X-FEM) will be used for the computations. The X-FEM [4] is an extension of the finite element method (FEM) that was developed from the need to improve the FEM approach for problems with complex geometrical features (cracks [4], material interfaces [18], free surfaces [18,19]). In contrast to classical finite elements, the X-FEM does not require the mesh to conform the geometry. Instead, a regular mesh is constructed for the domain of interest and the presence of internal boundaries is taken into account in the formulation of the finite elements at the corresponding locations by means of the partition of unity method [11]. The X-FEM approximation of the displacement field, $\mathbf{u}$, over an element Ω_e is given by:

$$\mathbf{u}(\mathbf{x})|_{\Omega_e} = \sum_{\alpha=1}^{n} \mathbf{N}^{\alpha} \left(u^{\alpha} + \sum_{\beta=1}^{n_e} a^{\alpha}_{\beta}\, \varphi_{\beta}(\mathbf{x}) \right) \tag{14}$$

where the approximation can be divided into a classical one that depends only on the vectorial shape functions $\mathbf{N}^{\alpha}(x)$[a] and classical degrees of freedom (dofs in the following) u^{α}, and an enriched one that depends on enrichment functions $\varphi_{\beta}(x)$ and enriched dofs a^{α}_{β}. Those functions prevent poor rates of convergence due to the non-conformity of the approximation or the singularity of the solution. The additional degrees of freedom are only added at the nodes whose support is split by the interface, which means that typically only a few of them are added. More precisely, if an element is fully enriched, then this number of enriched dofs is equal to $n \times n_e$. On the contrary, no enrichment is used in the case of a non-conforming approximation: the weak form is just integrated selectively in the domain. A level-set representation of the geometry is typically used: in this case, the level-set is interpolated on the approximation mesh. This couples the geometrical representation to the approximation [20], and prevents the use of higher order approximations due to an insufficient geometrical accuracy. A so-called sub-grid level-set approach has been proposed in [6,7] in order to uncouple geometry and approximation, and thus allow the use of high-order approximations. Alternatively, the use of the so-called Nurbs-Enhanced X-FEM [10] allows to consider the exact geometrical representation independently of the approximation mesh, see Figure 5 for a comparison of both approaches. In this case, geometries such as the one depicted in Figure 6 could be represented using a mesh whose characteristic length is large with respect to the geometrical details. However, the following question arise: what is the influence of the size mismatch between the mesh and the geometrical details on the convergence of the finite element approximation?

In this contribution, both low and high-order X-FEM will be considered. The latter case makes sense, as one can deal with meshes composed of big elements with simple shape. In this case, the enrichment scheme presented in (14) can lead to conditioning issues when a so-called geometrical enrichment is used [21]. Geometrical enrichment states that the size of the enriched region remains unchanged during refinement (it is also called "fixed area" enrichment in [22]). It is not related to the concept of "geometric mesh" which is commonly used in the p-fem community. In order to improve this issue, several strategies have been proposed [21-24]. In this contribution, the strategy proposed by Duarte et al. [25] and further studied in Chevaugeon et al. [26] is considered. It consists in using a vectorial enrichment, rather than a scalar one as in eqn. (14):

$$\mathbf{u}(\mathbf{x})|_{\Omega_e} = \sum_{\alpha=1}^{n} \mathbf{N}^{\alpha} u^{\alpha} + \sum_{\alpha=1}^{\tilde{n}} \tilde{N}^{\alpha} \left(\sum_{\beta=1}^{n_e} a_{\beta}^{\alpha} \, \boldsymbol{\varphi}_{\beta}(\mathbf{x}) \right) \quad (15)$$

in this expression, the first term corresponds to the classical finite element approximation while the second one corresponds to the enrichment. It involves $\tilde{N}^{\alpha}$, the *scalar* shape

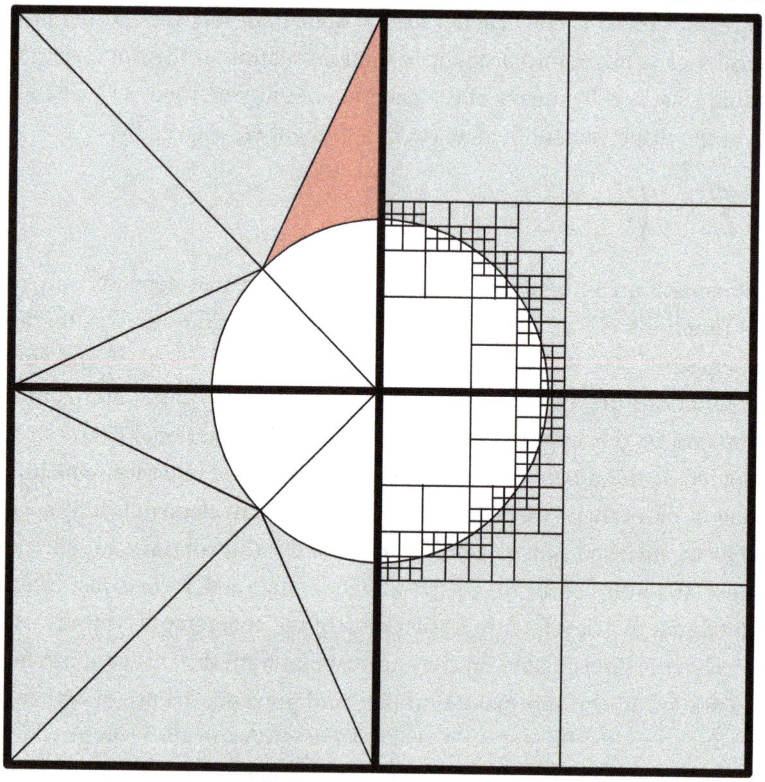

Figure 5 Left: Nurbs-Enhanced X-FEM (in red, a typical integration cell); Right: Sub-Grid level-set. Thick lines indicate elements boundaries, and thin lines integration cells boundaries.

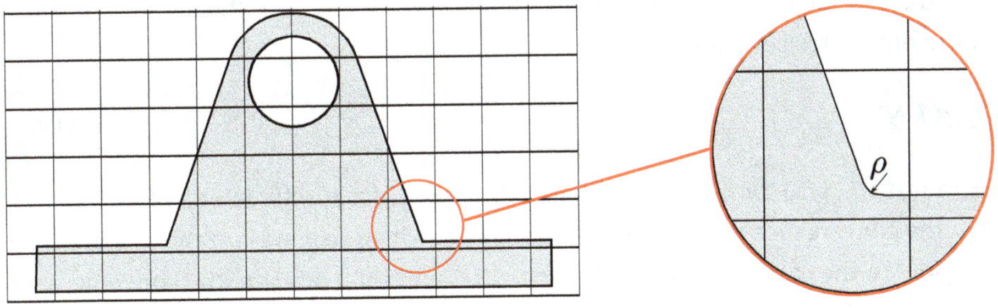

Figure 6 Geometrical details in an element.

function associated with the partition of unity, a_β^α the scalar enriched dof and $\boldsymbol{\varphi}_\beta(\mathbf{x})$ the β^{est} *vectorial* enrichment function. Note that the number of vectorial shape functions n remains unchanged with respect to (14), and that the number of scalar shape functions $\tilde{n}$ is smaller than n. More precisely, in the case where $\tilde{N}$ and $\mathbf{N}$ share the same polynomial order, we have $\tilde{n} = \frac{n}{d}$ with d the spatial dimension of the problem. It reflects the different nature of these shape functions (vectorial and scalar). This difference has also an influence on the number of enriched dofs: it is reduced by a factor d if (15) is used ($\tilde{n} \times n_e$ rather than $n \times n_e$).

In [26], the resulting conditioning number evolution was shown to increase in $O(1/h^2)$ for a model problem, which is the same rate as classical linear finite elements. This improvement in the conditioning number is of great interest in practice, as is allows to use the so-called geometrical enrichment which has been proved to be optimal in term of convergence. This aspect becomes fundamental when high-order shape functions are used, as the conditioning number increases with the polynomial order.

Methods

1D model problem

The behaviour of the finite element approximation is studied on a simple 1D model problem which is representative of the solution near the fillet:

$$\frac{d^2u}{dx^2} + f = 0 \quad x \in]0,1[\tag{16}$$

$$u(0) = u(1) = 0 \tag{17}$$

f is chosen such that the exact solution is:

$$u_0(x) = x^\alpha - x \tag{18}$$

for $\alpha > 1/2$. One can see that this solution is singular, and that the center of the singularity is for $x = 0$. Such a solution can be compared to typical 2D solutions near re-entrant corners: $u = r^\beta \phi(\theta)$ for $\alpha = \beta + 1/2$. The problem is solved using h and p finite element approximations, with both homogeneous or geometric meshes [27]. The finite element shape functions are based on integrated Legendre polynomials, as presented in [1]. The problem is solved for $x \in [\varepsilon, 1[$, $\varepsilon \in [0, 1[$. Remark that a conforming mesh is used for this simple problem: the conclusions can be extended to the non-conforming case. We first consider the case where $\varepsilon = 0$: the singularity emanates from the boundary of the domain. In the case of a quasi-uniform mesh, the estimates given in [27] states that the energy norm of the error evolves as:

$$\|e\|_E \leq k N^{-\beta} \tag{19}$$

with $\beta = \min(p, \alpha - 1/2)$ for h-refinement, and $\beta = 2\alpha - 1$ for p-refinement. Note that in both cases the convergence is algebraic, with an order which depends on the singularity of the solution. In the case where the singularity lies out of the domain, the convergence remains algebraic for h-refinement, whereas it becomes exponential for p-refinement [28]:

$$\|e\|_E \leq k \exp^{-\gamma N^{\theta}} \tag{20}$$

In this expression, k, γ and θ are positive constants that depend on the exact solution.

Convergence for nearly singular problems

Equations (19) and (20) are checked on the model problem (16) with $\alpha = 0.55$, and for $\varepsilon = 10^{-k}, k \in [1, 2, 3, 4, 5]$. Both h and p-convergence are considered, with a regular nodal distribution. Note that all the figures in this section represent absolute errors. The results are presented in Figures 7 and 8: it can be seen that ε has a visible effect on the behaviour of the two approaches. Large ε values correspond to very smooth solutions in the domain, so that estimates (19) with $\beta = p$ and (20) are verified. On the contrary, small values for ε correspond to nearly-singular solutions in spite of the absence of any singularity in the domain and on its boundary. The convergence tends to be algebraic with a rate that depends on the singularity ($\alpha - 1/2$ for h-convergence, and $2\alpha - 1$ for p-convergence). In particular, two regimes can be observed in Figure 7: for low h the convergence is driven by the singularity ($\beta = \alpha - 0.5$), whereas at some point the asymptotic $\beta = p$ convergence is observed. The smaller ε, the later this asymptotic convergence is recovered. Note that

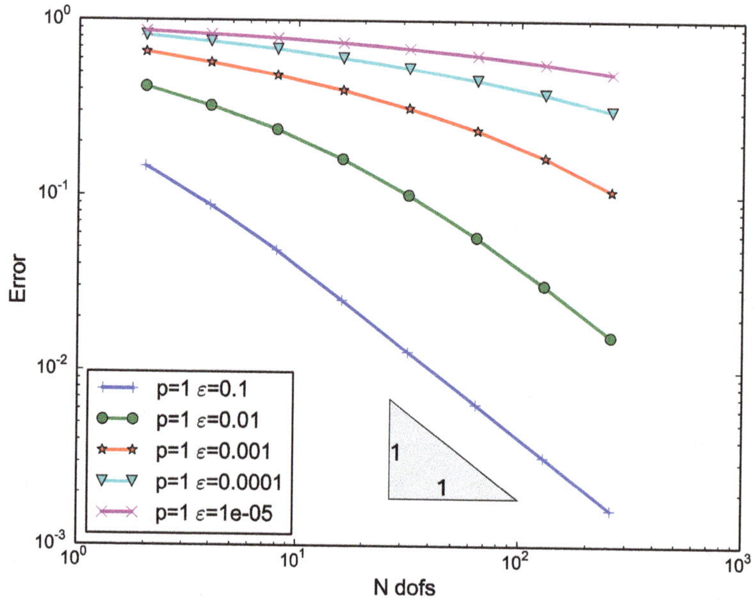

Figure 7 Influence of the distance to the singularity, *h*-convergence.

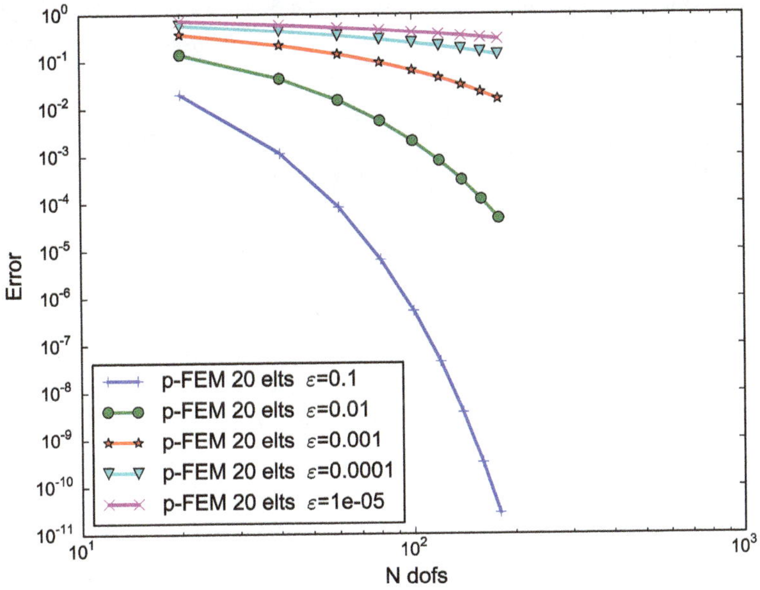

Figure 8 Influence of the distance to the singularity, *p*-convergence.

this regular convergences can always be obtained, but not necessarily for "engineering" meshes (i.e. with moderate element size).

The reason for this loss of convergence is now discussed more precisely in the p-Fem case: Consider the contribution of the first four elements to the global error which is presented in Figure 9(a) (for $\alpha = 0.55$, $\varepsilon = 10^{-5}$ and 20 elements). It can be seen that only the first element has an algebraic convergence, whereas the remaining elements converge exponentially. This behaviour is typical of the singular case, where this phenomenon is well known. For bigger ε, an exponential convergence is obtained for all the elements (see Figure 9(b)).

Following [27] (in the case where $\varepsilon = 0$), the contribution η_1 of the first element (that touches the center of the singularity) to the error writes:

$$\eta_1 \simeq C_0(\alpha)\frac{h_1^{\alpha-1/2}}{p^{2\alpha-1}}\left(1 + O\left(\frac{1}{p}\right)\right) \quad (p \to \infty) \tag{21}$$

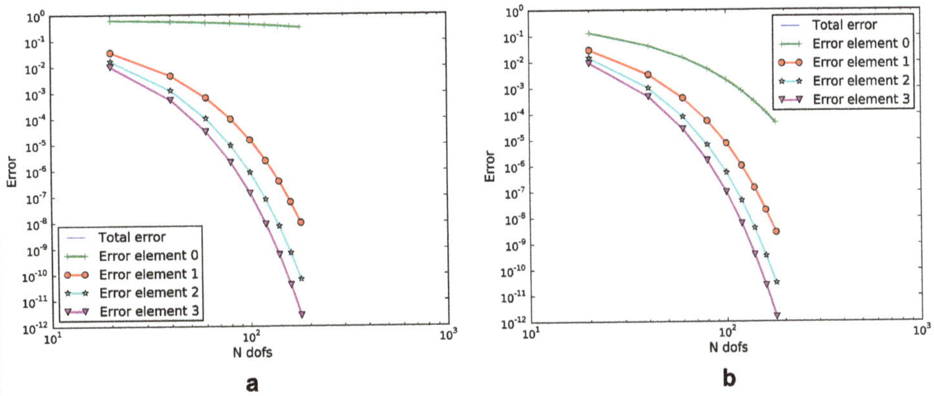

Figure 9 Contribution of the first elements to the global error: (a) $\varepsilon = 10^{-5}$; (b) $\varepsilon = 10^{-2}$.

Where h_1 is the length of the element, and $C_0(\alpha)$ is a constant. This shows that the error decrease in the first element is algebraic. On the contrary, the estimate for the remaining elements is:

$$\eta_{i,\,i>1} \simeq C_1(\alpha)\, h_i^{\alpha-1/2} \left(\frac{1-r_i^2}{2r_i}\right)^{\alpha-1} \frac{r_i^p}{p^\alpha}\left(1+o\left(\frac{1}{p^\sigma}\right)\right) \tag{22}$$

$$r_i = \frac{\sqrt{x_i}-\sqrt{x_{i-1}}}{\sqrt{x_i}+\sqrt{x_{i+1}}}$$

Where h_i is the length of the element, $C_1(\alpha)$ is a constant, and $\sigma > 0$. This equations states that the convergence is exponential in all the elements but the first one. This is why exponential convergence should be expected in the nearly-singular case, as all the elements should follow estimate (22) (none of them touches the center of the singularity). Although, Figure 9(a) seems to contradict eqn. (22), this is not the case. Indeed, eqn. (22) holds only if $0 < r_i^2 < 1 - 1/p$, which is not the case for Figure 9(a), as r^2 ranges from 0.999 to $6.6\,10^{-4}$ (see Table 1). It can be seen that only the first element has a r^2 greater than $1 - 1/p\; \forall p$, hence the algebraic convergence. When $1 - 1/p < r_i^2 < 1$, following [28], the estimates becomes:

$$E(I_i) \simeq h_i^{\alpha-1/2} \frac{r_i^{p+1-\alpha}}{p^{\alpha-1/2}} \left(1/p^{\alpha-1/2} + (1-r_i^2)^{\alpha-1/2}\right) \tag{23}$$

When r_i^2 is close to 1, one obtain the same estimate as eqn. (21), which is consistent with the numerical results (Figure 9(a))[b].

Strategies for recovering optimal convergence

Two strategies are proposed in order to recover a proper convergence in the case of nearly-singular problems. The first one is based on an enrichment of the approximation, using the Partition of Unity method [11], see eqn. (14). The second one is based on a proper mesh design, which is close to the approaches that are classically used in the context of p-Fem.

Enrichment of the approximation

The idea consists in the enrichment of the approximation in order to capture the steep gradients of the exact solution. The enrichment function considered is x^α as only this term is singular in (18), and a "geometrical" enrichment strategy is considered, as it has been shown in practice that it was leading to better convergence properties [21,22]. Such an approach can be used for both h and p Fem.

Table 1 Evolution of r^2 for the elements of a regular mesh ($\varepsilon = 10^{-5}$, 20 elements)

Element	r^2
1	$9.99\,10^{-1}$
2	$1.11\,10^{-1}$
3	$4.00\,10^{-2}$
4	$2.04\,10^{-2}$
20	$6.57\,10^{-4}$

Suitable mesh design

A second possibility, based on the construction of a suitable mesh is investigated in the case when p-refinement is considered. It has been highlighted in the previous section that in the case of p-Fem, the contribution of the first element was preventing any exponential convergence of the approximation. A sufficient condition for recovering the exponential convergence consists in ensuring that this first element converges exponentially. The objective is thus to prescribe $r^2 < 1 - \frac{1}{p}$ in the first element. In the context of p-refinement, we have chosen to verify this condition for small orders, i.e. $p = 2$. So, if $r^2 < 1/2$ exponential convergence is expected for all the elements and for any polynomial order greater than two. First, consider the case of a regular mesh depicted in Figure 10. We are interested in the first element which is highlighted in red. This element is located at a distance ε from the center of the singularity. For this first element, condition $r^2 < 1/2$ can be written as:

$$\left(\frac{\sqrt{\varepsilon+h}-\sqrt{\varepsilon}}{\sqrt{\varepsilon+h}+\sqrt{\varepsilon}}\right)^2 \leq \frac{1}{2} \tag{24}$$

Solving this equation for h gives the maximum length of the first element allowing for an exponential convergence:

$$\begin{aligned} h &\in \left[0,\ 4\left(3\sqrt{2}+4\right)\varepsilon\right] \\ h &\in [0,\ 33\,\varepsilon] \end{aligned} \tag{25}$$

Equation (25) states that in the case of a quasi-uniform mesh, the length of the first elements must have the same order of magnitude as ε. This condition is very restrictive in practice, as a quasi-uniform mesh of this type is unusable for real problems. Note that the numerical results from Figure 8 are consistent with this estimate, as exponential convergence is noticeable for $\varepsilon < 10^{-3}$ which is close to $h/33 \simeq 1.6\,10^{-3}$.

We now consider the use of a geometrical mesh. Indeed, it has been shown in the case of singular problems [1] that the use of a finite element mesh with geometrical progression of power 0.15 near the center of the singularity could lead to exponential rates of convergence for both p and $h-p$ fem (only in the pre-asymptotic range for the former case, while this rate can be maintained asymptotically in the latter). It is interesting to note that the geometric progression is independent of the order of the singularity. In this case, factor r is seen to be constant, and the condition $r^2 < 1/2$ becomes:

$$\left(\frac{1-\sqrt{q}}{1+\sqrt{q}}\right)^2 \leq \frac{1}{2} \tag{26}$$

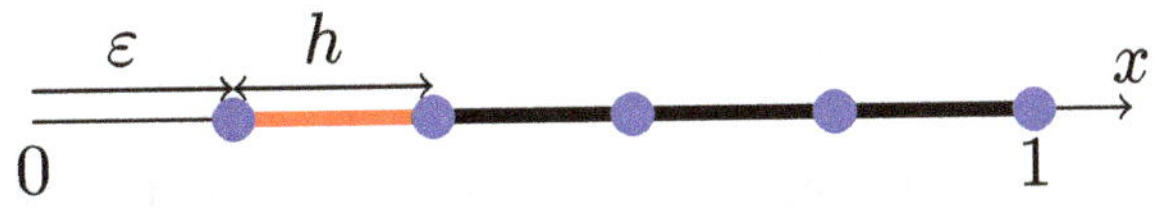

Figure 10 Regular mesh with singularity out of the domain ($x = 0$). In red, the element of interest.

Solving this equation for $0 < q < 1$ gives the range of geometrical progression that allows for an exponential convergence:

$$q \in \left[\frac{1}{17 + 12\sqrt{2}}, 1\right] \tag{27}$$

$$q \in [0.029, 1] \tag{28}$$

From this study one can see that in practice (i.e. $q \geq 0.15$), the exponential convergence is ensured no matter the value of the geometrical progression.

Results and discussion

1D Numerical examples

The two strategies presented above are now appraised considering $\alpha = 0.55$ and $\varepsilon = 10^{-5}$ (i.e. for the most unfavourable case). The energy norm of the error is monitored with respect to the number of degrees of freedom.

h-convergence

Only the enrichment strategy is considered in this case. The length of the enriched zone is set to 0.25, and h varies from 0.0625 to 0.00390625. The selection of the enriched nodes upon mesh refinement is illustrated is Figure 11, where the enriched nodes are depicted as squares. The results are presented in Figure 12, and it can be seen that (regular) optimal order of convergence are obtained for linear, quadratic and cubic approximations. Note the small loss in the convergence for P2 and small h which stems from the accuracy of the integration of the weak formulation.

p-convergence

In this section, both enrichment and mesh-based strategies are considered. In the case of the enrichment strategy, a four elements mesh is considered and only the first one is enriched (note that in this case, the enrichment strategy behaves like a "geometrical" enrichment). The polynomial order ranges from 1 to 10, and the evolution of the error in the energy norm is depicted in Figure 13. One can see that the exponential rate of convergence is recovered. This exponential convergence makes it possible to obtain a 10^{-8} absolute error level with ten times less dofs, compared to enriched h-convergence.

Finally, the use of a geometrical mesh is considered. Various geometrical progressions are used, ranging from 0.0032 to 0.24: a typical mesh with four elements (0.0562 progression) is depicted in Figure 14 to illustrate the strong grading near the singularity. It can be

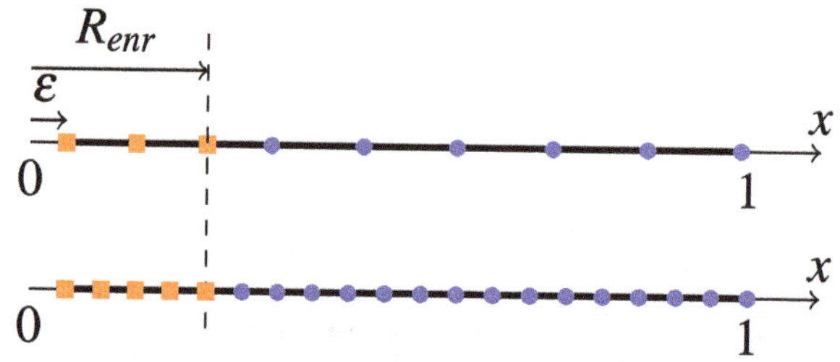

Figure 11 "Geometrical" enrichment strategy for two mesh size: R_{enr} corresponds to the size of the enriched zone, and square marks represent enriched nodes.

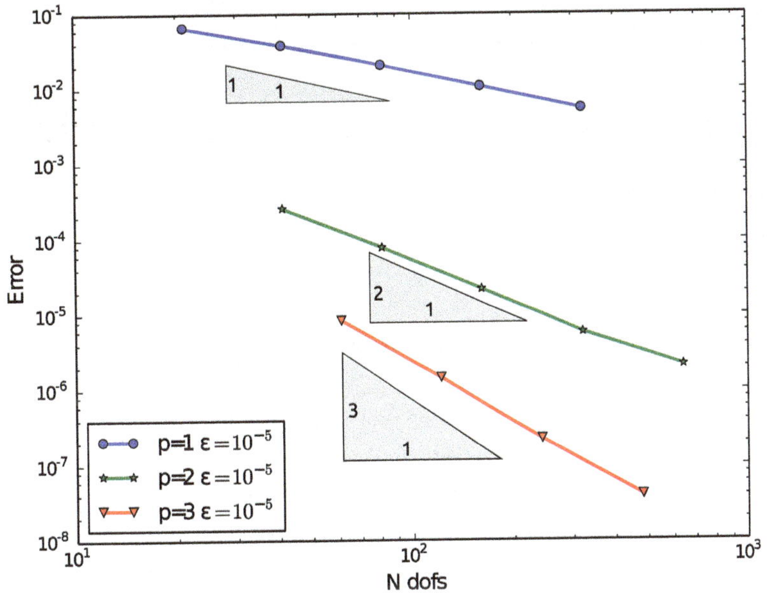

Figure 12 ***h*-convergence, enriched approximation.**

seen in Figure 15 that, as expected, exponential convergences are obtained for $q = 0.056$ and 0.24. The performances are similar for both meshes, although being less efficient than the enriched approximation. The small exponential convergence for $q = 0.0032$ may seem surprising, as this value is out of the range from eqn. (28). In fact, condition (28) was obtained for an exponential convergence with quadratic polynomials. If one accepts to shift the exponential convergence to higher polynomial orders, smaller progressions

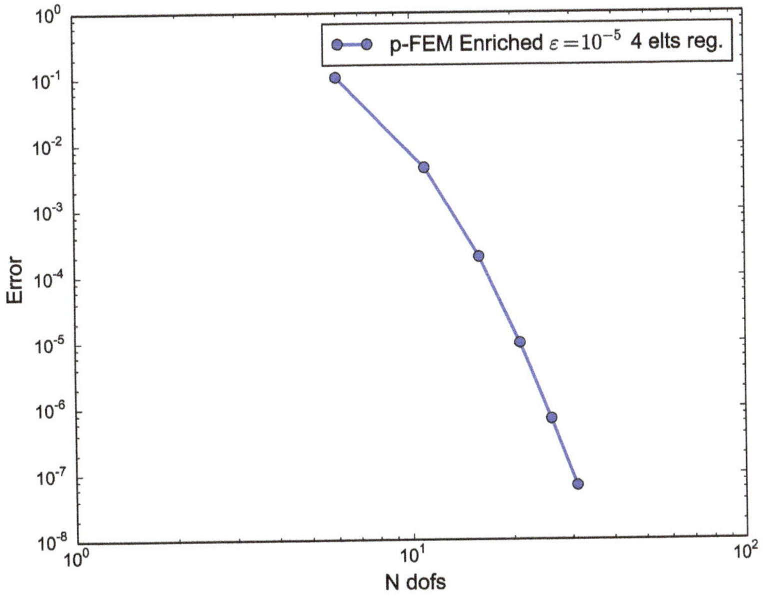

Figure 13 ***p*-convergence, enriched approximation.**

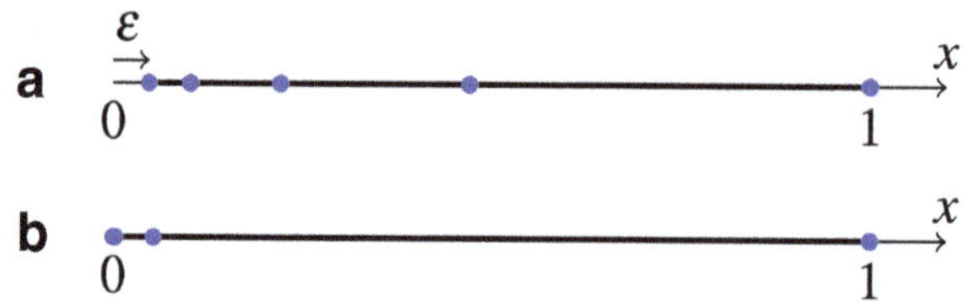

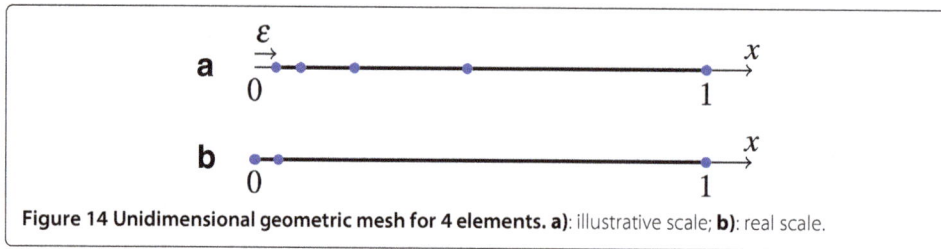

Figure 14 Unidimensional geometric mesh for 4 elements. a): illustrative scale; **b)**: real scale.

can be considered. In particular, $q = 0.0032$ is associated to an exponential convergence for a P5 approximation.

As a conclusion for this section, one can see that it is possible to recover regular rates of convergence in the case of nearly-singular solutions if one of the proposed strategies is used. The enrichment of the approximation seems to be the more versatile approach, as it can be applied for both h and p refinement. In addition, it has been demonstrated on the proposed example that it was performing more efficiently than the mesh-based approach. However, note that the geometrical mesh approach is less prone to conditioning and integration issues, and that is remains applicable even when the asymptotic behaviour is not precisely known.

Extension to 2D

We now discuss the extension to 2D of the proposed strategies.

Enrichment of the approximation

The adaptation is straightforward in this case. The vectorial formulation presented in eqn. (15) is considered, and a straight study of the asymptotic mechanical fields (13) can be

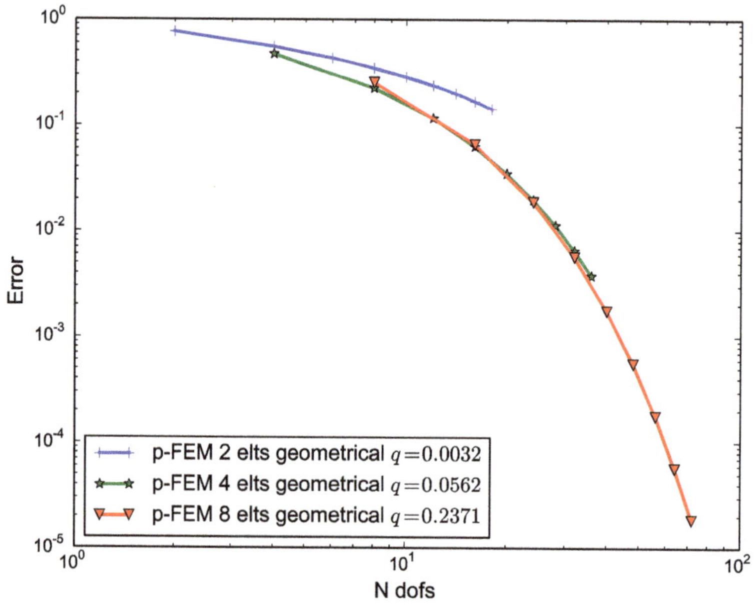

Figure 15 ***p*-convergence, geometrical meshes.**

used to build the enrichment functions. In this case, four vectorial enrichment functions $\boldsymbol{\varphi}_i$ will be used:

$$\begin{aligned}
\boldsymbol{\varphi}_1 &= u_r^{s,1}(r,\theta)\,\mathbf{e}_r + u_\theta^{s,1}(r,\theta)\,\mathbf{e}_\theta \\
\boldsymbol{\varphi}_2 &= u_r^{s,2}(r,\theta)\,\mathbf{e}_r + u_\theta^{s,2}(r,\theta)\,\mathbf{e}_\theta \\
\boldsymbol{\varphi}_3 &= u_r^{b,1}(r,\theta)\,\mathbf{e}_r + u_\theta^{b,1}(r,\theta)\,\mathbf{e}_\theta \\
\boldsymbol{\varphi}_4 &= u_r^{b,2}(r,\theta)\,\mathbf{e}_r + u_\theta^{b,2}(r,\theta)\,\mathbf{e}_\theta
\end{aligned} \tag{29}$$

where $u_r^{s,i}$ and $u_r^{b,i}$ are given in equations (30)–(33).

Geometrical mesh

In the case of the construction of an adapted mesh, the objective is to be able to blend a geometrical mesh into an existing finite element grid. The implementation is easy as the computational mesh does not need to conform to the geometry, and follows the three steps depicted in Figure 16. (i) the element containing the center of the singularity is removed from the mesh together with its neighbours. (ii) a geometrical nodes patch is inserted in the vacant space (Figure 16b)). The geometrical progression of this patch is obtained taking into account to the requested number of elements made by the user, and prescribing the length of the smallest elements to ρ (note that the value of the progression has a limited influence on the numerical efficiency, as shown in the following). Finally (iii), a local Delaunay algorithm is used in order to fill the space with finite elements, and build a transition to the existing elements (Figure 16c)). An example output of this procedure is given in Figure 17. Note that the center of the progression is not the center of the radius of curvature, but the point from which the singularity emanates (see section 'Rounded corner').

2D numerical examples

Consider the model problem depicted in Figure 18. It represents a plane domain whose behaviour is assumed linear elastic and containing a re-entrant corner with a fillet of radius ρ. Young's modulus is assumed to have a unit value, and Poisson's ratio is 0.3. Exact tractions whose expression are given in section 'Rounded corner' are applied on the boundaries of the domain (note that because of the geometrical approximations used to derive the solution, exact tractions have to be applied even on the free surface, see Figure 2). In the following, ρ will range from 0.0625 to 0.1, and both h and

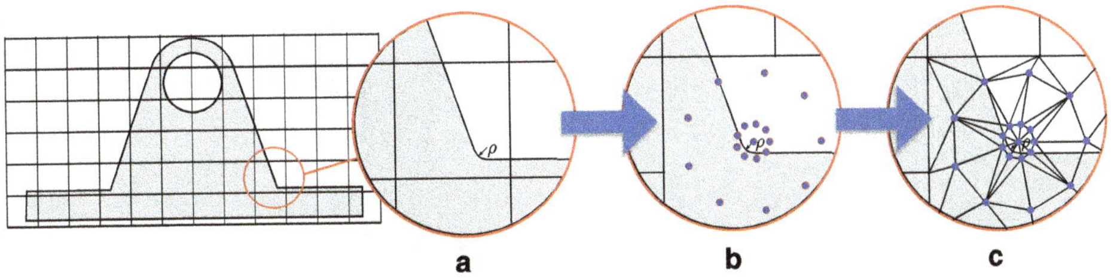

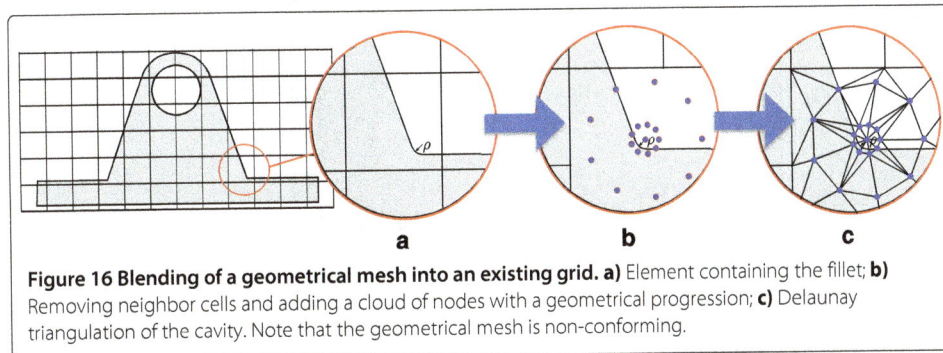

Figure 16 Blending of a geometrical mesh into an existing grid. a) Element containing the fillet; **b)** Removing neighbor cells and adding a cloud of nodes with a geometrical progression; **c)** Delaunay triangulation of the cavity. Note that the geometrical mesh is non-conforming.

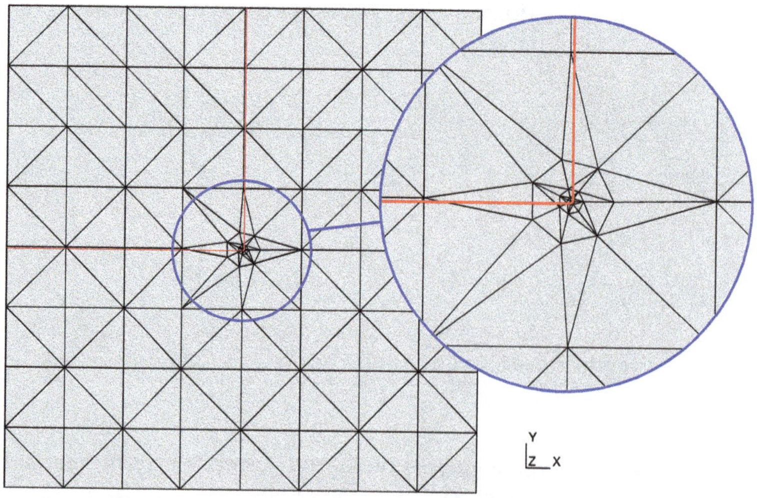

Figure 17 Resulting mesh for the L-shaped panel with a fillet (red: free-surface).

p-convergence are considered. In the former, only linear, quadratic and cubic approximations will be considered. In the case where an enrichment is considered, a "geometrical" enrichment strategy will be used by enriching all the elements lying in a circle of fixed radius centered on the center of the singularity. In order to integrate correctly the weak formulation in the enriched zone, the number of integration points is simply increased in the enriched elements, but not in the remaining part of the mesh. The objective of this section is to compare the performance of the 2D extension of the strategies considered in section 'Strategies for recovering optimal convergence', and propose good practice rules. All the numerical examples are conducted on a geometrical domain which consists in the bounding-box of the physical domain shown in Figure 18, and the X-FEM is used thanks to the definition of the geometry in terms of a level-set function. A sub-grid level-set [6,7] approach is used in order to be able to represent the geometry accurately on coarse

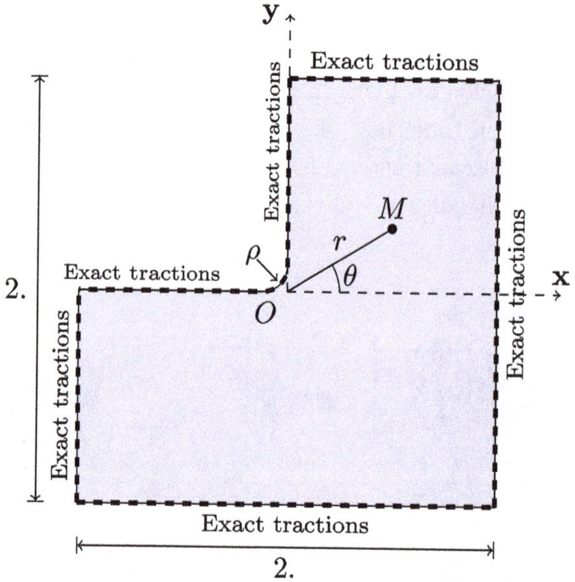

Figure 18 Blunt re-entrant corner convergence problem.

computational meshes. Unless mentioned, the h-convergence studies are conducted on regular triangular meshes composed of 4×4 to 128×128 elements per side, and all errors correspond to relative errors.

No enrichment

In a first step, the case with no enrichment is considered with different values for ρ. The results obtained for h-convergence, and mode I are presented in Figure 19. The behaviour highlighted in the 1D model problem is also observed here: small ρ prevent an $O(h^p)$ convergence, especially for low polynomial orders. It can be seen that the higher the polynomial order, the faster the convergence is recovered. The moment when this optimal convergence is recovered is now discussed: The convergence curves are now plotted with respect to h/ρ rather than h (see Figure 20). Looking at these curves, one can see that the change of regime in the convergence occurs in the shaded area. This means that the asymptotic convergence is recovered for $h \simeq \rho$ with linear elements. Moreover, this behaviour is consistent with the theoretical previsions from the last section (eqn. (25) for instance) Thus, the conclusion is the following: if no enrichment is used, one should consider meshes with element of size close to ρ in the fillet area. This condition is more easily verified when conforming low order finite elements are used, as the construction of a conforming mesh already requires this amount of refinement for a proper geometrical representation. This is not the case with non-conforming methods, and p-Fem meshes using blending mapping as elements far bigger than the radius of the fillet can be considered. When quadratic or cubic approximations are used, the same behaviour occurs: the only difference comes from the fact that the optimal convergence is recovered faster than

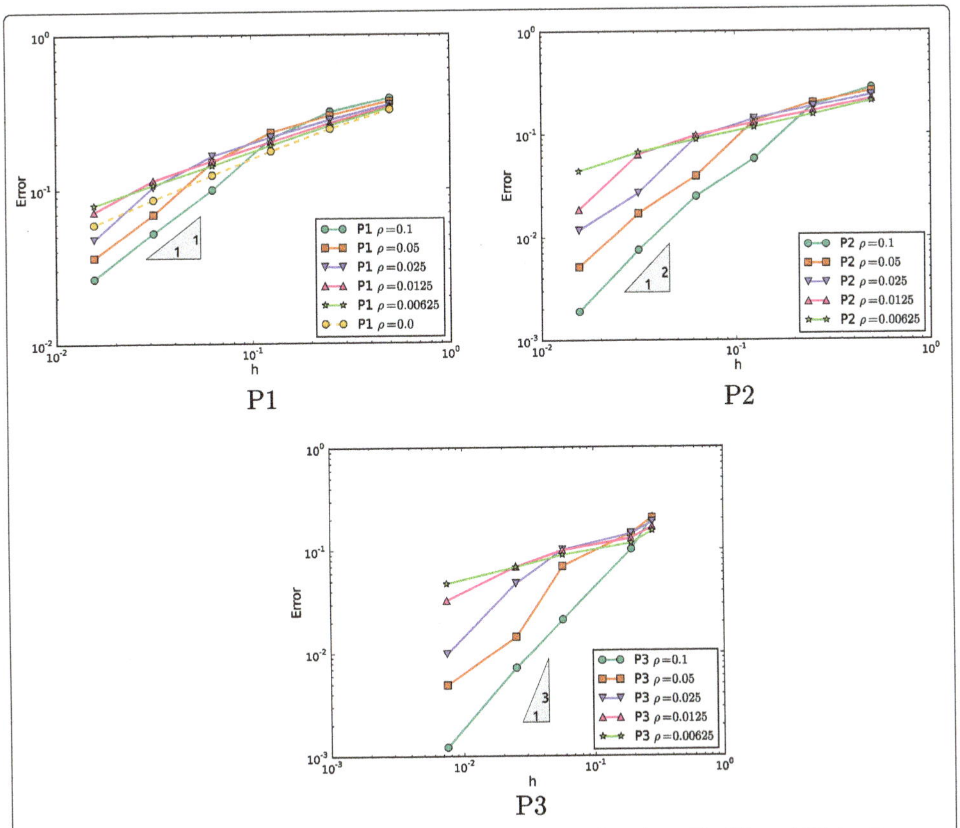

Figure 19 ***h*** **convergence for linear, quadratic and cubic approximation with different fillet radius.**

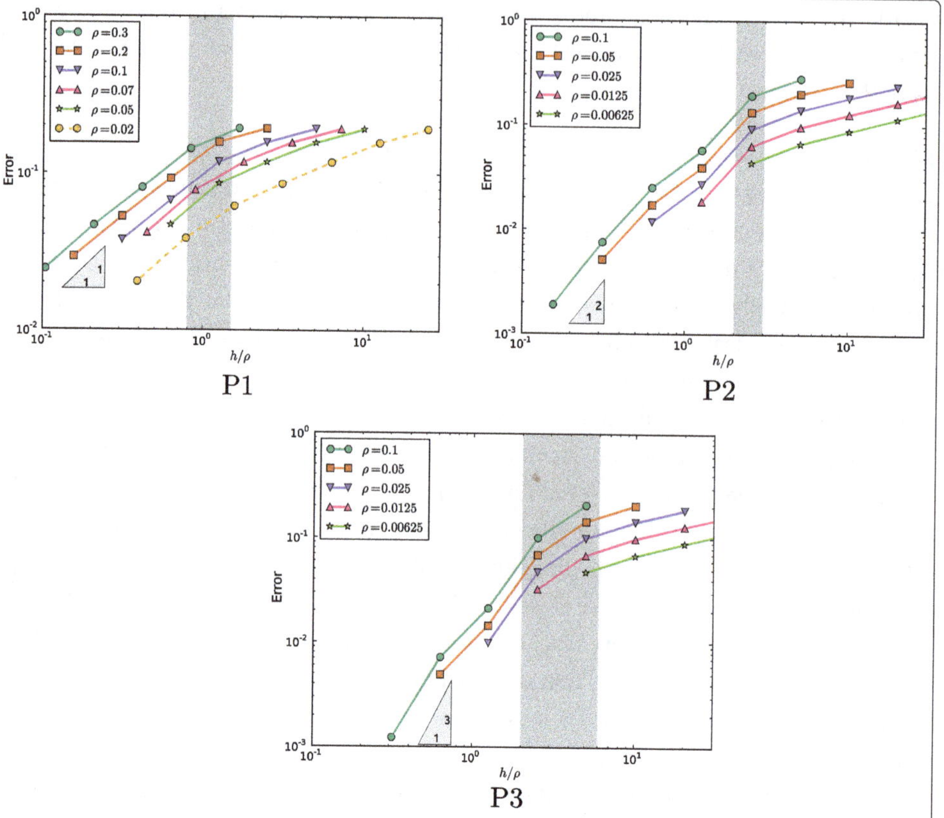

Figure 20 Convergence regime change with respect to ρ. The shaded zone represents the zone where the convergence regime is changing.

in the linear case. Yet, a very fine mesh has still to be considered ($h \simeq 2 - 3 \times \rho$ for P2 and $h \simeq 3 - 6 \times \rho$ for P3). In addition, optimal convergence couldn't be recovered in our tests for small radii.

Finally, p-refinement is considered with $\rho = 0.00625$ (smaller radius) in Figure 21. One can see that an algebraic rather than exponential convergence is obtained (as in the 1D case).

Effect of the enrichment

The approximation is now enriched by means of the enrichment functions presented in (29). The convergence study is done only for the smaller radius ($\rho = 0.00625$), mode 1 loading, and the enriched zone has a radius of 0.3. The results are presented in Figure 22. Regular convergence rates are recovered for linear, quadratic and cubic (not shown here) approximations, even for large element size.

Next, a p-convergence is performed on 16×16 and 8×8 meshes. The enriched zone is still a circular region of radius 0.30 and the results are given in Figure 23. Exponential convergence is recovered for both meshes with very accurate error levels (10^{-4}) and moderate number of dofs. Note that if the desired accuracy can be obtained with a reasonable polynomial order (below P8), then the coarse mesh is more efficient. Otherwise, conditioning issues with higher polynomial orders may prevent to get a meaningful solution. In this case a finer mesh or a preconditioner such as [21] should be used. Another alternative could be to use a non-homogeneous p distribution (smaller p in the enriched area).

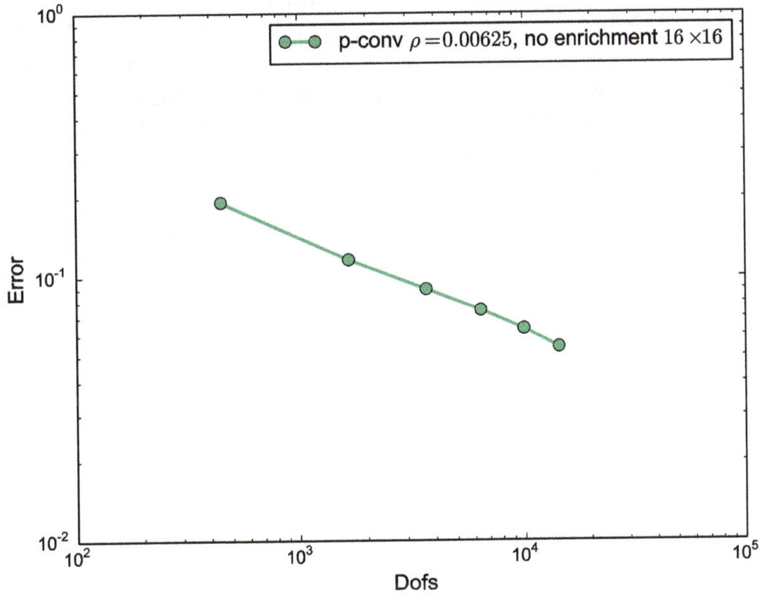

Figure 21 ***p*** **convergence, no enrichment.**

Effect of mesh refinement

Finally, the mesh-refinement strategy is evaluated. A 8 × 8 base mesh is considered (see Figure 17), and 4 or 5 layers of geometrical elements are focused near the center of the singularity (which is not the center of the radius of curvature). The results presented in Figure 24 show that an exponential convergence is obtained, and that the number of layers has very little influence on the performances of the approach.

Comparison of the two strategies

To conclude, the performances of the enrichment and geometrical mesh strategies are compared for $\rho = 0.00625$. In the case of the use of an enrichment, the computational

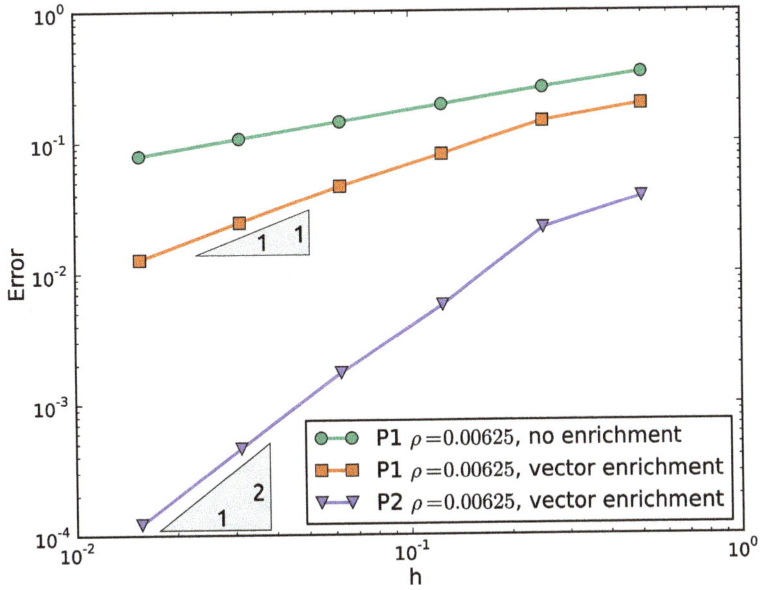

Figure 22 ***h*** **convergence for linear, quadratic approximation with and without enrichment.**

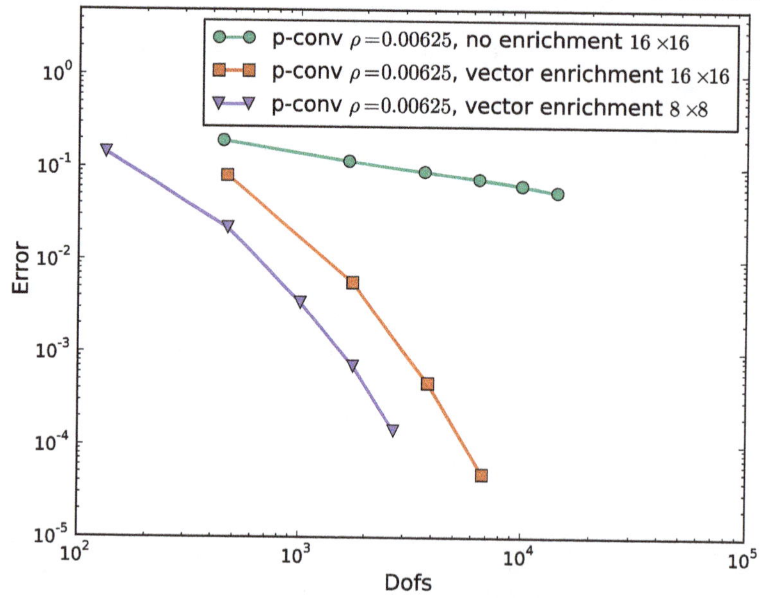

Figure 23 ***p*** **convergence with and without enrichment, and using two different meshes.**

setup is the same as in section 'Effect of the enrichment', whereas the number of layers is fixed for the geometrical mesh (but 4 × 4 and 8 × 8 base meshes are considered). The convergence curves are presented in Figure 25: it is shown that the behaviour of the different methods is quite similar, but that the enrichment strategy seems more efficient. However, it requires the derivation of the asymptotic fields for the problem of interest. Indeed, changing the boundary conditions on the free-surface of the corner has an influence on the order the singularity, and the trigonometric behaviour of the solution. On the contrary, the use of a geometrical mesh is more versatile.

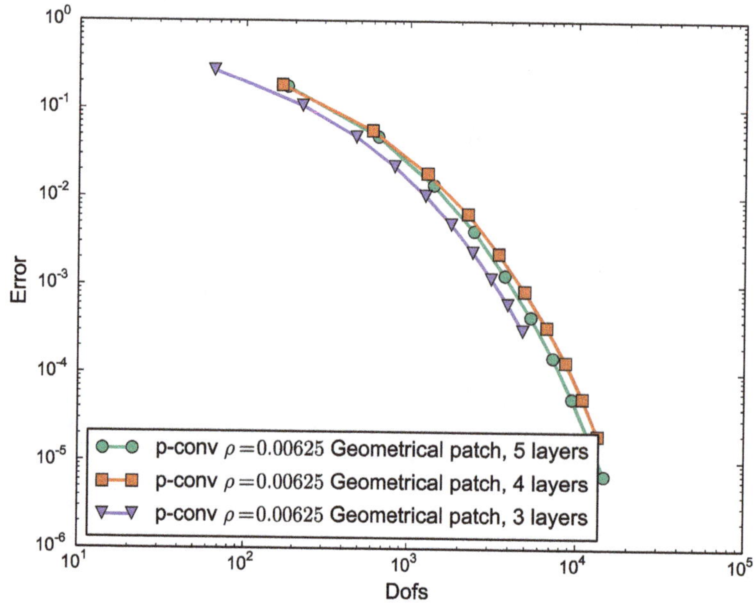

Figure 24 ***p*** **convergence, geometrical mesh (8 × 8 base mesh).**

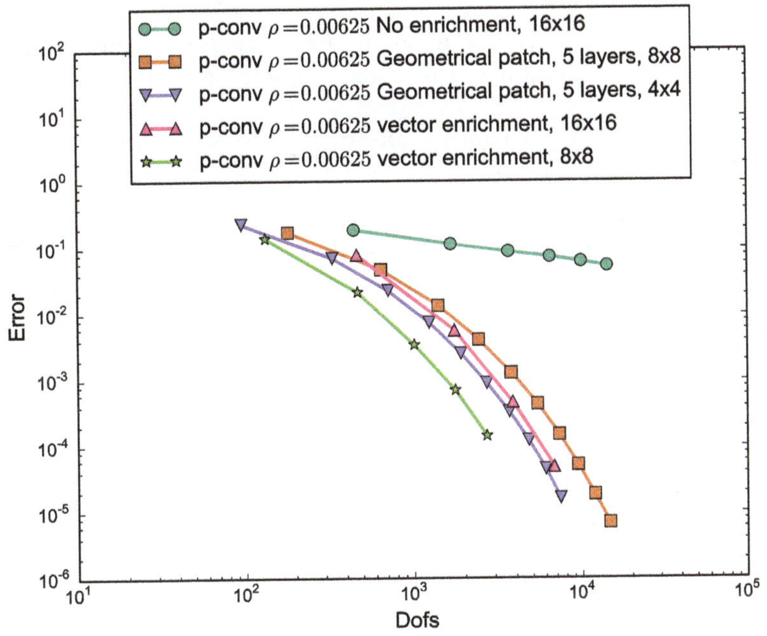

Figure 25 Comparison of two strategies, p convergence.

Conclusions

In this contribution, the behaviour of non-conforming h and p finite elements has been studied in the case of nearly-singular solutions (re-entrant corners with a fillet here). In particular, it has been shown in both 1D and 2D that despite being regular, the convergence rate was algebraic and limited by the order of the singularity. Therefore, it is not possible to use fictitious domain methods such as the X-FEM without enrichment or the Finite-Cell Method if high accuracy is needed in near the fillet. Thanks to the study of a 1D model problem, it has been possible to highlight the reasons for such a behaviour. Two strategies have been proposed in order to overcome this convergence bound. The first one is based on the enrichment of the approximation near the fillet, and is usable for both h and p methods. The second one is based on the use of a mesh with a geometrical progression towards the center of the singularity, and is restricted to p methods. The performances of these two strategies have been compared in both 1D and 2D: the enrichment method is the more efficient, but can lead to conditioning issues with high-order bases unless proper preconditioning strategies are used [21-24]. Moreover, the enrichment function is problem dependent, and can be tedious to obtain. In the present application, the enrichment functions are limited to stress-free bidimensional corners, and may be not be valid in the case where Dirichlet or non-homogeneous Neumann boundary conditions are applied. This kind of limitations was also present in the work of Wagner et al. [29] who considered rigid particles in Stokes flow, and used an enrichment function which is only valid for rigid and not too-close particles. The actual improvement with such not-fully adapted enrichment functions should be investigated further. On the contrary, exponential convergence is ensured in the case of the use of a geometrical mesh, no matter the progression, which makes it more versatile in case of high-order methods (only). The penalty of constructing such a mesh is greatly alleviated as this mesh does not need to conform the geometry, which is a contribution of the paper.

Endnotes

[a] In (14), the vectorial nature of the field is handled by the shape functions, and not the dofs that are just coefficients. This notation facilitates the writing of the discrete operators.

[b] For $\varepsilon = 10^{-2}$ (Figure 9(b)), the maximum value of r^2 is 0.51, which means that p convergence is obtained for the first element for any $p > 2$.

Appendix A: Asymptotic displacement fields

The asymptotic displacement fields associated to the asymptotic stress fields (4), (5), (9) and (10) are obtained by proper integration of the strain equations. Their expressions are given below:

$$u_r^{s,i} = \frac{r^{\lambda_i}}{2G}\Big[a_1\,(\kappa-\lambda_i)\cos\left[(\lambda_i-1)\,\theta\right] - a_2\,(\kappa-\lambda_i)\sin\left[(\lambda_i-1)\,\theta\right] - b_1\cos\left[(\lambda_i+1)\,\theta\right] + b_2\sin\left[(\lambda_i+1)\,\theta\right] \Big] \quad (30)$$

$$u_r^{b,i} = \frac{r^{\mu_i}}{2G}\Big[d_1\,(\kappa-\mu_i)\cos\left[(\mu_i-1)\,\theta\right] - d_2\,(\kappa-\mu_i)\sin\left[(\mu_i-1)\,\theta\right] - c_1\cos\left[(\mu_i+1)\,\theta\right] + c_2\sin\left[(\mu_i+1)\,\theta\right] \Big] \quad (31)$$

$$u_\theta^{s,i} = \frac{r^{\lambda_i}}{2G}\Big[a_1\,(\kappa+\lambda_i)\sin\left[(\lambda_i-1)\,\theta\right] + a_2\,(\kappa+\lambda_i)\cos\left[(\lambda_i-1)\,\theta\right] + b_1\sin\left[(\lambda_i+1)\,\theta\right] + b_2\cos\left[(\lambda_i+1)\,\theta\right] \Big] \quad (32)$$

$$u_\theta^{b,i} = \frac{r^{\mu_i}}{2G}\Big[d_1\,(\kappa+\mu_i)\sin\left[(\mu_i-1)\,\theta\right] + d_2\,(\kappa+\mu_i)\cos\left[(\mu_i-1)\,\theta\right] + c_1\sin\left[(\mu_i+1)\,\theta\right] + c_2\cos\left[(\mu_i+1)\,\theta\right] \Big] \quad (33)$$

The expression of a_1, b_1, c_1 as a function of a_1 and b_2, c_2, d_2 as a function of a_2 are also given in [14]:

$$b_1 = \chi_{b1}(1-\lambda_1)\,a_1 \quad (34)$$

$$c_1 = \frac{q\lambda_1 r_0^{\lambda_1-\mu_1}}{4\mu_1(q-1)}\chi_{c1}\,a_1 \quad (35)$$

$$d_1 = \frac{q\lambda_1 r_0^{\lambda_1-\mu_1}}{4\mu_1(q-1)}\chi_{d1}\,a_1 \quad (36)$$

$$b_2 = -\chi_{b2}(1-\lambda_2)\,a_2 \quad (37)$$

$$c_2 = \frac{q\lambda_2 r_0^{\lambda_2-\mu_2}}{4\mu_2\,(\mu_2-1)}\chi_{c2}\,a_2 \quad (38)$$

$$d_2 = \frac{q\lambda_2 r_0^{\lambda_2-\mu_2}}{4\mu_2\,(\mu_2-1)}\chi_{d2}\,a_2 \quad (39)$$

Competing interests

The authors declare that they have no competing interests.

Authors' contributions

GL worked on the algorithms, performed all the computations and drafted the manuscript. NM worked on the algorithms and carried out detailed revision. All authors read and approved the final manuscript.

Acknowledgements

The support of the ERC Advanced Grant XLS no 291102 is gratefully acknowledged

References

1. Szabó B, Babuška I (1991) Finite element analysis. 1st edition. Wiley
2. Sevilla R, Fernández-méndez S (2008) NURBS-Enhanced Finite Element Method (NEFEM). Int J Numer Meth Eng 76:56–83
3. Sevilla R, Fernández-méndez S, Huerta S (2011) Comparison of high-order curved finite elements. Int J Numer Meth Eng 87:719–734
4. Moës N, Dolbow JE, Belytschko T (1999) A finite element method for crack growth without remeshing. Int J Numer Meth Eng 46:131–150
5. Saulèv VK (1963) On the solution of some boundary value problems on high performance computers by fictitious domain method. Siberian Math J 4:912–925
6. Dréau K, Chevaugeon N, Moës N (2010) Studied X-FEM enrichment to handle material interfaces with higher order finite element. Comput Meth Appl Mech Eng 199(29–32):1922–1936
7. Legrain G, Chevaugeon N, Dréau K (2012) High order X-FEM and levelsets for complex microstructures: uncoupling geometry and approximation. Comput Meth Appl Mech Eng 241–244:172–189
8. Groß S, Reusken A (2007) An extended pressure finite element space for two-phase incompressible flows with surface tension. J Comput Phys 224(1):40–58
9. Düster A, Rank E (2001) The p-version of the finite element method compared to an adaptive h-version for the deformation theory of plasticity. Comput Meth Appl Mech Eng 190:1925–1935
10. Legrain G (2013) A NURBS enhanced extended finite element approach for unfitted CAD analysis. Comput Mech 52(4):913–929
11. Melenk JM, Babuška I, Babuskab I (1996) The partition of unity finite element method: basic theory and applications. Comput Meth Appl Mech Eng 139:289–314
12. Williams ML (1952) Stress singularities resulting from various boundary conditions in angular corners of plates in tension. ASME J Appl Mech 19:526–528
13. Lazzarin P, Tovo R (1996) A unified approach to the evaluation of linear elastic stress fields in the neighborhood of cracks and notches. Int J Fract 78(1):3–19
14. Filippi S, Lazzarin P, Tovo R (2002) Developments of some explicit formulas useful to describe elastic stress fields ahead of notches in plates. Int J Solid Struct 39(17):4543–4565
15. Dini D, Hills DA (2004) Asymptotic characterisation of nearly-sharp notch root stress fields. Int J Fract 130(3):651–666
16. Creager M, Paris PC (1967) Elastic field equations for blunt cracks with reference to stress corrosion cracking. Int J Fract Mech 3(4):247–252
17. Glinka G (1985) Calculation of inelastic notch-tip strain- stress histories under cyclic loading. Eng Fract Mech 22(5):839–854
18. Sukumar N, Chopp DL, Moës N, Belytschko T (2001) Modeling holes and inclusions by level sets in the extended finite element method. Comp Meth Appl Mech Eng 190:6183–6200
19. Lian WD, Legrain G, Cartraud P (2012) Image-based computational homogenization and localization: comparison between X-FEM/levelset and voxel-based approaches. Comput Mech 51(3):279–293
20. Legrain G, Allais R, Cartraud P (2011) On the use of the extended finite element method with quadtree/octree meshes. Int J Numer Meth Eng 86:717–743
21. Béchet É, Minnebo H, Moës N, Burgardt B (2005) Improved implementation and robustness study of the X-FEM for stress analysis around cracks. Int J Numer Meth Eng 64(8):1033–1056
22. Laborde P, Pommier J, Renard Y, Salaün M (2005) High-order extended finite element method for cracked domains. Int J Numer Meth Eng 64(3):354–381
23. Babuška I, Banerjee U, Osborn JE (2004) Generalized finite element methods – main ideas, results and perspective. Int J Comput Methods 1(1):67-103
24. Ndeffo M, Massin P, Moës N (2013) Crack propagation modelisation using XFEM With 2D and 3D quadratic elements. In: International conference on Extended Finite Element Methods - XFEM 2013
25. Duarte CA, Babuška I, Oden JT (2000) Generalized finite element methods for three-dimensional structural mechanics problems. Comput Struct 77:215–232
26. Chevaugeon N, Moës N, Minnebo H (2013) Improved crack tip enrichment functions and integration for crack modeling using the extended finite element method. Int J Multiscale Comput Eng 11(6):597–631
27. Babuška I, Guo B (1992) The h, p and h-p version of the finite element method; basis theory and applications. Adv Eng Softw 15:159–174
28. Gui W, Babuška I (1986) The p andh-p versions of the finite element method in 1 dimension Part I: the basic approximation results. Numerische Mathematik 612:577–612
29. Wagner GJ, Moës N, Liu WK, Belytschko T (2001) The extended finite element method for stokes flow past rigid cylinders. Int J Numer Meth Eng 51:293–313

8

Membrane wrinkling revisited from a multi-scale point of view

Noureddine Damil[1], Michel Potier-Ferry[2,3*] and Heng Hu[4]

* Correspondence: michel.potier-ferry@univ-lorraine.fr
[2]LEM3, Laboratoire d'Etudes des Microstructures et de Mécanique des Matériaux, UMR CNRS 7239, Université de Lorraine, Ile du Saulcy, 57045, Metz Cedex 01, France
[3]Laboratory of Excellence on Design of Alloy Metals for low-mAss Structures (DAMAS), Université de Lorraine, Lorraine, France
Full list of author information is available at the end of the article

Abstract

Background: Membrane modeling in the presence of wrinkling is revisited from a multi-scale point of view. In the engineering literature, wrinkling is generally accounted at a macroscopic level by nonlinear constitutive laws without compressive stiffness, but these models ignore the properties of wrinkles, such as their wavelength, their size and spatial distribution.

Methods: A new multi-scale approach is discussed that belongs to the family of Ginzburg- Landau bifurcation equations. By using the method of Fourier series with variable coefficients, several nonlinear macroscopic models are derived that couple the membrane response with equations governing the evolution of the wrinkles.

Results: Contrary to previous approaches, these macroscopic models are completely deduced from the "microscopic" shell model without any phenomenological assumptions. Some analytical and numerical solutions are discussed that prove the relevance of the presented modeling.

Conclusions: A new class of models has been established. It permits to predict the characteristics of the wrinkles and their influence on membrane behavior.

Keywords: Wrinkling; Buckling; Bifurcation; Membrane; Slowly variable Fourier coefficients; Multi-scale

Background

This paper deals with macroscopic modeling of very thin shells, also called membranes in the scientific literature and in everyday life. Membranes have increasing application fields such as spacecraft structures (antenna, telescope lenses, Gossamer structures...), civil structures, life jackets, electronics, biological tissues, textile composites [1-6]. The appearance of wrinkles is a major feature in the mechanical behavior of membrane, due to their vanishing bending stiffness.

Membrane wrinkling is a multi-scale phenomenon, but it is generally described by one-scale models that can be "microscopic" or "macroscopic". Microscopic models are simply based on elastic shell theory; see for instance [7-11]. Nowadays many commercial finite element codes permit to carry out such nonlinear shell computations. Shell analyses are able to describe the details of the membrane response: size, wavelength, orientation of the wrinkles, instability threshold, etc., but this leads to large scale numerical models that are moreover very difficult to be controlled in cases with a large number of wrinkles and therefore with a large number of equilibrium solutions. Indeed wrinkling can be seen as a small wavelength buckling phenomenon and, in such a case, there are several neighbor bifurcation modes and it is well known that this leads to compound response curves involving secondary bifurcations, see for instance [12,13] among an extensive literature. Even if the complexity of wrinkling patterns requires numerical treatments, one can mention several papers by Coman et al. [14-16] that were able to find analytically wrinkling modes in cases with inhomogeneous pre-bifurcation stresses.

The second group of numerical descriptions are pure membrane models that account for the effect of wrinkling on membrane behaviour. The bending stiffness is neglected and the compressive stresses are dropped: this yields nonlinear constitutive laws that are of unilateral type, the tensile behaviour being about linear and the compressive states being forbidden. This has led to many numerical studies; see for instance [17-22]. In most of these papers, the constitutive law relates simply membrane stress and strain and it distinguishes three states: slack, wrinkled and taut. Some authors prefer the method of Roddeman [23,24] that splits the deformation gradient into consistent membrane part and wrinkling part [25-28]. The partial differential equations deduced from these membrane models are not elliptic (or hyperbolic in the dynamical case) in the presence of a non-positive principal stress so that this problem is mathematically ill-posed. A well-posed problem can be obtained if the macroscopic model includes an internal length, for instance within Cosserat theory [29,30]. However this regularisation may be not necessary for an explicit dynamic computation. All these membrane models have a phenomenological character, the only mention to full shell models being the vanishing of the membrane compressive stress. A variant has been introduced by [31] and used for buckling problems of sheets under residual stresses generated by manufacturing processes [32].

The aim of this paper is to revisit membrane modelling from a multi-scale point of view. It will be clearly established that a consistent wrinkling model depends both on microscopic quantities (wavelength, orientation of the wrinkles...) and macroscopic data like boundary conditions, size and shape of the structure... In this respect, a new multi-scale membrane model was briefly presented [33] that couple a nonlinear 2D membrane model with an amplitude equation governing the evolution of wrinkles. Amplitude equations to describe spatial pattern formation follow generally from the asymptotic bifurcation analyses of Ginzburg-Landau type [34-37]. Here a slightly different method will be applied, where the nearly periodic fields are represented by Fourier series with slowly varying coefficients [38-41]. In other terms, we use a multi-scale modelling method whose result is a generalised continuum including an internal length and where the macroscopic stresses are Fourier coefficients of the microscopic stress.

In this paper, the new membrane model of [33] is discussed in detail. Some additional variants, several analytical and numerical solutions and the corresponding numerical schemes will be presented. The new membrane models including wrinkling will be deduced from Föppl-Von Karman plate theory. These analyses will be carried out by applying standard techniques of multi-scale bifurcation analyses: asymptotic multi-scale method for the linear bifurcation analysis and multi-scale Fourier approach leading to nonlinear models valid away from the wrinkling threshold. Note that, contrarily to most of macroscopic membrane models, the present models are deduced from the shell equations without any phenomenological assumptions. The bending stiffness effects are included not only to define the wrinkling wavelength, but also to predict the macroscopic evolution of the buckling pattern.

The paper is organized as follows. In section "About the origin of short wavelength instabilities", the origin of the wrinkling instability is briefly recalled, in order to underline its multi-scale origin and the generic character of this instability mechanism. section "Multi-scale nonlinear models for membrane wrinkling" is devoted to deducing new nonlinear wrinkling models from Föppl-Von Karman plate theory. Then two analytical solutions of these models will be presented (section "Two analytical solutions for clamped rectangular membranes"), the first one for defining the bifurcation features (bifurcation stress, wavelength and shape of the envelope) and the second one for a post-bifurcation analysis characterizing the evolution of the size of the wrinkles as a function of the applied load. The numerical analysis is presented in section "Numerical implementation" (techniques of implementation) and section Two numerical solutions where two examples of plate wrinkling will be solved. Finally, in section "Linear wrinkling analysis revisited by a double scale asymptotic analysis", a more classical asymptotic multi-scale analysis will be performed in a special case and the results will be compared with those arising from the multi-scale Fourier approach.

About the origin of short wavelength instabilities

Spatially periodic instability patterns are encountered in a lot of physical problems. The instability often occurs in domains whose dimension is smaller than the others, typically in rectangles or parallelepipeds with a small aspect ratio. The instability wavelength can be of the same order as the smallest dimension of the domain as in Rayleigh-Bénard convection [42,43].

The instability wavelength is not always directly related to a structural dimension. A well known typical case of periodic patterns is the flexible beam resting on a Winkler foundation that is also referred as Swift-Hohenberg equation [44-50]. The governing equation has the following form in 1D linear case:

$$A\frac{d^4v(X)}{dX^4} + \lambda\frac{d^2v(X)}{dX^2} + Bv(X) = 0, \qquad X \in IR \tag{1}$$

The wrinkling modes $v(X)$ and the wrinkling load λ are then given by:

$$v(X) = \cos(QX + \varphi), \qquad \lambda(Q) = AQ^2 + \frac{B}{Q^2}. \tag{2}$$

The instability wavenumber Q or the instability wavelength $\ell = 2\pi/Q$ corresponds to the minimum of the neutral stability curve $Q \rightarrow \lambda(Q)$:

$$\frac{1}{Q} = \frac{\ell}{2\pi} = \sqrt[4]{\frac{A}{B}}, \qquad \lambda = 2AQ^2 = 2\sqrt{AB}. \tag{3}$$

A physical example of such instability is the symmetric microbuckling mode of long fiber composite materials that are represented by an infinite stack of hard and soft layers (fiber, matrix), see Figure 1 and Rosen [51].

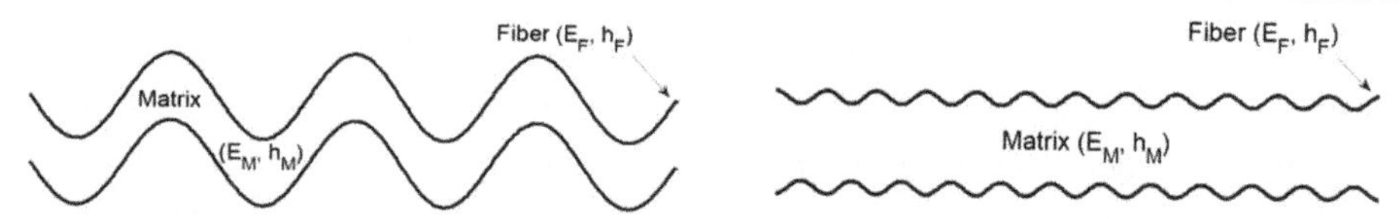

Figure 1 Antisymmetric and symmetric microbuckling modes of long fiber composite materials.

In this problem, the coefficients of (1) are related to Young moduli and thickness's of the layers ($A \cong E_F h_F^3$, $B \cong E_M / h_M$), the microbuckling wavelength is related to the width of the fiber and of the matrix by the following formula: $\ell \cong h_F^{3/4} h_M^{1/4} (E_F/E_M)^{1/4}$. Quite similar wrinkling phenomena with a small wavelength can be found in studies about thin films coupled with compliant substrate, see for instance [52-55], or in the buckling of sheets due to residual stresses and forming processes [56,57]. In this second class of instability problems, the wavelength is larger than the microscopic lengths $h_B\, h_M$ because of a large stiffness ratio. Here is the origin of the multi-scale behavior: the characteristic distance in the width is smaller than the longitudinal wavelength and moreover these longitudinal waves are generally modulated on larger distances.

The wrinkling of membrane belongs to a third class of instability problems that can be represented by a partial differential equation in a 2D domain [3]:

$$A \frac{\partial^4 v(X,Y)}{\partial X^4} + \lambda \frac{\partial^2 v(X,Y)}{\partial X^2} - C \frac{\partial^2 v(X,Y)}{\partial Y^2} = 0, \quad X \in IR \quad Y \in IR. \tag{4}$$

The instability modes are in the form $V(Y)\cos(QX+\varphi)$ and their instability wavelength $2\pi/Q$ is generally small with respect to the characteristic length in the transverse direction. The Equations (1) and (4) can be related by assuming that the coefficient B is an eigenvalue of the differential operator $-C\partial^2/\partial Y^2$, or equivalently $B = C/L_Y^2$, where L_Y is a macroscopic characteristic length in the transverse direction.

In the problem of a plane membrane of thickness h submitted to a tensile stress σ_Y in the transverse direction, $A \cong Eh^3$, $C = \sigma_Y h$ and $\sigma_X = -\lambda/h$ is a compressive stress in the X-direction so that the wrinkling wavelength is $\ell_x \cong (E/\sigma_Y)^{1/4} h^{1/2} L_Y^{1/2}$ [3]. Hence this wavelength is much smaller than the transverse length L_Y because a membrane is very thin ($h << L_Y$) and it decreases when the tensile stress increases. From Equation (3), one can deduce the critical compressive stress $|\sigma_X| \cong \sqrt{E\sigma_Y} h/L_Y$ that is much smaller than the prescribed tensile stress σ_y if this stress σ_y is not too small. It is worth to mention that the wrinkling stress depends both on the microscopic length h and on the macroscopic transverse one L_Y.

Note that the Equation (4) seems to be more or less generic. It has also been established in the antisymmetric microbuckling of long fiber composite [51,58], with $A \cong E_F h_F^3$, $C \cong E_M h_F$ (this antisymmetrical mode occurs when the thickness of the fiber and of the matrix are of the same order $h_M \cong h_F$). Hence the instability wavelength is $\ell_x \cong \left(\frac{E_F}{E_M}\right)^{1/4} h_F^{1/2} L_Y^{1/2}$ where the transverse length L_Y can be associated with the composite plate thickness or to the ply thickness [58]. Note that the nonlinear versions of these fiber microbuckling models permit to predict the compression failure of long fiber composite materials [59-61].

Hence a wrinkling mode follows from two stiffness's: first the bending stiffness A, that is very small for a thin membrane or a fiber in a composite material, second a transverse stiffness C that can be purely elastic in the antisymmetrical microbuckling or the so called geometric stiffness due to the tensile stress in the wrinkling of a membrane. Clearly, all these instability problems involve several length scales. That is why we propose several multi-scale approaches to analyze the behavior of membranes in the presence of wrinkling.

Methods

A family of nonlinear membrane models is presented that is based on a multi-scale approach. Unlike macroscopic models of the literature, there is no phenomenological assumption and the models are completely deduced from the full plate model. The deduction method relies on the concept of Fourier series with slowly varying coefficients [38,40], whose principle is to work with envelopes of the spatial oscillations. These envelopes are solutions of systems of nonlinear partial differential equations established in the next sections "Multi-scale nonlinear models for membrane wrinkling". These equations can be solved, sometimes analytically and in general numerically, as sketched in the section "numerical implementation".

Multi-scale nonlinear models for membrane wrinkling

The method of Fourier series with slowly variable coefficients

A new multiple scale approach: the method of Fourier series with slowly variable coefficients

We present here the methodology that will be used to deduce macroscopic nonlinear models of membrane wrinkling. For simplicity, this first discussion is limited to one-dimensional case (space variable $X \in IR$) and to a one-dimensional beam model of Von Karman type that was studied in numerous papers [38-41,44-49]:

$$\frac{dn}{dx} + f = 0 \qquad (5-\text{a})$$

$$\frac{n}{ES} = \gamma = \frac{du}{dx} + \frac{1}{2}\left(\frac{dv}{dx}\right)^2 \qquad (5-\text{b})$$

$$EI\frac{d^4v}{dx^4} - \frac{d}{dx}\left(n\frac{dv}{dx}\right) + cv + c_3v^3 = 0. \qquad (5-\text{c})$$

Let us suppose that the instability wavenumber Q is known. Within this method, the unknown field $U(X) = (u(X) \quad v(X) \quad n(X) \quad K(X) \quad \gamma(X))$, whose components are axial displacement, transverse displacement, resultant stress, curvature and membrane strains, is written in the following form:

$$U(X) = \sum_{m=-\infty}^{+\infty} U_m(X)\exp(imQX) \qquad (6)$$

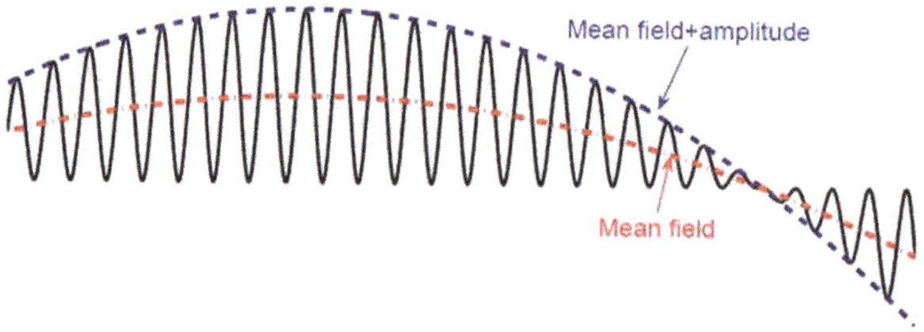

Figure 2 At least two macroscopic fields are necessary to describe a nearly periodic response: the mean field and the amplitude of the fluctuation.

The new macroscopic unknown fields $U_m(X)$ vary slowly on a single period $\left[X, X+\frac{2\pi}{Q}\right]$ of the pattern oscillation. As pictured in Figure 2, at least two functions $U_0(X)$ and $U_1(X)$ are necessary to describe nearly periodic patterns: $U_0(X)$ can be identified to the mean value and $U_1(X)$ represents the envelope or the amplitude of the spatial oscillations.

The first envelope $U_0(X)$ is real-valued while the next ones are complex. In the whole paper, we limit ourselves to two envelopes $U_0(X)$ and $U_1(X)$. Most of the time, we shall assume that they are real.

The derivation operators are calculated exactly according to the following rule:

$$\left(\frac{\mathrm{da}}{\mathrm{dX}}\right)_m = \frac{da_m}{dX} + miQa_m = \left(\frac{d}{dX} + miQ\right)a_m \tag{7}$$

$$\left(\frac{\mathrm{d}^2\mathrm{a}}{\mathrm{dX}^2}\right)_m = \frac{d^2a_m}{dX^2} - m^2Q^2a_m + 2imQ\frac{da_m}{dX}. \tag{8}$$

It was established [38,40] that it is necessary to keep some spatial derivatives of the envelopes, despite of the assumption of slowly varying envelopes $d/dX < < Q$. This will be re-discussed in the membrane case. At first all the derivatives are kept as in (7) (8), but some ones can be dropped later.

Application to the membrane constitutive law

The application of the Fourier method to simple nonlinear equation is straightforward and it can follow from a simple identification of Fourier coefficients. For instance let us consider the 1-D membrane constitutive law (5-b) that relates the membrane stress n, the membrane strain γ and the displacement (u,v). From (5-b), one can deduce a macroscopic constitutive law for the m-th Fourier envelope:

$$\frac{n_m}{ES} = \gamma_m = \left(\frac{d}{dX} + imQ\right)u_m + \frac{1}{2}\sum_{k=-\infty}^{+\infty}\left(\frac{d}{dX} + ikQ\right)v_k\left(\frac{d}{dX} + i(m-k)Q\right)v_{m-k}. \tag{9}$$

Especially the constitutive law for the mean stress $n_0(X)$ will be very useful. In the case of two real envelopes $(u_0, v_0, n_0, \gamma_0, K_0)$ and $(u_1, v_1, n_1, \gamma_1, K_1)$, it can be expressed as:

$$\frac{n_0}{ES} = \gamma_0 = \frac{du_0}{dX} + \frac{1}{2}\left(\frac{dv_0}{dX}\right)^2 + \left(\frac{dv_1}{dX}\right)^2 + Q^2v_1^2. \tag{10}$$

Let us mention the last two terms are positive, what means that the membrane stretches out when it wrinkles.

Energetic approach

There is another manner to derive the full macroscopic model [38,40] by starting from the potential energy. In the elastic beam problem (5), the initial potential is given by:

$$P(u,v) = \frac{1}{2}\int\left(EI\,K^2 + ES\,\gamma^2 + cv^2 + \frac{c_3v^4}{2}\right)dX, \qquad K = \frac{d^2v}{dX^2}. \tag{11}$$

As a consequence of the assumptions of slowly varying envelopes, these envelopes are assumed to be constant on each period so that only terms corresponding to the harmonic zero of K^2,γ^2,v^2 and v^4contribute to the approximated values of the potential energy. According to Parseval formula, the potential energy of the macroscopic model can be written as (the contribution of v^4 is omitted here for simplicity, full reduced models can be found in [40]):

$$\begin{aligned} P(u_j,v_j) &= \frac{1}{2}\int\left(EI\,\left(K_0^2 + 2\sum_{j=1}^{\infty}|K_j|^2\right) + ES\,\left(\gamma_0^2 + 2\sum_{j=1}^{\infty}|\gamma_j|^2\right) + c\left(v_0^2 + 2\sum_{j=1}^{\infty}|v_j|^2\right) + \ldots\right) dX \\ &\approx \frac{1}{2}\int\left(EI\,\left(K_0^2 + 2|K_1|^2\right) + ES\left(\gamma_0^2 + 2|\gamma_1|^2\right) + c\left(v_0^2 + 2|v_1|^2\right) + \ldots\right) dX. \end{aligned} \tag{12}$$

Further simplifications can be introduced in this model if one looks only at local bending instabilities from one-dimensional elastic states. In this respect, we can drop the mean deflection $v_0(X)$ and the envelope of the axial displacement $u_1(X)$. In this framework, the curvature and membrane strains of the wrinkled beam are approximated by (see (7) (9) and suppress the imaginary terms):

$$\begin{cases} K_0 = 0 & |K_1|^2 = \left(\frac{d^2v_1}{dX^2} - Q^2v_1\right)^2 + 4\left(\frac{dv_1}{dX}\right)^2 \\ \gamma_0 = \frac{du_0}{dX} + \left(\frac{dv_1}{dX}\right)^2 + Q^2v_1^2 & \gamma_1 = 0. \end{cases} \tag{13}$$

A last approximation can be introduced by disregarding $\frac{d^2v_1}{dX^2}$ in the potential energy. This latter approximation has been done in previous works [40] and it is justified by the slow spatial variations of the envelope. Finally, the potential energy can be approximated in the following form:

$$P(u_0,v_1) = \int\left(6EIQ^2\left(\frac{dv_1}{dX}\right)^2 + EIQ^4v_1^2 + \frac{ES}{2}\left(\frac{du_0}{dX} + \left(\frac{dv_1}{dX}\right)^2 + Q^2v_1^2\right)^2 + cv_1^2 + 3c_3v_1^4/2\right) dX \tag{14}$$

Thus the extrema of the macroscopic energy (14) are solutions of the following system:

$$\begin{cases} \dfrac{dn_0}{dX} + f_0 = 0 & (15-a) \\ \dfrac{n_0}{ES} = \gamma_0 = \dfrac{du_0}{dX} + \dfrac{1}{2}\left(\dfrac{dv_1}{dX}\right)^2 + Q^2 v_1^2 & (15-b) \\ -6Q^2 \dfrac{d^2 v_1}{dX^2} + \left(EIQ^4 + n_0 Q^2 + c\right) v_1 - \dfrac{d}{dX}\left(n_0 \dfrac{dv_1}{dX}\right) = 0 & (15-c). \end{cases}$$

Comments

Hence the membrane law (15-b) accounts for the wrinkling oscillation $v_1(X)$ that is governed by a sort of bifurcation Equation (15-c) looking like a Ginzburg-Landau equation. The full model is a nonlinear system coupling membrane behavior and evolution of wrinkles. It has been established [38,40] that the Fourier approach generalizes the asymptotic Ginzburg- Landau method, but it is not limited to the neighborhood of the bifurcation.

This modeling by few Fourier envelopes can be applied with several levels of approximation. The system (15) is the simplest possible with an internal length. It can still be simplified by neglecting the derivatives of v_1 in (15-c): this leads to a nonlinear relation between membrane stress and strain, see [40] § 4.4. This pure membrane model will be extended in the 2D case in the following. It is also possible to include more harmonics, see [40] § 3.2, with obviously the cost of a greater complexity. The model with a complex envelope $v_1(X)$ has been evaluated in [62]: it permits for instance a better account of the phase in the bulk and improves the response near the boundary.

The previous approach can be considered as a multi-scale method or a computational homogenization technique. The account of the local behavior is very simplified by the assumption of a harmonic local variation and it is described only by the instability wavenumber Q. If the Equation (15-c) is reduced to its linear version without modulation of the envelope, ones recovers the classical approach (2) (3) that gives the wavenumber at the first bifurcation by minimizing the critical load. When solving a nonlinear macroscopic problem such as (15), the wavenumber Q has to be prescribed and this could be considered as a weak point of the present macroscopic approach. Nevertheless it is known that, in a cellular bifurcation problem, many solutions can exist [45,46,63], each one being characterized by its wavelength. Furthermore a model with a complex envelope permits to predict a wavelength slightly deviating from the one a priori prescribed [62]. Higher order harmonics could also be accounted: for instance, it was established in [40] that a rough account of the second harmonic is sometimes necessary to recover a consistent post-bifurcation behavior. But this will be not necessary in the case of membrane wrinkling.

In the next parts, the interaction membrane-wrinkling will be modelised in a bi-dimensional case within a framework similar to (14) (15), the starting model being the Föppl-Von Karman equations.

Föppl-Von Karman plate equations as a microscopic model

The main objective of the paper is to obtain nonlinear membrane models where the wrinkling is described by a bifurcation equation deduced from the initial plate model.

The method of Fourier series with slowly variable coefficients is applied to deduce the sought model from the well known Föppl-Von Karman equations for elastic isotropic plates that will be considered here as the reference model:

$$\begin{cases} D\Delta^2 w - div(\mathbf{N}.\nabla w) = 0 \\ \mathbf{N} = \mathbf{L^m}.\boldsymbol{\gamma} \\ 2\boldsymbol{\gamma} = \nabla\mathbf{u} + \nabla^t\mathbf{u} + \nabla w \otimes \nabla w \\ div\mathbf{N} = 0 \end{cases} \tag{16}$$

where $\mathbf{u} = (u,v) \in IR^2$ is the in-plane displacement, w is the deflection, $\mathbf{N}$ and $\boldsymbol{\gamma}$ are the membrane stress and strain. With the vectorial notations ($\mathbf{N} \rightarrow {}^t(N_X N_Y N_{XY})$, $\gamma \rightarrow {}^t(\gamma_X \gamma_Y 2\gamma_{XY})$), the membrane elasticity tensor is represented by the matrix $\frac{Eh}{1-\nu^2}\begin{bmatrix} 1 & \nu & 0 \\ \nu & 1 & 0 \\ 0 & 0 & \frac{1-\nu}{2} \end{bmatrix}$. The corresponding energy $\mathcal{E}$ can be split into a membrane part $\mathcal{E}_{mem}$ and a bending part $\mathcal{E}_{ben}$, as follows:

$$\begin{cases} \mathcal{E}(\mathbf{u}, w) = \mathcal{E}_{ben}(w) + \mathcal{E}_{mem}(\mathbf{u}, w) \\ 2\mathcal{E}_{ben}(w) = D \iint \left((\Delta w)^2 - 2(1-\nu)\left(\frac{\partial^2 w}{\partial X^2}\frac{\partial^2 w}{\partial Y^2} - \left(\frac{\partial^2 w}{\partial X \partial Y} \right)^2 \right) \right) d\omega \\ 2\mathcal{E}_{mem}(\mathbf{u}, w) = \iint {}^t\boldsymbol{\gamma}.\mathbf{L}^m.\boldsymbol{\gamma} dw = \frac{Eh}{1-\nu^2} \iint (\gamma_X^2 + \gamma_Y^2 + 2(1-\nu)\gamma_{XY}^2 + 2\nu\gamma_X\gamma_Y) d\omega. \end{cases} \tag{17}$$

Macroscopic modeling of membrane strain

As explained previously, the unknown fields $U(X,Y)$ are expressed in terms of two harmonics: the mean field $U_0(X,Y)$ and the first order harmonicsr, $\mathbf{U}_1(X, Y)e^{iQX}$, $\overline{\mathbf{U}}_1(X, Y)e^{-iQX}$. The second harmonic should be taken into account to recover the results of the asymptotic Ginzburg-Landau bifurcation approach, see [40]. Nevertheless the second harmonic does not contribute to the membrane energy in the present case, because the rapid one-dimensional oscillations e^{iQX} are inextensional so that $\mathbf{N}_2 = 0$, $w_2 = 0$. Hence the second harmonic does not influence the simplest macroscopic models. For simplicity, the details of this calculation are omitted.

A unique direction OX for the wrinkling oscillations is chosen in the whole domain. Of course this assumption is a bit restrictive and should be removed in the future. A true multi-scale approach should include two levels of modelisation as in the FE2 method [64], but with a basic cell that is not a priori known [65], what should require a rather intricate management. Thus a more realistic goal is to first discuss the multi-scale approach with a given wrinkling wavelength and a given direction of the wrinkles.

The derivation rules (7) (8) can be extended easily in a bi-dimensional framework. For instance, the first Fourier coefficient of the gradient and the 0th order coefficient of the strain (i.e. its mean value on a period) are given by:

$$\{(\nabla w)_1\} = \begin{Bmatrix} \frac{\partial w_1}{\partial X} + iQw_1 \\ \frac{\partial w_1}{\partial Y} \end{Bmatrix} \tag{18}$$

$$\{\gamma_0\} = \begin{Bmatrix} \gamma_{X0} \\ \gamma_{Y0} \\ 2\gamma_{XY0} \end{Bmatrix} = \{\gamma^{FK}\} + \{\gamma^{wr}\} \tag{19}$$

$$\{\gamma^{FK}\} = \begin{Bmatrix} \dfrac{\partial u_0}{\partial X} + \dfrac{1}{2}\left(\dfrac{\partial w_0}{\partial X}\right)^2 \\ \dfrac{\partial v_0}{\partial Y} + \dfrac{1}{2}\left(\dfrac{\partial w_0}{\partial Y}\right)^2 \\ \dfrac{\partial u_0}{\partial Y} + \dfrac{\partial v_0}{\partial X} + \dfrac{\partial w_0}{\partial X}\dfrac{\partial w_0}{\partial Y} \end{Bmatrix}$$

$$\{\gamma^{wr}\} = \begin{Bmatrix} \left|\dfrac{\partial w_1}{\partial X} + iQw_1\right|^2 \\ \left|\dfrac{\partial w_1}{\partial Y}\right|^2 \\ \left(\dfrac{\partial w_1}{\partial X} + iQw_1\right)\dfrac{\partial \bar{w}_1}{\partial Y} + \left(\dfrac{\partial \bar{w}_1}{\partial X} - iQ\bar{w}_1\right)\dfrac{\partial w_1}{\partial Y} \end{Bmatrix} \tag{20}$$

As in formula (10), the strain is split into a classical part γ^{FK}that has the same form as the original Föppl-Von Karman model (16) and a wrinkling part γ^{wr} which depends only on the envelope of the deflection w_1.

Let us now turn to a simplified version of the displacement–strain law (19) (20), in the same spirit as for the 1D model (15). First the displacement field is reduced to a membrane mean displacement and to a bending wrinkling, i.e. $\mathbf{u}_1 = 0$, $w_0 = 0$. This means that we account for the influence of the local buckling on the membrane behavior, but not for the coupling between local and global buckling as in [66,67]. Second the deflection envelope $w_1(X, Y)$ is assumed to be real, which disregards the phase modulation of the wrinkling pattern. Thus the envelope of the displacement has only three components $\mathbf{u}_0 = (u_0, v_0)$ and w_1 that will be rewritten for simplicity as $(u, v, w) \stackrel{def}{=} (u_0, v_0, w_1)$. The simplified version of the strain field becomes:

$$\{\gamma\} \stackrel{def}{=} \{\gamma_0\} = \{\varepsilon(\mathbf{u})\} + \{\gamma^{wr}\} \tag{21}$$

$$\{\varepsilon(\mathbf{u})\} = \begin{Bmatrix} \dfrac{\partial u}{\partial X} \\ \dfrac{\partial v}{\partial Y} \\ \dfrac{\partial u}{\partial Y} + \dfrac{\partial v}{\partial X} \end{Bmatrix}, \qquad \{\gamma^{wr}\} = \begin{Bmatrix} \left(\dfrac{\partial w}{\partial X}\right)^2 + Q^2 w^2 \\ \left(\dfrac{\partial w}{\partial Y}\right)^2 \\ 2\dfrac{\partial w}{\partial X}\dfrac{\partial w}{\partial Y} \end{Bmatrix} \tag{22}$$

The simplified membrane strain formula (21) (22) is quite similar to that of the initial Von Karman model. It is split, first in a linear part $\varepsilon(\mathbf{u})$ that is the symmetric part of the mean displacement gradient corresponding to the pure membrane linear strain, second in a nonlinear part $\gamma^{wr}(w)$ more or less equivalent to wrinkling stain of [23]. The main difference with the initial Föppl-Von Karman strain (16) is the extension Q^2w^2 in the direction of the wrinkles. This wrinkling strain is always positive and this corresponds to a stretching. In the case of a compressive membrane strain, this wrinkling term leads to a decrease of the true strain.

As for the 1D model (14), we limit ourselves to the 0th order harmonic to compute the reduced membrane energy:

$$2\mathcal{E}_{mem}(\mathbf{u},w)=\frac{Eh}{1-\nu^2}\iint\left\{\begin{array}{l}\left(\frac{\partial u}{\partial X}+\left(\frac{\partial w}{\partial X}\right)^2+Q^2w^2\right)^2+\left(\frac{\partial v}{\partial Y}+\left(\frac{\partial w}{\partial Y}\right)^2\right)^2+\\ 2(1-\nu)\left(\frac{1}{2}\left(\frac{\partial u}{\partial X}+\frac{\partial v}{\partial Y}\right)+\frac{\partial w}{\partial X}\frac{\partial w}{\partial Y}\right)^2+\\ 2\nu\left(\frac{\partial u}{\partial X}+\left(\frac{\partial w}{\partial X}\right)^2+Q^2w^2\right)\left(\frac{\partial v}{\partial Y}+\left(\frac{\partial w}{\partial Y}\right)^2\right)\end{array}\right\}d\omega. \tag{23}$$

Macroscopic bending energy

We saw two possible ways to get a reduced-order model via the technique of Fourier series with slowly variable coefficients: either identify the Fourier coefficient as in (9), or simplify the energy by keeping only the 0th order term in the energy, as in (12). Here we shall prefer the second approach which permits to provide a formulation easier to be managed for the numerical discretisation. The computation of the energy is based on the fact that only the 0th order harmonic φ_0 of a function φ has a non zero mean value:

$$\iint_\omega \varphi d\omega=\iint_\omega \varphi_0 d\omega. \tag{24}$$

The identity (24) is applied to the two terms of the bending energy in the same framework as in (15) (21) (22), i.e. $\mathbf{u}_1=(u_1,v_1)=(0,0)$, $w_0=0$, $w_1\in IR$:

$$\varphi=(\Delta w)^2-2(1-\nu)\left[\frac{\partial^2 w}{\partial X^2}\frac{\partial^2 w}{\partial Y^2}-\left(\frac{\partial^2 w}{\partial X\partial Y}\right)^2\right]=\varphi^A-2(1-\nu)\varphi^B$$

$$\begin{aligned}\varphi_0^A&=(\Delta w)^2=\sum_{n=-\infty}^{+\infty}(\Delta w)_n(\Delta w)_{-n}=2(\Delta w)_1(\Delta w)_{-1}=2\left|(\Delta w)_1\right|^2\\ &=2\left|\Delta w_1-Q^2w_1+2iQ\frac{\partial w_1}{\partial X}\right|^2.\end{aligned}$$

Because of the assumption of a real envelope $w=w_1$, the first term of the bending energy is obtained:

$$\varphi_0^A=2\left(\Delta w-Q^2w\right)^2+8Q^2\left(\frac{\partial w}{\partial X}\right)^2. \tag{25}$$

The second term of the bending energy φ_0^B is computed in the same way:

$$\varphi_0^B=2\left(\frac{\partial^2 w}{\partial X^2}-Q^2w\right)\frac{\partial^2 w}{\partial Y^2}-2\left(\frac{\partial^2 w}{\partial X\partial Y}\right)^2-2Q^2\left(\frac{\partial w}{\partial Y}\right)^2. \tag{26}$$

To be consistent with the approach in the previous 1D case, the derivatives of order three or four in the differential equations are neglected. This has been justified in [39]: indeed spurious oscillations can appear in the response of the macroscopic model if these high order derivatives are kept. Finally the bending energy in the simplified framework is reduced as:

$$\mathcal{E}_{ben}(w) = D \iint \left\{ Q^4 w^2 - 2Q^2 w \Delta w + 4Q^2 \left(\frac{\partial w}{\partial X}\right)^2 + 2(1-\nu^2)Q^2 \left[w \frac{\partial^2 w}{\partial Y^2} + \left(\frac{\partial w}{\partial Y}\right)^2 \right] \right\} d\omega. \tag{27}$$

Three macroscopic membrane wrinkling models

In this section, one establishes the Partial Differential Equations of the macroscopic models associated with the strain energies previously presented. In fact, we have not defined a closed model, but a family of models that can depend on the number of harmonics and various other assumptions, such as the assumption of a real wrinkling envelope or the hypothesis $w_0 = 0$, which means no bending before wrinkling. Three cases will be considered. First the model presented in [33] couples a linear membrane behavior with a real envelope governed by a sort of Ginzburg-Landau equation. It corresponds to (15) in the 1D case and can be considered as a reference model because it is simple and able to account for the influence of wrinkling on membrane behavior. Next a pure membrane model will be derived by neglecting the dependence with respect to the derivatives of the envelopes. Last we shortly describe a model with a complex envelope, as the one studied in [62] in the 1D case that is a relatively simple improvement of the reference model.

A reference macroscopic membrane model

The first proposed macroscopic membrane model follows from the minimum of the potential energy. In the absence of body forces, the sum of the membrane energy (23) and of the bending energy (27) is stationary at equilibrium. Let us recall that these energies associate 0th order harmonic for membrane quantities and a real first order harmonics for the deflection, as in the 1D model (15). In other words, it permits to couple a spatially modulated wrinkling with a linear membrane model:

$$\delta\mathcal{E}_{ben} + \delta\mathcal{E}_{mem} = 0,$$

for any virtual displacement that is zero at the boundary. This gives:

$$\delta\mathcal{E}_{\text{ben}} + \iint_\omega \mathbf{N} : \delta\boldsymbol{\gamma}^{wr} d\omega = 0 \tag{28}$$

$$\iint_\omega \mathbf{N} : \delta\boldsymbol{\varepsilon} d\omega = 0. \tag{29}$$

After straightforward calculations, one obtains the partial differential equations of the macroscopic problems in the following form, the wrinkling membrane strain $\gamma^{wr}(w)$ being given in (22):

$$div\mathbf{N} = 0 \tag{30}$$

$$\mathbf{N} = \mathbf{L}^m : [\boldsymbol{\varepsilon}(u) + \boldsymbol{\gamma}^{wr}(w)] \tag{31}$$

$$-6DQ^2 \frac{\partial^2 w}{\partial X^2} - 2DQ^2 \frac{\partial^2 w}{\partial Y^2} + \left(DQ^4 + N_X Q^2\right) w - div(\mathbf{N}.\nabla w) = 0. \tag{32}$$

The nonlinear model (30) (31) (32) couples nonlinear membrane equations with a bifurcation Equation (32) satisfied by the envelope of wrinkling pattern. It extends the previous analysis of the 1D case that couples a beam membrane with a one-dimensional Ginzburg-Landau equation. Hence the bifurcation Equation (32) is a sort of bi-dimensional Ginzburg-Landau equation, but it differs from the amplitude equation of Segel-Newell-Whitehead [42,43] who consider cases where the pre-bifurcation state is invariant under rotation. We shall see that finite element discretisation of (30) (31) (31) is straightforward. Two analytical solutions will be also presented in the section "Two analytical solutions for clamped rectangular membranes".

Within the nonlinear model (30) (31) (32), one recovers two ideas of classical macroscopic membrane models. First the splitting between membrane and wrinkling strain of Roddeman theory [23] has been deduced from Föppl-Von Karman Equation, see (31), without any phenomenological assumption. Then the final bifurcation Equation (32) includes an internal length, which permits to retrieve the multi-scale instability analysis of Part 2. Finally this model can be qualitatively compared with the models of Banerjee et al. [29,30], where an internal length is introduced via Cosserat theory. Body forces and boundary forces can be introduced easily by the same procedure. This requires that these forces can also be put in the form of Fourier series with slowly variable coefficients. For instance if these forces vary slowly at the scale of the wrinkles, this leads to a classical body force in the membrane Equation (30), as well as in the corresponding boundary conditions. In the same way, transverse forces can also be accounted for within the more complete model entitled "A more sophisticated macroscopic membrane model with a complex envelope".

A pure membrane model

The reference membrane model (30) (31) (32) is not a pure membrane model because the Equation (32) includes spatial derivatives of the envelope of the wrinkles. It is natural to try to recover a pure membrane model, where the kinematic unknown is only the in-plane displacement, as suggested in [40], § 4.4. By dropping all the spatial derivatives in (32), one gets a bifurcation equation $(N_X + DQ^2)w = 0$ that can be transformed in a perturbed bifurcation equation as:

$$(N_X + DQ^2)w = \delta \tag{33}$$

From (33), one can obtain the deflection as a function of one component of the membrane stress. In (33), δ is a small perturbation parameter that transforms the perfect bifurcation equation into a perturbed bifurcation, what is more convenient for numerical path following calculations. If one simplifies the wrinkling strain (22) $\gamma^{wr}(w) = Q^2 w^2 e_x \otimes e_x$ and if one combines (31) and (33), one can drop the deflection and deduce a nonlinear relation between membrane strain $\boldsymbol{\varepsilon}(\mathbf{u})$ and membrane stress $\mathbf{N}$. With account of the balance of membrane forces, the obtained full model is restricted to:

$$\begin{cases} div\mathbf{N} = 0 \\ \boldsymbol{\varepsilon}(u) + \dfrac{Q^2\delta^2}{(N_X + DQ^2)^2}\mathbf{e}_X \otimes \mathbf{e}_X = (\mathbf{L}^m)^{-1}: \mathbf{N}. \end{cases} \tag{34}$$

The model (34) is consistent with the pure membrane theories of the literature, see for instance [23], because N_X cannot be lower than the wrinkling stress $-DQ^2$.Generally, this wrinkling threshold is approximated by zero so that the membrane does not admit compressive stresses. In sections "Two analytical solutions for clamped rectangular membranes" and "Two numerical solutions", this model will be compared with our reference model (30) (31) (32).

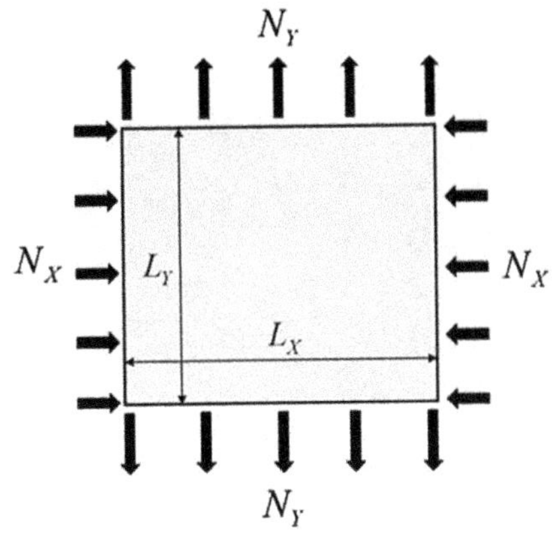

Figure 3 Rectangular membrane under biaxial load.

A more sophisticated macroscopic membrane model with a complex envelope

More sophisticated models can be introduced in a rather easy way. For instance models with five envelopes (harmonics 0, ±1, ±2) have been presented in [39,40], starting models being respectively 2D hyperelasticity and the beam model of section The method of Fourier series with slowly variable coefficients. The reference model of section A reference macroscopic membrane model can be improved at least in two ways. First one can reintroduce the mean deflection w_0 to account for a coupling between local wrinkling and global buckling, as in [66]: in this manner, one will associate an envelope equation with the full Föppl-Von Karman plate equations. Moreover, this envelope can be a complex one for a better account of the phase field and of the boundary behavior [62]. To keep a relatively simple model, a more questionable hypothesis will be done: the membrane behavior will be described only by the 0th order harmonic, i.e. $\mathbf{u}_1 = 0, \mathbf{N}_1 = 0, \boldsymbol{\gamma}_1 = 0$. Hence the remaining unknown fields will be:

$$\mathbf{u} \overset{def}{=} \mathbf{u}_0,\ \mathbf{N} \overset{def}{=} \mathbf{N}_0, \boldsymbol{\gamma} \overset{def}{=} \boldsymbol{\gamma}_0, w_0 \in IR,\ w_1 \in C, \tag{35}$$

i.e. the same variables as in the basic Föppl-Von Karman equations completed by the complex envelope w_1 of the wrinkles. This leads to the following system:

$$div\mathbf{N} = 0 \tag{36}$$

$$\mathbf{N} = \mathbf{L}^m : \left[\boldsymbol{\varepsilon}(u) + \frac{1}{2}\nabla w_0 \otimes \nabla w_0 + \boldsymbol{\gamma}^{wr}(w_1)\right] \tag{37}$$

$$D\Delta^2 w_0 - div(\mathbf{N}.\nabla w_0) = 0 \tag{38}$$

$$-6DQ^2\frac{\partial^2 w_1}{\partial X^2} - 2DQ^2\frac{\partial^2 w_1}{\partial Y^2} + (DQ^4 + N_X Q^2)w_1 - iQ((\mathbf{N}.e_X).\nabla w_1 + div(w_1\mathbf{N}.e_X)) - div(\mathbf{N}.\nabla w_1) = 0. \tag{39}$$

This system is only presented as an example of the multi-scale procedure and it will not be discussed further in this paper.

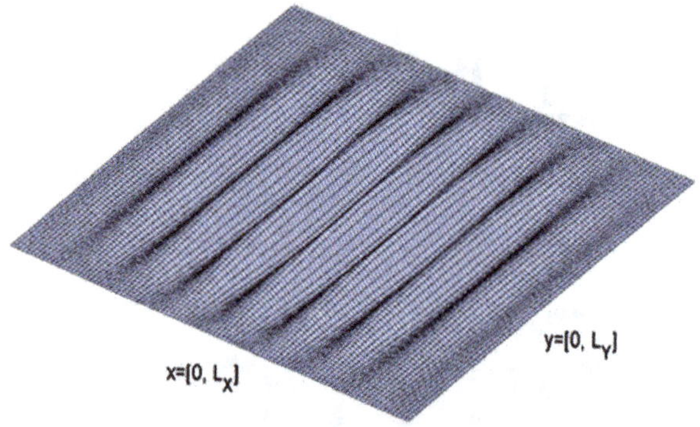

Figure 4 Sketch of wrinkling phenomena of a rectangular membrane under biaxial load.

Two analytical solutions for clamped rectangular membranes

In this Section, the reference macroscopic model (22) (30) (31) (32) is studied by seeking closed-form solutions in the case of clamped rectangular membranes: $\omega = [0, L_X] \times [0, L_Y]$ as pictured in Figure 3. The membrane is submitted to a large uniform tensile stress $N_Y = h\sigma_Y > 0$ and to a small uniform compressive stress loading $N_X = h\sigma_X = -\lambda < 0$. If the plate is clamped, it is known [40] that the envelope w vanishes on the boundary. So the corresponding boundary conditions will be

$$\begin{cases} \mathbf{N}.\mathbf{e}_X = -\lambda \mathbf{e}_X & \text{on the sides} \quad X = 0, X = L_X \\ \mathbf{N}.\mathbf{e}_Y = N_Y \mathbf{e}_Y & \text{on the sides} \quad Y = 0, Y = L_Y \\ w(X, Y) = 0 & \text{on } \partial\omega. \end{cases} \tag{40}$$

Classically, we start with a "linear stability analysis" that is quite simple due to the stress state depending linearly on the applied force. This will establish unambiguously the multi-scale character of membrane wrinkling. Next a classical nonlinear bifurcation analysis will be done, in the same manner as for a classical post-buckling computation [68,69]. This will provide a relation between the applied compressive stress and the size of the wrinkles.

An analytical solution for wrinkling initiation

The linearised version of the envelope Equation (32) is rewritten as

$$-6DQ^2 \frac{\partial^2 w}{\partial X^2} - \left(2DQ^2 + h\sigma_Y\right) \frac{\partial^2 w}{\partial Y^2} + DQ^4 w = h|\sigma_X| \left(Q^2 w - \frac{\partial^2 w}{\partial Y^2} \right). \tag{41}$$

Since the envelope vanishes at the boundary, the smallest eigenfunction of (41) is in the form: $w(X, Y) = \sin(\pi X/L_x)\sin(\pi Y/L_Y)$, as shown in Figure 4. This leads to a classical relation between compressive stress and wavenumber.

$$h|\sigma_X|(Q) = \frac{6DQ^2 \frac{\pi^2}{L_X^2} + \left(2DQ^2 + h\sigma_Y\right) \frac{\pi^2}{L_Y^2} + DQ^4}{Q^2 + \frac{\pi^2}{L_X^2}} \tag{42}$$

Classically, the instability wavenumber Q is chosen by minimizing the stress as a function of the wavenumber. For simplicity, we take into account the orders of magnitude $1 << QL_X$, $2DQ^2/h \approx |\sigma_X| << \sigma_Y$ to simplify (42) in the following manner:

$$|\sigma_X|(Q) = \frac{\sigma_Y \pi^2}{Q^2 L_Y^2} + \frac{DQ^2}{t} \tag{43}$$

The minimum of the latter yields values of the wavenumber and of the critical compressive stress that are consistent with the results of the literature [2,7]:

$$Q^{wr} = \sqrt[4]{12\pi^2(1-\nu^2)}\frac{1}{\sqrt{hL_Y}}\sqrt[4]{\frac{\sigma_Y}{E}} \approx 3.2\frac{1}{\sqrt{hL_Y}}\sqrt[4]{\frac{\sigma_Y}{E}}\ell^{wr} = \sqrt{hL_Y}\sqrt[4]{\frac{E}{\sigma_Y}}|\sigma_X^{wr}| = \frac{\pi}{\sqrt{3(1-\nu^2)}}\sqrt{E\sigma_Y}\frac{h}{L_Y} \tag{44}$$

This simple calculation brings out the multiple scale character of the wrinkling phenomenon: indeed, the wrinkling threshold depends on the wavelength that is a microscopic quantity, but this wavelength depends of the width of the plate that is a macroscopic length. Thus a full wrinkling analysis has to associate micro and macro scales.

Recent experimental results [9] have established that the wavelength increases when the thickness h increases and decreases when the tension σ_Y increases. The scaling law $l \approx \sqrt{h}\,\sigma_Y^{-\frac{1}{4}}$ seems qualitatively consistent with experimental results.

The wrinkling stress can also be written in terms of the wavenumber by $N_X = -2DQ^2$, to be compared with the bifurcation load $N_X = -DQ^2$ of the pure membrane model (34), but of course the pure membrane model is not able to predict the wavenumber.

Initial post-bifurcation analysis

Now we try to connect the applied loads and the size of the wrinkles. As in any plate buckling, the deflection is proportional to the square root of the difference between the current load and its critical value. This post-bifurcation analysis will be done by starting from the new reference membrane model (22) (30) (31) (32) in order to illustrate its ability to characterize the evolution of the wrinkles beyond the instability threshold.

The membrane model (22) (30) (31) (32) is a macroscopic one already simplified by a multi-scale analysis. Therefore there is no need to come back to a multiple scale bifurcation analysis of Ginzburg-Landau type. A standard post-bifurcation analysis as in [68,69] will be sufficient to predict the amplitude of wrinkles.

According to [69], the solution of the symmetric bifurcation problem (22) (30) (31) (32) can be solved by seeking the unknowns and the compression load as Taylor series with respect to a scalar bifurcation parameter "a":

$$\begin{Bmatrix} \mathbf{u}(X,Y) \\ \mathbf{N}(X,Y) \\ w(X,Y) \\ \lambda \end{Bmatrix} = \begin{Bmatrix} \mathbf{u}^{(0)} \\ \mathbf{N}^{(0)} \\ 0 \\ \lambda^{(0)} \end{Bmatrix} + a\begin{Bmatrix} 0 \\ 0 \\ w^{(1)} \\ 0 \end{Bmatrix} + a^2\begin{Bmatrix} \mathbf{u}^{(2)} \\ \mathbf{N}^{(2)} \\ 0 \\ \lambda^{(2)} \end{Bmatrix} + a^3\begin{Bmatrix} 0 \\ 0 \\ w^{(3)} \\ 0 \end{Bmatrix} + \ldots \tag{45}$$

To avoid any ambiguity with Fourier expansion (U_i), the terms of the Taylor expansions have been denoted with a superscript $U^{(i)}$. This straightforward bifurcation analysis leads to Partial Differential Equations at each order.

At order 0, the solution is a membrane state:

$$\mathbf{N}^{(0)}(X,Y) = -\lambda^{(0)}\mathbf{e}_X \otimes \mathbf{e}_X + N_y \mathbf{e}_Y \otimes \mathbf{e}_Y. \tag{46}$$

By introducing the linear operator

$$\mathcal{L}(w) = -6DQ^2\frac{\partial^2 w}{\partial X^2} - 2DQ^2\frac{\partial^2 w}{\partial Y^2} + DQ^4 w - N_Y\frac{\partial^2 w}{\partial Y^2}. \tag{47}$$

the PDE's at orders 1, 2, 3 are:

$$\begin{cases} \mathcal{L}(w^{(1)}) - \lambda^{(0)}Q^2 w^{(1)} + \lambda^{(0)}\dfrac{\partial^2 w^{(1)}}{\partial X^2} = 0 \\ w^{(1)} = 0 \qquad \text{on } \partial\omega \end{cases} \tag{48}$$

$$\begin{cases} div\mathbf{N}^{(2)} = 0 & (49-a) \\ \varepsilon(\mathbf{u}^{(2)}) + \gamma^{wr}(w^{(1)}) = \mathbf{L}^{m^{-1}} : \mathbf{N}^{(2)} & (49-b) \\ N_X^{(2)} = -\lambda^{(2)} \qquad \text{on the sides } X = 0,\ X = L_X & (49-c) \\ N_Y^{(2)} = 0 \qquad \text{on the sides } Y = 0,\ Y = L_Y & (49-d) \end{cases}$$

$$\begin{cases} \mathcal{L}(w^{(3)}) - \lambda^{(0)}Q^2 w^{(3)} + \lambda^{(0)}\dfrac{\partial^2 w^{(3)}}{\partial X^2} = -N_x Q^2 w^{(1)} + div(\mathbf{N}^{(2)}.\nabla w^{(1)}) & (50-a) \\ w^{(3)} = 0 \qquad \text{on } \partial\omega & (50-b) \end{cases}$$

Of course the first order Equation (48) corresponds exactly to the bifurcation Equation (41) that has been solved previously. It is not necessary to re-discuss this point. The smallest buckling mode is

$$w^{(1)}(X,Y) = \sin\left(\frac{\pi X}{L_X}\right)\sin\left(\frac{\pi Y}{L_Y}\right). \tag{51}$$

Let us solve the second order problem (49). The equilibrium Equation (49-a) can be solved by introducing a stress function:

$$N_X^{(2)} = \frac{\partial^2\varphi}{\partial Y^2},\quad N_Y^{(2)} = \frac{\partial^2\varphi}{\partial X^2},\quad N_{XY}^{(2)} = -\frac{\partial^2\varphi}{\partial X\partial Y}. \tag{52}$$

The membrane constitutive law (49-b) can be detailed as:

$$\begin{cases} \dfrac{\partial u^{(2)}}{\partial X} + \left(\dfrac{\partial w^{(1)}}{\partial X}\right)^2 + Q^2\left(w^{(1)}\right)^2 = \dfrac{1}{Eh}\left[\dfrac{\partial^2 \varphi}{\partial Y^2} - \nu\dfrac{\partial^2 \varphi}{\partial X^2}\right] \\ \dfrac{\partial v^{(2)}}{\partial Y} + \left(\dfrac{\partial w^{(1)}}{\partial Y}\right)^2 = \dfrac{1}{Eh}\left[\dfrac{\partial^2 \varphi}{\partial X^2} - \nu\dfrac{\partial^2 \varphi}{\partial Y^2}\right] \\ \dfrac{\partial u^{(2)}}{\partial Y} + \dfrac{\partial v^{(2)}}{\partial X} + 2\dfrac{\partial w^{(1)}}{\partial X}\dfrac{\partial w^{(1)}}{\partial Y} = -2\dfrac{1+\nu}{Eh}\dfrac{\partial^2 \varphi}{\partial X \partial Y}. \end{cases} \tag{53}$$

The in-plane displacement $u^{(2)}, v^{(2)}$ can be eliminated from the combination $\frac{\partial^2}{\partial Y^2}(53\text{–}a) + \frac{\partial^2}{\partial X^2}(53\text{–}b) - \frac{\partial^2}{\partial X \partial Y}(53\text{–}c)$. The resulting equation is:

$$\frac{1}{Eh}\Delta^2\varphi = 2\left\{\left(\frac{\partial^2 w^{(1)}}{\partial X \partial Y}\right)^2 - \frac{\partial^2 w^{(1)}}{\partial X^2}\frac{\partial^2 w^{(1)}}{\partial Y^2}\right\} + 2Q^2\left\{\left(\frac{\partial w^{(1)}}{\partial Y}\right)^2 + w^{(1)}\frac{\partial^2 w^{(1)}}{\partial Y^2}\right\}. \tag{54}$$

It looks like the in-plane equation of the starting Föppl-Von Karman model, with a new second term in the r.h.s. due to the rapid oscillations. This second term is of the order $O\left(\frac{1}{\ell^2 L^2}\right)$ while the first term of the r.h.s. involves only the macroscopic length and is $O\left(\frac{1}{L^4}\right)$. In what follows, this first term will be neglected. To account for the second order term $\lambda^{(2)}$ in (52-c), one can introduce a new stress function $\psi(X, Y)$ by:

$$\varphi(X, Y) = -\lambda^{(2)}\frac{Y^2}{2} + \psi(X, Y),$$

so that $\psi(X, Y)$ is solution of the following boundary value problem:

$$\begin{cases} \dfrac{1}{Eh}\Delta^2\psi = 2Q^2\left\{\left(\dfrac{\partial w^{(1)}}{\partial Y}\right)^2 + w^{(1)}\dfrac{\partial^2 w^{(1)}}{\partial Y^2}\right\} = \dfrac{\pi^2 Q^2}{L_Y^2}\left[1 - \cos(\dfrac{2\pi X}{L_X})\right]\cos(\dfrac{2\pi Y}{L_Y}) & (55-a) \\ \psi = 0 \quad \text{on } \partial\omega & (55-b) \\ \dfrac{\partial \psi}{\partial X} = 0 \quad \text{on the sides } X = 0, L_X & (55-c) \\ \dfrac{\partial \psi}{\partial Y} = 0 \quad \text{on the sides } Y = 0, L_Y & (55-d) \end{cases}$$

The Dirichlet boundary conditions in (55) are deduced from (49-c) (49-d) by integration along the boundary, by a quite well known calculation when dealing with a stress function.

Because of the boundary conditions, it seems difficult to get a closed form solution of (55). We propose to build an approximate equation by using a Galerkin approximation. Let us transform the domain $\omega = [0, L_X] \times [0, L_Y]$ into a reference domain $\omega_{ref} = [-1, +1] \times [-1, +1]$ by:

$$X = \frac{1+r}{2}L_X, \quad Y = \frac{1+s}{2}L_Y$$

$$\Rightarrow w^{(1)}(r, s) = \cos\left(\frac{\pi r}{2}\right)\cos\left(\frac{\pi s}{2}\right).$$

Let us consider a shape function that satisfies the boundary conditions (55-b) (55-c) (55-d), for instance

$$W(r,s) = \left(1-r^2\right)^2\left(1-s^2\right)^2. \tag{56}$$

The PDE (55-a) can be rewritten as (with $\alpha = \frac{L_Y}{L_X}$):

$$\left(\alpha^2\frac{\partial^2}{\partial r^2}+\frac{\partial^2}{\partial s^2}\right)^2\psi = \pi^2 EhQ^2\frac{L_Y^2}{16}(1-\cos\pi(1+r))\cos\pi(1+s).$$

Its Galerkin approximated solution is:

$$\begin{aligned}
\psi &= \pi^2 EhQ^2\frac{L_Y^2}{16}\frac{I_2}{I_1}W(r,s)\\
I_1 &= \iint_{\omega_{ref}}\left[\left(\alpha^2\frac{\partial^2}{\partial r^2}+\frac{\partial^2}{\partial s^2}\right)^2 W\right]Wdrds\\
I_2 &= \iint_{\omega_{ref}}[1-\cos\pi(1+r)]\cos\pi(1+s)Wdrds.
\end{aligned} \tag{57}$$

The corresponding axial stress field is:

$$N_X^{(2)} = -\lambda^{(2)} + \frac{\pi^2}{4}EhQ^2\frac{I_2}{I_1}\frac{\partial^2 W}{\partial s^2}. \tag{58}$$

Finally we take into account the third order Equation (50-a), but we only need to write an orthogonality condition between the mode $w^{(1)}$ and the r.h.s. of (50-a):

$$\iint_{\omega_{ref}}\left\{N_X^{(2)}Q^2\left(w^{(1)}\right)^2-div\left[\mathbf{N}^{(2)}.\nabla w^{(1)}\right]w^{(1)}\right\}d\omega = 0. \tag{59}$$

For simplicity we disregard the second term of (59) that is of order $O\left(\frac{1}{L^2}\right)$ while the first one is of order $O(Q^2) = O\left(\frac{1}{\ell^2}\right)$ (ℓ is the small instability wavelength), which leads to a simpler solvability condition:

$$\iint_{\omega_{ref}} N_X^{(2)}\left(w^{(1)}\right)^2 d\omega = 0.$$

This leads to the value of the second order load $\lambda^{(2)}$, that is written in terms of four adimensional integrals I_1, I_2, I_3, I_4

$$\lambda^{(2)} = \frac{\pi^2}{4}EhQ^2\frac{I_2 I_4}{I_1(\alpha)I_3} \tag{60}$$

Table 1 Membrane response for different tensile loads as a function of the tensile stress: wrinkling compressive load, wrinkling wavenumber, size of the wrinkles when the compressive load is twice the wrinkling load

N_Y (N/mm)	N_X^{wr} (N/mm)	$\frac{L_X}{\ell^{wr}}$	$\frac{a}{h}$ for $N_X = 2N_X^{wr}$
5	−0.064	6.28	1.66
10	−0.090	7.48	1.66
15	−0.110	8.28	1.66
20	−0.127	8.90	1.66

Data: E = 70000 MPa, ν=0; $L_X = L_Y$ = 200 mm, h = 0.05 mm.

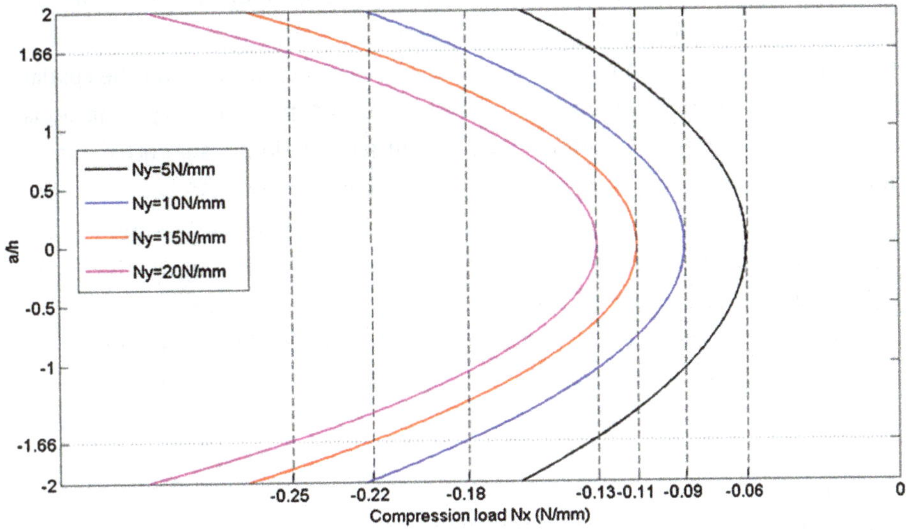

Figure 5 **Amplitude of the wrinkles normalised by membrane thickness versus compression load for different tensile loads.**

$$I_3 = \iint_{\omega_{ref}} \left(w^{(1)}\right)^2 drds = 1$$

$$I_4 = \iint_{\omega_{ref}} \frac{\partial^2 W}{\partial s^2} \left(w^{(1)}\right)^2 drds.$$

This identity (60) permits to evaluate the increment of the load due to the wrinkles. It is convenient to compare this load to the approximated critical load $\lambda^{(0)} = 2DQ^2$:

$$\frac{\lambda}{\lambda^{(0)}} = 1 + \frac{\lambda^{(2)}}{\lambda^{(0)}} a^2 = 1 + \frac{3}{2}\pi^2(1-\nu^2)\frac{I_2 I_4}{I_1(\alpha) I_3}\left(\frac{a}{h}\right)^2. \tag{61}$$

Curiously, this ratio depends rather little on the macroscopic lengths L_X, L_Y and the wrinkling wavelength ℓ, except via the aspect ratio $\alpha = \frac{L_Y}{L_X}$. Obviously the critical load depends strongly on ℓ or on L_Y by $\frac{\lambda^{(0)}}{Eh} \approx \left(\frac{h}{l}\right)^2 \approx \frac{h}{L_Y}$.

This means that for a given aspect ratio $\alpha = \frac{L_Y}{L_X}$, there is an universal law connecting the ratio "amplitude of wrinkles over the thickness" to the ratio "loading over critical wrinkling load", but clearly the critical wrinkling load depends on the traction load. Finally by noting;

$$C(\alpha) = \sqrt{\frac{2}{3\pi^2(1-\nu^2)} \frac{I_1(\alpha) I_3}{I_2 I_4}},$$

and by accounting for (44), we obtain the law connecting the wrinkling amplitude as a function of the compressive stress:

$$\frac{a}{h} = C(\alpha)\sqrt{\frac{|\sigma_X|}{\sqrt{E\sigma_Y}} \frac{L_Y}{h} \sqrt{\frac{3(1-\nu^2)}{\pi}} - 1}. \tag{62}$$

Note that this analytical formula comes from a structural analysis. Hence the results depend on the domain via the coefficient $\alpha = \frac{L_Y}{L_X}$ and on the boundary conditions. The formula (62) is illustrated by numerical values in the case of a square membrane in Table 1 and Figure 4, for four values of the tensile force.

Let us underline that the previous bifurcation analysis depends strongly on the spatial derivatives in the envelope Equations (32) and therefore on the boundary conditions satisfied by the envelope w. For instance, if one considers a simply supported membrane, the macroscopic boundary conditions are of Newman type ($\frac{\partial w}{\partial n} = 0$) and the mode is $w^{(1)} = 1$ so that the bifurcation analysis from (32) should imply a bifurcation branch with a constant load, as in straight beam buckling. In other terms, the stable postbuckling response obtained in (62) is strongly related to the spatial evolution of the wrinkling mode. The non-linearity of this response is not negligible, as shown by the last column of Table 1 and Figure 5, which means that a significant wrinkling amplitude requires a significant increasing of the compression.

Numerical implementation

We describe now the discretisation techniques and the solving algorithm for the reference model (30) (31) (32). Because this system is a second order partial differential system, any classical C^0 finite element is acceptable. In the applications, quadratic quadrilateral finite elements (Q8) will be used. The resulting nonlinear discrete equations will be solved by the Asymptotic Numerical Method [70]. Some numerical tests will be performed with the pure membrane model (34); its discretisation being straightforward, the details will be omitted.

Matrix form of the bending virtual work

The bending energy (27) depends only on the envelope of the deflection that can be represented by the vector:

$$\{\boldsymbol{\beta}\} = {}^t\left\langle w, \frac{\partial w}{\partial X}, \frac{\partial w}{\partial Y} \right\rangle \tag{63}$$

After introducing the matrix

$$\left[\mathbf{L}^f\right] = 2D \begin{bmatrix} Q^4 & 0 & 0 \\ 0 & 6Q^2 & 0 \\ 0 & 0 & 2Q^2 \end{bmatrix} \tag{64}$$

The bending energy can be written as:

$$\delta \mathcal{I}_{ben} = \iint \langle \delta \boldsymbol{\beta} \rangle \left[\mathbf{L}^f\right] \{\boldsymbol{\beta}\} \, d\omega \tag{65}$$

Matrix form of the membrane virtual work

The membrane energy (23) is a quadratic function of the membrane strain $\boldsymbol{\gamma}$ given in (21) (22) that is a quadratic function of the displacement. The displacement field and its useful derivatives are described by the vector:

$$\{\boldsymbol{\theta}\} = {}^t\left\langle \frac{\partial u}{\partial X}, \frac{\partial u}{\partial Y}, \frac{\partial v}{\partial X}, w, \frac{\partial w}{\partial X}, \frac{\partial w}{\partial Y} \right\rangle \tag{66}$$

Note that the two displacement vectors (63) (66) are connected by a transform matrix $[\mathbf{T}_\beta]$:

$$\{\boldsymbol{\beta}\} = \mathbf{T}_\beta\{\boldsymbol{\theta}\}$$

$$[\mathbf{T}_\beta] = \begin{bmatrix} 0 & 0 & 0 & 0 & 1 & 0 & 0 \\ 0 & 0 & 0 & 0 & 0 & 1 & 0 \\ 0 & 0 & 0 & 0 & 0 & 0 & 1 \end{bmatrix} \quad (67)$$

The strain $\{\gamma\}$ can be expressed by a constant matrix $[\mathbf{H}]$ and by a matrix $[\mathbf{A}(\boldsymbol{\theta})]$ that depends linearly of the displacement vector:

$$\{\gamma\} = \left([\mathbf{H}] + \frac{1}{2}[\mathbf{A}(\boldsymbol{\theta})]\right)\{\boldsymbol{\theta}\} \quad (68)$$

$$[\mathbf{H}] = \begin{bmatrix} 1\,0\,0\,0\,0\,0\,0 \\ 0\,0\,0\,1\,0\,0\,0 \\ 0\,1\,1\,0\,0\,0\,0 \end{bmatrix} \quad (69)$$

$$[\mathbf{A}(\boldsymbol{\theta})] = 2\begin{bmatrix} 0\,0\,0\,0 & Q^2 w & \dfrac{\partial w}{\partial X} & 0 \\ 0\,0\,0\,0 & 0 & 0 & \dfrac{\partial w}{\partial Y} \\ 0\,0\,0\,0 & 0 & \dfrac{\partial w}{\partial Y} & \dfrac{\partial w}{\partial X} \end{bmatrix} \quad (70)$$

With these notations, the membrane constitutive laws and the membrane energy can be written as:

$$\{\mathbf{N}\} = [\mathbf{L}^m]\left([\mathbf{H}] + \frac{1}{2}[\mathbf{A}(\boldsymbol{\theta})]\{\boldsymbol{\theta}\}\right) \quad (71)$$

$$\delta E_{mem} = \iint \langle\delta\boldsymbol{\theta}\rangle \left({}^t[\mathbf{H}] + {}^t[\mathbf{A}(\boldsymbol{\theta})]\right)\{\mathbf{N}\}\, d\omega \quad (72)$$

Discretisation

A classical 2D-Q8 finite element that is defined by eight nodes and three degrees of freedom per node is used. The displacement (**u**,w) and the full vector on each element are discretised in the form:

$$\begin{Bmatrix} \mathbf{u} \\ w \end{Bmatrix} = [\mathcal{N}]\{\mathbf{q}\}^e \quad (73)$$

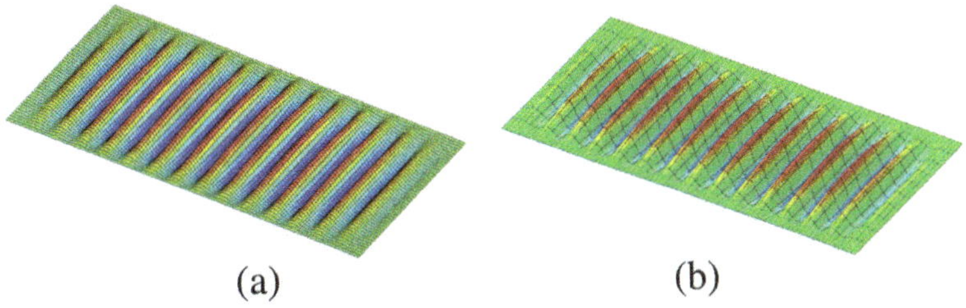

Figure 6 Rectangular membrane submitted to tension and compression. Post-bifurcation patterns just after the bifurcation, with the new reduced model **(a)** and the full shell model **(b)**. $N_Y = 10$ N/mm, $N_X = -0.09$ N/mm.

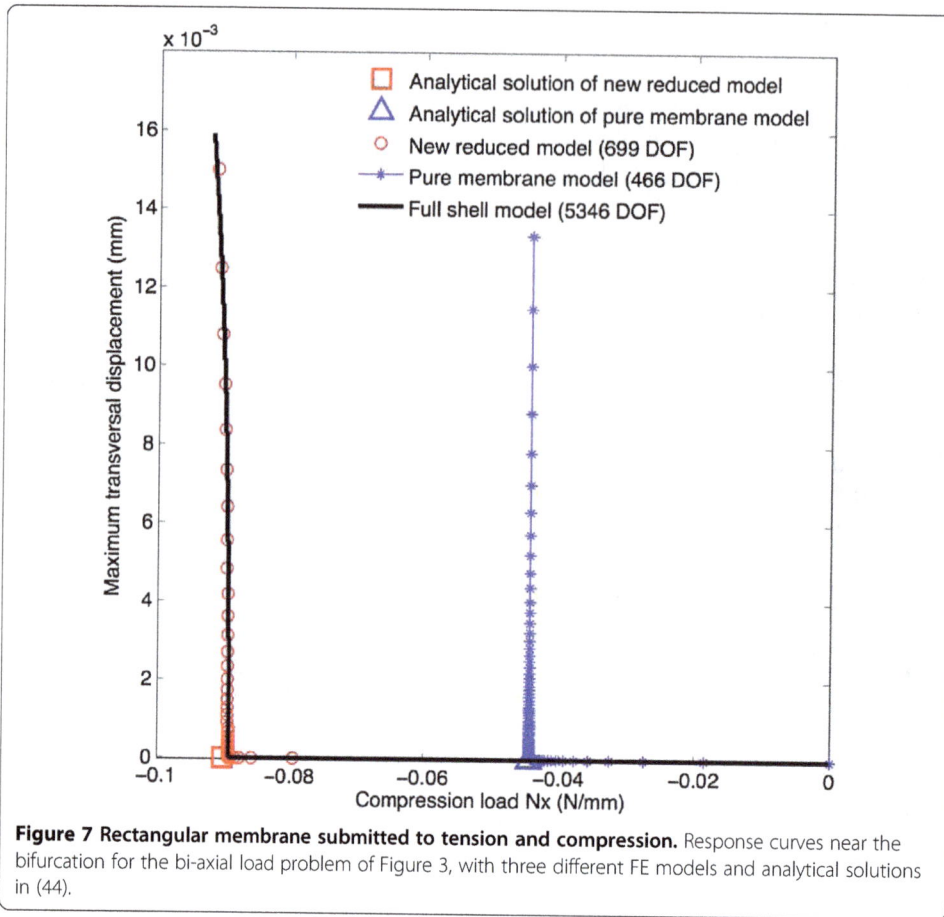

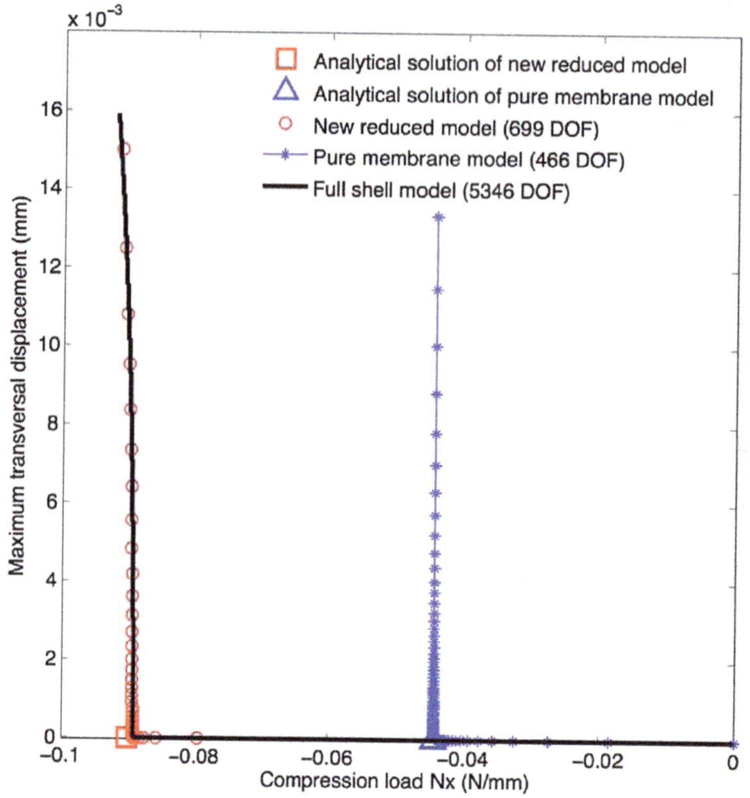

Figure 7 Rectangular membrane submitted to tension and compression. Response curves near the bifurcation for the bi-axial load problem of Figure 3, with three different FE models and analytical solutions in (44).

$$\{\boldsymbol{\theta}\}^e = [\mathbf{G}]\{\mathbf{q}\}^e \tag{74}$$

We need also few matrices to connect the discrete displacement and the membrane strain:

$$[\mathbf{B}(\mathbf{q})] = [\mathbf{H} + \mathbf{A}(\boldsymbol{\theta}(\mathbf{q}))]\ [\mathbf{G}] \tag{75}$$

$$[\mathbf{B}_l] = [\mathbf{H}]\ [\mathbf{G}] \quad [\mathbf{B}_{nl}(\mathbf{q})] = [\mathbf{A}(\boldsymbol{\theta}(\mathbf{q}))]\ [\mathbf{G}] \tag{76}$$

Hence we obtain the discrete form of the full internal virtual work:

$$\delta E_{ben} + \delta E_{mem} = \sum_e \langle\delta\mathbf{q}\rangle^e \iint {}^t[\mathbf{B}(\mathbf{q})]\{\mathbf{N}\} + {}^t[\mathbf{G}]^t[\mathbf{T}_\beta]\left[\mathbf{L}^f\right][\mathbf{T}_\beta][\mathbf{G}]\{\mathbf{q}\}^e d\omega \tag{77}$$

$$\{\mathbf{N}\} = [\mathbf{L}^m]\left([\mathbf{B}_l] + \frac{1}{2}[\mathbf{B}_{nl}(\mathbf{q})]\right)\{\mathbf{q}\}^e \tag{78}$$

The external virtual work is quite classical and the details will be omitted.

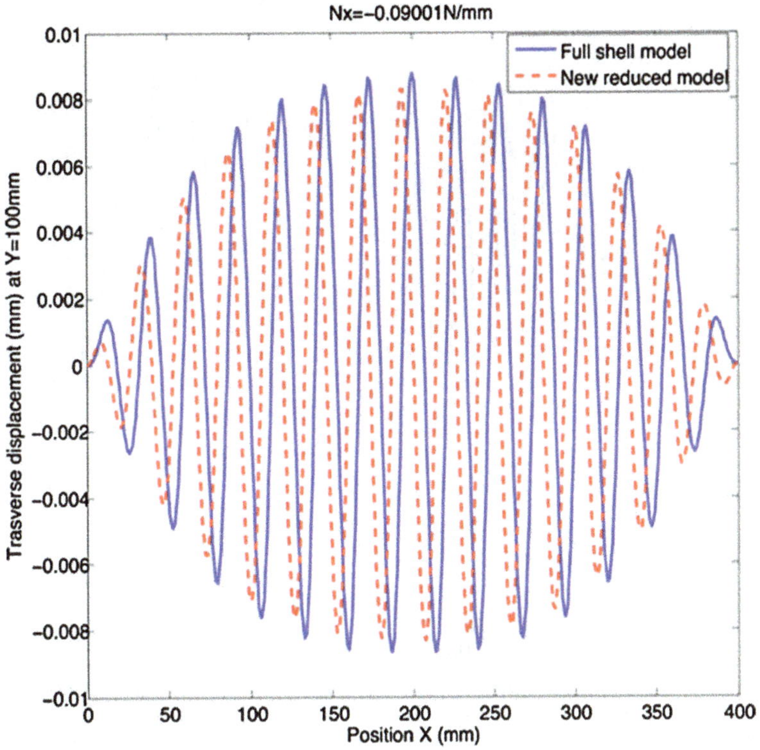

Figure 8 Rectangular membrane submitted to tension and compression. Post-bifurcation profile along the length, at the beginning of the bifurcation curve. $N_X = -0.09001$ N/mm.

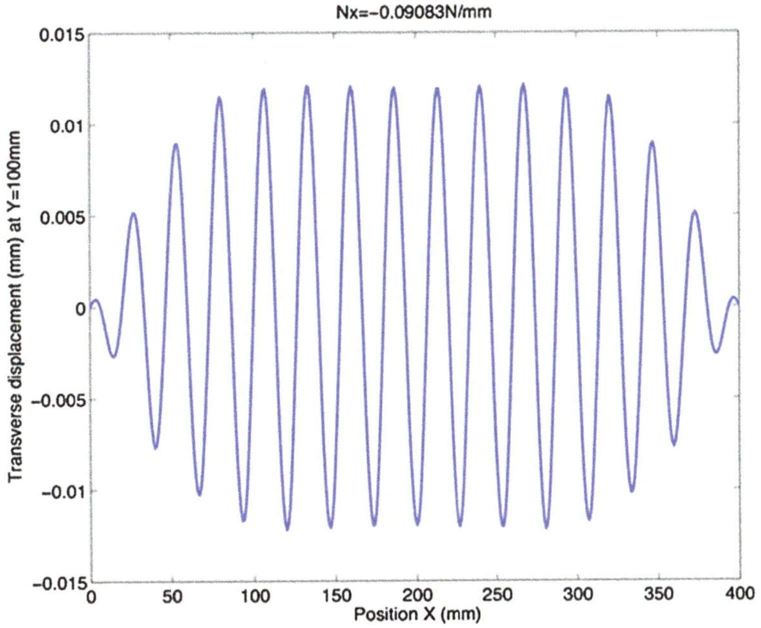

Figure 9 Rectangular membrane submitted to tension and compression. Post-bifurcation profile along the length in the medium range predicted by the new reduced (22) (30) (31) (32). $N_X = -0.090528$ N/mm.

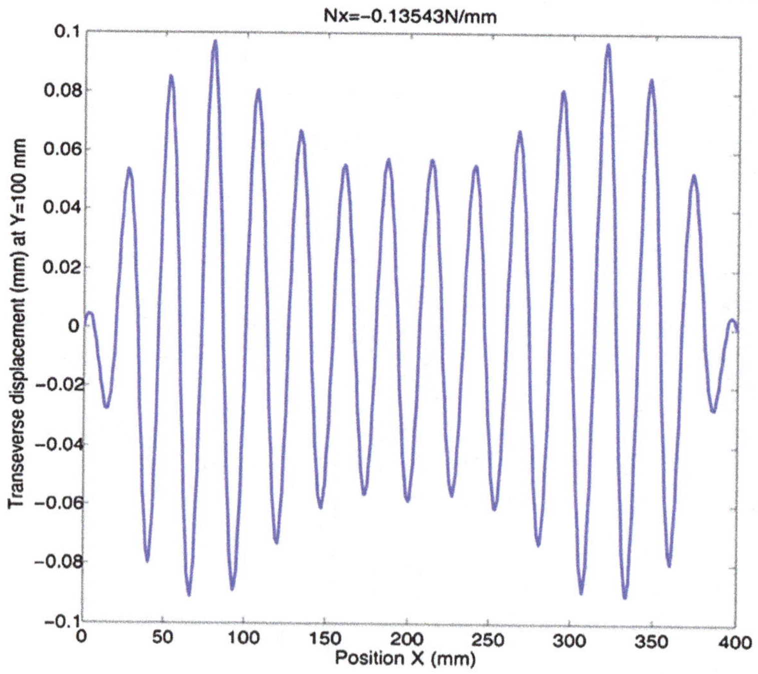

Figure 10 Rectangular membrane submitted to tension and compression. Post-bifurcation profile along the length in the large predicted by the new reduced (22) (30) (31) (32). $N_X = -0.13543$ N/mm.

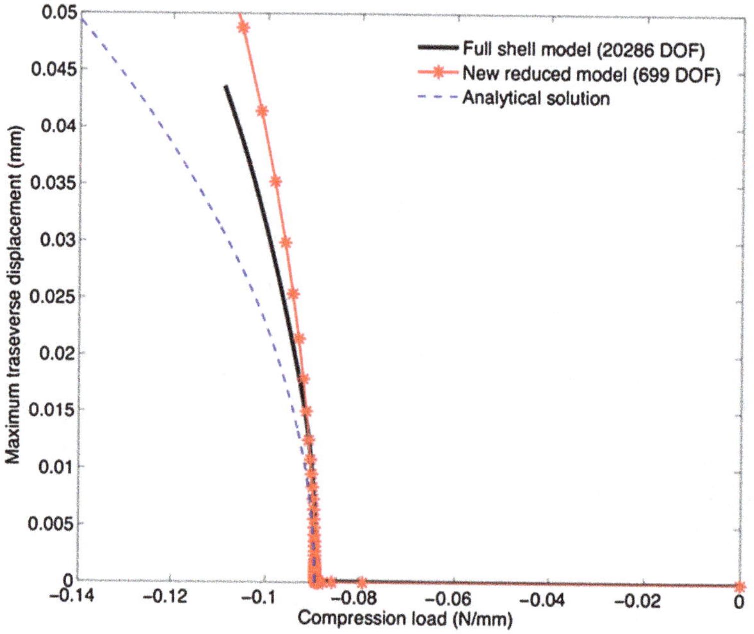

Figure 11 Rectangular membrane submitted to tension and compression. Bifurcation curves obtained with two numerical models and the analytical solution of § 4.2.

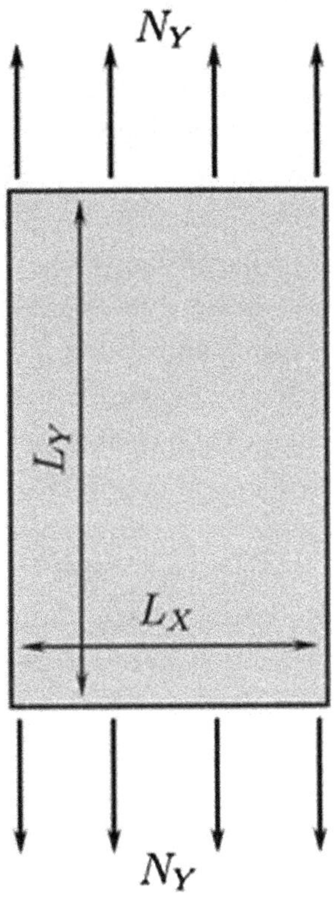

Figure 12 A rectangular membrane under uniaxial load.

Continuation procedure

The full discrete problem (54) (55) is solved by a continuation procedure called Asymptotic Numerical Method [70] that is a step by step continuation algorithm. In each step, the unknowns are expanded into series with respect to a real path parameter "a":

$$\begin{Bmatrix} \mathrm{N} \\ \mathrm{q} \end{Bmatrix} = \sum_{n=0}^{N_{order}} a^n \begin{Bmatrix} \mathrm{N}_n \\ \mathrm{q}_n \end{Bmatrix} \tag{79}$$

To compute the term of order "n" in (56), we followed a procedure described in numerous papers of the literature [66,70], which is not developed in this paper.

Two numerical solutions

The procedure described in the previous section has been applied to two numerical tests: the wrinkling of a rectangular membrane submitted to a compressive-tensile loading studied previously in section "Clamped rectangular membrane submitted to tension and compression" and the wrinkling of long rectangular membrane submitted to a uniaxial traction that was studied by several authors in the literature. The main question is to evaluate the validity range of the reduced model (30-32) and (34).

Clamped rectangular membrane submitted to tension and compression

In this section, we study again the example of a clamped rectangular membrane under bi-axial tension-compression load (see Figure 3). The side lengths L_X and L_Y are respectively 400 mm and 200 mm, and the thickness h is 0.05 mm. The applied tension is $N_Y = 10$ N/mm and the compression force N_X increases in the path following procedure. In the macroscopic model, a wavenumber Q has to be chosen and we take the one predicted by the analytic formula (44). The Figure 6 presents the wrinkling pattern just after the bifurcation. Clearly the envelope is nearly sinusoidal in the two directions OX and OY, as predicted by the analytical solutions. It appears also that the envelope model is able to describe correctly the wrinkling shape predicted by the full shell model. A more quantitative picture is provided by Figure 7, where the results of the pure membrane model (34) are also reported. First the finite element study corroborates the analytical bifurcation study. Second the pure membrane model (34) underestimates very much the wrinkling load. Last the macroscopic model (30) (31) (32) is able to describe quite perfectly the initial post-bifurcation response. Let us underline that the macroscopic model requires less degrees of freedom than the full shell model.

As it is well known [43,46] with the Ginzburg-Landau equation, the post-bifurcation profile evolves rapidly. Just after the bifurcation ($w/h = 0.16$), the envelope is nearly sinusoidal as predicted analytically, cf Figure 8. Next it changes rapidly for a hyperbolic tangent shape ($w/h = 0.2$, Figure 9), what is also predicted by Ginzburg-Landau theory. The numerical results show that the profile changes gradually and becomes more and more localized, with two peaks at the end of the boundary layers ($w/h = 2$, Figure 10).

The Figure 11 evaluates the ability of the macroscopic model to predict large sizes of wrinkles. It appears that correct amplitudes can be obtained for rather large wrinkles ($w/h = 1$). Nevertheless one has to be careful in the analysis of the responses for large deflections because these problems can have many stable and unstable solutions and it is not obvious that a path following method with a full shell model provides always the most relevant solution. In the same Figure 11, we reported the result predicted by the approximate analytical solution (61). It gives correctly the beginning of the bifurcated branch up to about $w/h \approx 0.3$.

A rectangular membrane submitted to uniaxial traction

A thin rectangular membrane whose dimensions are $h = 0.05$ mm, $L_X = 1400$ mm, $L_Y = 200$ mm, is submitted to a uniaxial load (see Figure 12). The long sides are stress-free. Along the short sides, an increasing displacement is applied in the X-direction, the OY-displacement being locked. Linear bifurcation was studied in [56,71] with the same boundary conditions and a nonlinear bifurcation as here in [33], but with a prescribed stress at the boundary. In this case, the wrinkling occurs for rather

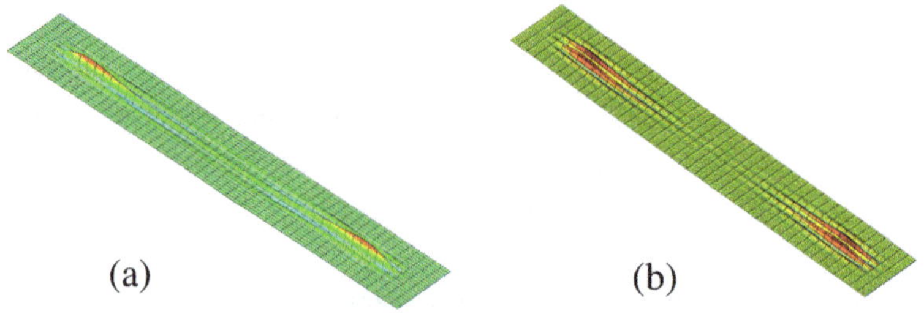

Figure 13 Long membrane in uniaxial tension. Post-bifurcation patterns obtained with the envelope model **(a)** and a full shell model **(b)**.

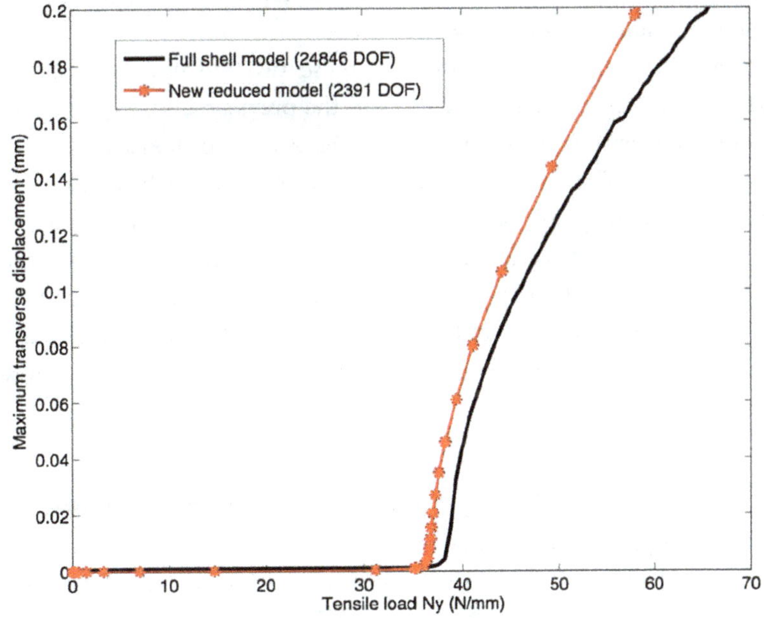

Figure 14 Long membrane in uniaxial tension. Bifurcation curves with two different models.

high axial stresses since it is caused by small transverse compressive stresses due to the boundary effects (for a finer analysis, see [71]). Comparing with the previous tests, it permits to evaluate the ability of the macroscopic model in a case of non uniform pre-buckling stresses.

In Figure 13, one sees that the post-bifurcation patterns obtained by the new reduced model (30) (31) (32) are quite similar as those provided by the full shell model. This

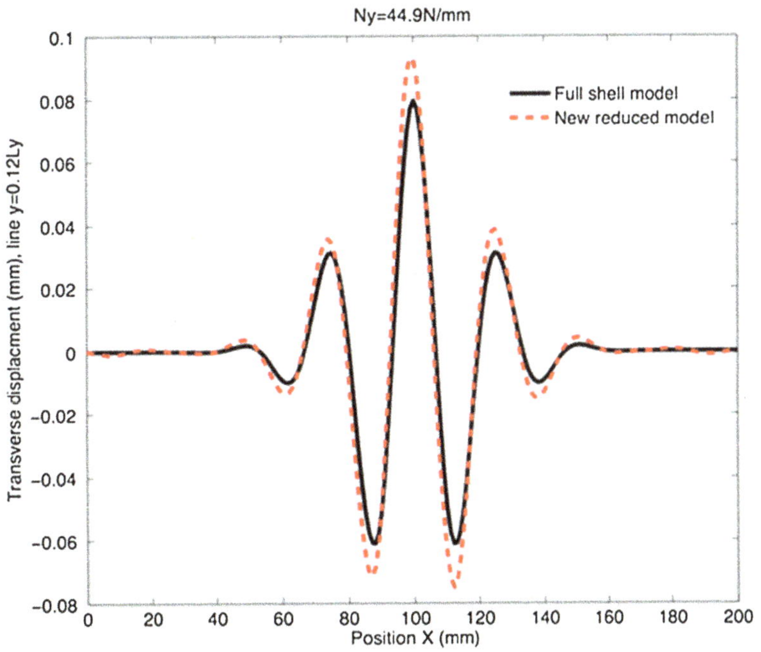

Figure 15 Long membrane in uniaxial tension. Post-bifurcation profile along the width.

establishes the relevance of this new reduced model to represent the wrinkling modes in a case with a non uniform pre-buckling stress field.

In Figure 14, a response curve is plotted for these two models: the macroscopic model (30) (31) (32) and the full shell model. As in the previous case, the new reduced model gives about the same bifurcation point as the reference model as well as the post-bifurcation response. The number of degrees of freedom is much smaller with the envelope models because they do not need to describe explicitly the full details of the wrinkles. Then a wrinkling pattern along the width is plotted in Figure 15 for the full and the reduced model. Globally they are quite similar but with slight differences for the amplitude and for large sizes of wrinkles (up to w/h = 4). One can wonder if these results could still be improved by keeping a complex envelope as in (36-39) or by keeping higher order harmonics.

Linear wrinkling analysis revisited by a double scale asymptotic analysis

Last we re-discuss the problem of membrane wrinkling from the point of view of the traditional asymptotic double scale method [72,73] that is slightly different from the technique of Fourier series with slowly variable coefficients. In both methods, one distinguishes two levels of spatial evolution. Within the asymptotic double scale method, the considered problem involves further small parameters and the solutions are also expanded with respect to these small parameters. It will be interesting to compare shortly the two approaches. This will done for the problem of wrinkling initiation in a rectangular membrane $0 < X < L_X, 0 < Y < L_Y$ studied previously. A uniform ompression-traction membrane field ($N_X = h\sigma_X$, $N_Y = h\sigma_Y$, $\sigma_X < 0 < \sigma_Y$, $\sigma_Y >> |\sigma_X|$) is applied. The wrinkling mode and the wrinkling compressive stress are assumed to solve the linearized Fôppl-Von Karman equation:

$$D\Delta^2 w - N_Y \frac{\partial^2 w}{\partial Y^2} = -|N_X| \frac{\partial^2 w}{\partial X^2} \tag{80}$$

A small parameter η is introduced that can be identified to the stress ratio:

$$\eta = \sqrt{\frac{|N_X|}{N_Y}} = \sqrt{\frac{|\sigma_X|}{\sigma_Y}} << 1. \tag{81}$$

The unknown wrinkling mode is assumed to depend on three independent spatial variables: the starting space coordinates X, Y and another rapidly varying coordinate $x = \frac{X}{\eta}$ in the compressive direction. The eigenpair $w(x, X, Y)$, N_X is sought in the form of an asymptotic expansion with respect to η. In agreement with the order of magnitude (81), we define an unknown control parameter λ and another fixed load parameter N_Y^{-4} by:

$$N_X = -\frac{\lambda}{\eta^2}, \qquad \lambda > 0, \qquad N_Y = \frac{N_Y^{-4}}{\eta^4}, \tag{82}$$

in such a way that λ and N_Y^{-4} will be $O(1)$ in the formal asymptotic expansion. The classical rules of the asymptotic expansion are as follows:

$$\lambda = \lambda_{(0)} + \eta\lambda_{(1)} + \eta^2\lambda_{(2)} + \ldots$$

$$w(x,X,Y) = w_{(0)}(x,X,Y) + \eta w_{(1)}(x,X,Y) + \eta^2 w_{(2)}(x,X,Y) + \ldots$$

$$\begin{aligned} &\frac{\partial}{\partial X} \to \frac{1}{\eta}\frac{\partial}{\partial x} + \frac{\partial}{\partial X} \\ &\frac{\partial}{\partial Y} \to \frac{\partial}{\partial Y} \\ &\Delta = \frac{\partial^2}{\partial X^2} + \frac{\partial^2}{\partial Y^2} \to \frac{1}{\eta^2}\frac{\partial^2}{\partial x^2} + \frac{2}{\eta}\frac{\partial}{\partial x}\frac{\partial}{\partial X} + \Delta \\ &\Delta^2 \to \frac{1}{\eta^4}\frac{\partial^4}{\partial x^4} + \frac{4}{\eta^3}\frac{\partial^3}{\partial x^3}\frac{\partial}{\partial X} + \frac{1}{\eta^2}\left[4\frac{\partial^2}{\partial x^2}\frac{\partial^2}{\partial X^2} + 2\frac{\partial^2}{\partial x^2}\Delta\right] + \frac{4}{\eta}\frac{\partial}{\partial x}\frac{\partial}{\partial X}\Delta + \Delta^2. \end{aligned} \tag{83}$$

To distinguish from the coefficients of the Fourier expansion (6) *Ui(X)*, the indices of the asymptotic expansions (83) have been denoted by $\lambda_{(i)}, w_{(i)}$.

Next the rules (82) (83) are inserted in the considered PDE (80) and a sequence of differential problems is deduced by identifying the PDE at the orders $O\left(\frac{1}{\eta^4}\right)$, $O\left(\frac{1}{\eta^3}\right)$, $O\left(\frac{1}{\eta^2}\right)$, etc. This leads to

$$\mathcal{L}(w_{(0)}) \stackrel{def}{=} D\frac{\partial^4 w_{(0)}}{\partial x^4} + \lambda_{(0)}\frac{\partial^2 w_{(0)}}{\partial x^2} - N_Y^{-4}\frac{\partial^2 w_{(0)}}{\partial Y^2} = 0 \tag{84}$$

$$\mathcal{L}(w_{(1)}) = -4D\frac{\partial^3}{\partial x^3}\frac{\partial w_{(0)}}{\partial X} - 2\lambda_{(0)}\frac{\partial}{\partial x}\frac{\partial w_{(0)}}{\partial X} - \lambda_{(1)}\frac{\partial^2 w_{(0)}}{\partial x^2} \stackrel{def}{=} -M(w_{(0)}) - \lambda_{(1)}\frac{\partial^2 w_{(0)}}{\partial x^2} \tag{85}$$

$$\begin{aligned} \mathcal{L}(w_{(2)}) = &-M(w_{(1)}) - 4D\frac{\partial^2}{\partial x^2}\frac{\partial^2 w_{(0)}}{\partial X^2} - 2D\frac{\partial^2}{\partial x^2}\Delta w_{(0)} \\ &-\lambda_{(0)}\frac{\partial^2 w_{(0)}}{\partial X^2} - 2\lambda_{(1)}\frac{\partial}{\partial x}\frac{\partial w_{(1)}}{\partial X} - \lambda_{(1)}\frac{\partial^2 w_{(1)}}{\partial x^2} - \lambda_{(2)}\frac{\partial^2 w_{(0)}}{\partial x^2}. \end{aligned} \tag{86}$$

As it is classical, the first term $w_{(0)}(x,X,Y)$ of the expansion of the mode will be defined by taking into account the equations at the three levels (84) (85) (86) with the associated boundary conditions. The first Equation (84) is exactly the same as Equation (4). It depends on two variables: the fast variable x in the direction of the wrinkles and the slow variable *Y* in the transverse direction; the dependence with respect to *X* will appear later. One recovers a main characteristic of the wrinkling: its modelisation requires two length scales: the scale of the wrinkles and a macroscopic length scale. Since the physical domain is very large with respect to the wavelength of the wrinkles, the mode $w_{(0)}(x,X,Y)$ may be assumed harmonic with respect to the fast variable x, the wavelength $\ell = \frac{2\pi}{q}$ being not prescribed at this stage:

$$w_{(0)}(x,X,Y) = A(X,Y)\cos(qx + \varphi). \tag{87}$$

The assumption (87) permits to define a slowly varying envelope *A(X,Y)* and (84) leads to a differential equation satisfied by this envelope:

$$-N_Y^{-4}\frac{\partial^2 A}{\partial Y^2} + \left(Dq^4 - \lambda_{(0)}q^2\right)A = 0. \tag{88}$$

As expected, (88) looks like a linearized Ginzburg-Landau equation. The solution of (88) requires boundary conditions on the sides of the rectangle that are parallel to OX.

If the deflection w and its Y-derivative are assumed to be zero on the sides $Y = 0$, $Y = L_Y$, the amplitude $A(X,Y)$ must satisfy the boundary conditions ([62] § 2.4):

$$A(X,0) = A(X,L_Y) = 0. \tag{89}$$

This leads to a first bifurcation pair:

$$\begin{aligned} A(X,Y) &= B(X)\sin\left(\frac{\pi Y}{L_Y}\right) \\ \lambda_{(0)}(q) &= Dq^2 + \frac{N_Y^{-4}}{q^2}\frac{\pi^2}{L_Y^2}. \end{aligned} \tag{90}$$

If one goes back to the corresponding initial quantities $Q = \frac{q}{\eta}$, $N_{X(0)} = -\frac{\lambda_{(0)}}{\eta^2}$ and to $\sigma_Y = \frac{N_Y}{h} = \frac{N_Y^{-4}}{h\eta^4}$, one recovers exactly the classical wrinkling stress and wavelength of (44). The wrinkling wavelength is defined at this level by minimizing the critical load $\lambda_{(2)}(q)$ with respect to the wavenumber.

Now Equation (85) corresponding to the second level $O\left(\frac{1}{\eta^3}\right)$ is discussed. Because of (87), (85) is rewritten as:

$$\mathcal{L}\left(w_{(1)}\right) = -2\sin(qx+\varphi)\left\{2Dq^2 - \lambda_{(0)}\right\} q\frac{\partial A}{\partial X} + \lambda_{(1)}A\cos(qx+\varphi). \tag{91}$$

The first term of the right hand side of (91) vanishes because $\lambda_{(0)} = 2Dq^2$. Since the operator $\mathcal{L}(.)$ is singular, the existence of a bounded solution $x \to w_{(1)}(x, X, Y)$ requires the elimination of "secular terms". This solvability condition leads to $\lambda_{(1)} = 0$. Then the solution of (91) has the same shape as the first order term:

$$w_{(1)}(x,X,Y) = A_{(1)}(X,Y)\cos\left(qx + \varphi_{(1)}\right). \tag{92}$$

Last we look at the Equation at the third level (86). By taking into account the previous results, we get

$$\mathcal{L}\left(w_{(2)}\right) = \cos(qx+\varphi)\left\{\left(6Dq^2 - \lambda_{(0)}\right)\frac{\partial^2 A}{\partial X^2} + 2Dq^2\frac{\partial^2 A}{\partial Y^2} + \lambda_{(2)}q^2 A\right\}. \tag{93}$$

The elimination of the secular terms leads to a PDE satisfied by the slowly varying envelope $A(X, Y) = B(X)\sin(\pi Y/L_Y)$:

$$-2D\left(2\frac{\partial^2 A}{\partial X^2} + \frac{\partial^2 A}{\partial Y^2}\right) = \lambda_{(2)}A \quad \Rightarrow \quad -4D\frac{\partial^2 B}{\partial X^2} + 2D\frac{\pi^2}{L_Y^2}B = \lambda_{(2)}A. \tag{94}$$

The resulting Equation (94) is a one-dimensional eigenvalue problem that permits to define a correction $\lambda_{(2)}\eta^2$ to the critical stress and the variation of the amplitude $A(X, Y) = B(X)sin(\pi Y/L_Y)$ in the compressive direction. In this respect, boundary conditions along the sides $X = 0, X = L_X$ are needed. For instance with Dirichlet boundary conditions, the smallest eigenvalue and eigenmode of (94) are:

$$A(X,Y) = \sin\left(\frac{\pi X}{L_X}\right)\sin\left(\frac{\pi Y}{L_Y}\right) \tag{95 – a}$$

$$\lambda_{(2)} = 2D\pi^2\left(\frac{2}{L_X^2}+\frac{1}{L_Y^2}\right). \tag{95 – b}$$

Hence the analysis of the corrective Equation (86) permits to define the evolution of the mode shape in the membrane, especially in the X-direction. This equally permits to correct the buckling stress according to (95-b). With account to this value of $\lambda_{(2)}$, the critical stress is corrected as $\sigma_X^{wr2} = -\frac{\lambda_{(0)}+\lambda_{(2)}\eta^2}{h\eta^2}$:

$$\begin{aligned}\frac{\left|\sigma_X^{wr2}\right|}{E} &= \frac{\pi}{\sqrt{3(1-\nu^2)}}\sqrt{\frac{\sigma_Y}{E}\frac{h}{L_Y}}+\frac{\pi^2}{6(1-\nu^2)}\left(2\frac{h^2}{L_X^2}+\frac{h^2}{L_Y^2}\right)\\ &= \frac{\left|\sigma_X^{wr1}\right|}{E}+\frac{\pi^2}{6(1-\nu^2)}\left(2\frac{h^2}{L_X^2}+\frac{h^2}{L_Y^2}\right).\end{aligned} \tag{96}$$

Nonetheless the correction is generally small for very thin membranes, i.e. $\frac{h}{L_X} = O(10^{-2})$, $\frac{h}{L_Y} = O(10^{-2})$. This correction depends mainly on the bending effects, as this can be seen from (95) or (96). It depends also on the boundary condition on the four sides, while the first order term σ_X^{wr1} depends only on boundary conditions on $X = 0, L$.

Last comments about the asymptotic modeling

The asymptotic modeling splits the modal analysis into several steps. First a harmonic fast evolution is obtained at the scale of the wrinkle wavelength, see (87). Then the differential Equation (88) yields a first approximation of the wrinkling load, the wavelength and the variation of the envelope in the transverse direction. Finally the mode is completely characterized if one takes into account its evolution along OX via (94), the latter equation being an eigenvalue problem that provides a small correction $\lambda_{(2)}$ to the wrinkling load.

This splitting of the governing envelope equation corresponds to a consistent account of the different levels of applied stresses and of length scales. This cumbersome multi-scale analysis does not exist within the Fourier series that only distinguishes the Fourier coefficients of the response. Unlike the purely asymptotic approach, Fourier method leads to consistent bi-dimensional models as (30) (31) (32) that are easier to handle numerically. Nevertheless the two approaches are not contradictory and an asymptotic procedure could be applied by starting from the reduced models (30-32)(34) or (36-39).

Conclusion

Membrane wrinkling has been re-discussed from multi-scale approaches. Basically, a local wrinkling wavelength is chosen and macroscopic models are deduced, whose unknowns are envelopes of the fastly varying oscillations of the wrinkles. This amounts to coupling a membrane model or a shell model with a sort of bifurcation equation that governs the appearance of wrinkles. This new class of models leads to well-posed partial differential systems that are easy to be solved numerically. These models have been studied in details, by computing several numerical and analytical solutions. This new approach is rather different from those classically used in the literature, but it permits to recover their main features: a quite different behavior of the membrane with or without wrinkling, a splitting into macroscopic strain and wrinkling strain and the presence of an internal length that proves to be necessary to get a consistent distribution of the wrinkles. A main originality is the ability to predict the main characteristics of the wrinkles from macroscopic models and coarse meshes. Likely the weakest point is the necessity to choose a priori the wavelength in the case of a nonlinear analysis, but this could be improved by coming back to a true multi-scaled approach as in FE^2 method.

Competing interests

The authors declare that they have no competing interests.

Authors' contributions

A new class of multi-scale nonlinear models for membrane wrinkling is proposed. They are deduced from a basic plate model without any phenomenological assumptions via the method of Fourier series with variable coefficients. It permits also to predict both the characteristics of the wrinkles and their influence on the membrane behavior. Two analytical solutions are provided that illustrate the multi-scale character of the wrinkling phenomenon. Two of these models are discretized by finite element method, giving relevant solutions with much less degrees of freedom than the full shell model. All authors read and approved the final manuscript.

Author details

[1]Laboratoire d'Ingénierie et Matériaux LIMAT, Faculté des Sciences Ben M'Sik, Université Hassan II Mohammedia Casablanca, Sidi Othman, Casablanca, Maroc. [2]LEM3, Laboratoire d'Etudes des Microstructures et de Mécanique des Matériaux, UMR CNRS 7239, Université de Lorraine, Ile du Saulcy, 57045, Metz Cedex 01, France. [3]Laboratory of Excellence on Design of Alloy Metals for low-mAss Structures (DAMAS), Université de Lorraine, Lorraine, France. [4]School of Civil Engineering, Wuhan University, 8 South Road of East Lake, 430072, Wuhan, PR China.

References

1. Jenkins CH (1996) Nonlinear dynamic response of membranes: state of the art–update. Applied Mechanics Review 49:S41–S48
2. Cerda E, Ravi-Chandar K, Mahadevan L (2000) Wrinkling of an elastic sheet under tension. Nature 419:579
3. Cerda E, Mahadevan L (2003) Geometry and physics of wrinkling. Phys Rev Lett 90:074302-1-4
4. Jenkins CHM, Korde UA (2006) Membrane vibration experiments: an historical review and recent results. J Sound Vib 295:602–613
5. Allaoui S, Boisse P, Chatel S, Hamila N, Hivet G, Soulat D, Vidal-Salle E (2011) Experimental and numerical analyses of textile reinforcement forming of a tetrahedral shape. Compos Part A 42:612–622
6. Li B, Cao YP, Feng XQ, Gao HJ (2012) Mechanics of morphological instabilities and surface wrinkling in soft materials: a review. Soft Matter 8:5728–5745
7. Wong YW, Pellegrino S (2006) Wrinkled membranes-Part1: experiments. J Mech Mater Struct 1:3–25
8. Wang CG, Du XW, Tan HF, He XD (2009) A new computational method for wrinkling analysis of gossamer space structures. Int J Solids Struct 46:1516–1526
9. Lecieux Y, Bouzidi R (2010) Experimentation analysis on membrane wrinkling under biaxial load – comparison with bifurcation analysis. Int J Solids Struct 47:2459–2475
10. Lecieux Y, Bouzidi R (2012) Numerical wrinkling prediction of thin hyperelastic structures by direct energy minimization. Adv Eng Softw 50:57–68
11. Healey TJ, Li Q, Cheng RB (2013) Wrinkling behavior of highly stretched rectangular elastic films via parametric global bifurcation. J Nonlinear Sc 23:777–805
12. Everall PR, Hunt G (1999) Arnold tongue predictions of secondary buckling in thin elastic plates. J Mech Phys Solids 47:2187–2206
13. Everall PR, Hunt G (2000) Mode jumping in the buckling of struts and plates: a comparative study. Int J Non-Linear Mech 35:1067–1079
14. Coman CD (2007) Edge-buckling in stretched thin films under in-plane bending. Z Angew Math Phys 58:510–525
15. Coman CD, Bassom AP (2007) On the wrinkling of a pre-stressed annular thin film in tension. J Mech Phys Solids 55:1601–1617
16. Coman CD, Liu X (2013) Semi-analytical approximations for a class of multi-parameter eigenvalue problems related to tensile buckling. Z Angew Math Phys 64:863–883
17. Diaby A, Le Van A, Wielgosz C (2006) Buckling and wrinkling of prestressed membranes. Finite Element Analysis and Design 42:992–1001
18. Liu XX, Jenkins CH, Schur WW (2001) Large deflection analysis of pneumatic envelopes using a penalty parameter modified material model. Finite Element Analysis and Design 37:233–251
19. Rossi R, Lazzari M, Vitaliani R, Onate E (2005) Simulation of light-weight membrane structures by wrinkling model. Int J Numer Methods Eng 62:2127–2153
20. Tabarrok B, Qin Z (1992) Nonlinear analysis of tension structures. Comput Struct 45:973–984
21. Troufflard J, Cadou JM, Rio G (2010) Recherche de forme des gilets de sauvetage gonflables. Mécanique et Industries 11:117–122

22. Rodriguez J, Rio G, Cadou JM, Troufflard J (2011) Numerical study of dynamic relaxation with kinetic damping applied to inflatable fabric structures with extensions for 3D solid element and non-linear behavior. Thin-Walled Struct 49:1468–1474
23. Roddeman DG, Oomens CWJ, Janssen JD, Drukker J (1987) The wrinkling of thin membranes. Part 1: theory. J Appl Mech 54:884–887
24. Roddeman DG, Oomens CWJ, Janssen JD, Drukker J (1987) The wrinkling of thin membranes. Part 2: numerical analysis. J Appl Mech 54:888–892
25. Lu K, Accorsi M, Leonard J (2001) Finite element analysis of membrane wrinkling. Int J Numer Methods Eng 50:1017–1038
26. Schoop H, Taenzer L, Hornig J (2002) Wrinkling of nonlinear membranes. Comput Mech 29:68–74
27. Miyazaki Y (2006) Wrinkle/slack model and finite element dynamics of membrane. Int J Numer Methods Eng 66:1179–1209
28. Akita T, Nakashino T, Natori MC, Park KC (2007) A simple computer implementation of membrane wrinkle behaviour via a projection technique. Int J Numer Methods Eng 71:1231–1259
29. Banerjee B, Shaw A, Roy D (2009) The theory of Cosserat points applied to the analyses of wrinkled and slack membranes. Comput Mech 43:415–429
30. Pimprikar NA, Banerjee B, Roy D, Vasu RM, Reid SR (2010) New computational approaches for wrinkled and slack membranes. Int J Solids Struct 47:2476–2486
31. Counhaye C (2000) Modélisation et contrôle industriel de la géométrie des aciers laminés à froid (modelling and industrial control of the geometry of cold rolled steels). Université de Liège, Dissertation (in French)
32. Abdelkhalek S, Montmitonnet P, Legrand N, Buessler P (2011) Coupled approach for flatness prediction in thin strip cold rolling. Int J Mech Sci 53:661–675
33. Damil N, Potier-Ferry M, Hu H (2013) New nonlinear multiscale models for membrane wrinkling. Comptes Rendus Mecanique 341:616–624
34. Wesfreid JE, Zaleski S (1984) Cellular Structures in Instabilities. In: Lecture Notes in Physics, vol 210. Springer-Verlag, Heidelberg
35. Iooss G, Mielke A, Demay Y (1989) Theory of steady Ginzburg-Landau equation in hydrodynamic stability problems. Eur J Mech B Fluids 8:229–268
36. Cross MC, Hohenberg PC (1993) Pattern formation out of equilibrium. Rev Mod Phys 65:851–1112
37. Hoyle R (2006) Pattern Formation, an Introduction to Methods. Cambrige University Press, Cambridge
38. Damil N, Potier-Ferry M (2006) A generalized continuum approach to describe instability pattern formation by a multiple scale analysis. Comptes Rendus Mecanique 334:674–678
39. Damil N, Potier-Ferry M (2008) A generalized continuum approach to predict local buckling patterns of thin structures. European Journal of Computational Mechanics 17:945–956
40. Damil N, Potier-Ferry M (2010) Influence of local wrinkling on membrane behaviour: a new approach by the technique of slowly variable Fourier coefficients. J Mech Phys Solids 58:1139–1153
41. Hu H, Damil N, Potier-Ferry M (2011) A bridging technique to analyze the influence of boundary conditions on instability patterns. J Comput Phys 230:3753–3764
42. Segel LA (1969) Distant side walls cause slow amplitude modulation of cellular convection. J Fluid Mech 38:203–224
43. Newell AC, Whitehead JA (1969) Finite band width, finite amplitude convection. J Fluid Mech 38:279–303
44. Amazigo JC, Budiansky B, Carrier GF (1970) Asymptotic analyses of the buckling of imperfect columns on non-linear elastic foundations. Int J Solids Struct 6:1341–1356
45. Pomeau Y, Zaleski S (1981) Wavelength selection in one-dimensional cellular structures. J Phys 42:515–528
46. Potier-Ferry M (1983) Amplitude Modulation, Phase Modulation and Localization of Buckling Patterns. In: Thompson JMT, Hunt GW (eds) Collapse: The Buckling of Structure in Theory and Practice. Cambridge University Press, Cambridge, pp 148–159
47. Hunt GW, Bolt HM, Thompson JMT (1989) Structural localization phenomena and the dynamical phase space analogy. Proc R Soc Lond A 425:245–267
48. Lee SH, Waas AM (1996) Initial post-buckling behavior of a finite beam on an elastic foundation. Int J Non Linear Mech 31:313–328
49. Hunt GW, Peletier MA, Champneys AR, Woods PD, Ahmer Wadee M, Budd CJ, Lord GJ (2000) Cellular buckling in long structures. Nonlinear Dynamics 21:3–29
50. Coman CD, Bassom AP, Wadee MK (2003) Elasto-plastic localised response in one-dimensional structural models. J Eng Math 47:83–100
51. Rosen BW (1965) Influence of Fiber and Matrix Characteristics on Mechanics of Deformation and Fracture of Fibrous Composites, in "Fiber Composite Materials, Papers Presented at a Seminar of the American Society for Metals, October 17 and 18, 1964". Metal Parks, Ohio, pp 37–75
52. Léotoing L, Drapier S, Vautrin A (2002) Nonlinear interaction of geometrical and material properties in sandwich beam instabilities. Int J Solids Struct 39:3717–3739
53. Chen X, Hutchinson JW (2004) Herringbone buckling patterns of compressed thin films on compliant substrates. J Appl Mech 71:597–603
54. Audoly B, Boudaoud A (2008) Buckling of a stiff film bound to a compliant substrate-Part II: A global scenario for the formation of herringbone pattern. J Mech Phys Solids 56:2422–2443
55. Wang S, Song J, Kim DH, Huang Y, Rogers JA (2008) Local versus global buckling of thin films on elastomeric substrates. Appl Phys Lett 93:023126
56. Fischer FD, Rammerstorfer FG, Friedl N, Wieser W (2000) Buckling phenomena related to rolling and levelling of sheet metal. Int J Mech Sci 42:1887–1910
57. Abdelkhalek S, Montmitonnet P, Potier-Ferry M, Zahrouni H, Legrand N, Buessler P (2010) Strip flatness modelling including buckling phenomena during thin strip cold rolling. Ironmak Steelmak 37:290–297
58. Grandidier JC, Ferron G, Potier-Ferry M (1992) Microbuckling and strength in long-fiber composites: theory and experiments. Int J Solids Struct 29:1753–1761

59. Kyriakides S, Arseculeratne R, Perry EJ, Liechti KM (1995) On the compressive failure of fiber reinforced composites. Int J Solids Struct 32:689–738
60. Waas AM, Schultheisz CR (1996) Compressive failure of composites. 2 Experimental studies. Prog Aerosp Sci 32:43–78
61. Drapier S, Grandidier JC, Potier-Ferry M (2001) A structural approach of plastic microbuckling in long fibre composites: comparison with theoretical and experimental results. Int J Solids Struct 38:3877–3904
62. Mhada K, Braikat B, Hu H, Damil N, Potier-Ferry M (2012) About macroscopic models of instability pattern formation. Int J Solids Struct 49:2978–2989
63. Boucif M, Wesfreid JE, Guyon E (1991) Experimental study of wavelength selection in the elastic buckling instability of thin plates. Eur J Mech A Solids 10:641–661
64. Feyel F, Chaboche JL (2000) FE2 multiscale approach for modeling the elastoviscoplastic behavior of long fiber SiC/Ti composite materials. Comput Methods Appl Mech Eng 183:309–330
65. Geymonat G, Muller S, Triantafyllidis N (1993) Homogenization of nonlinearly elastic materials: microscopic bifurcation and macroscopic loss of rank-one convexity. Archives for Rational Mechanics and Analysis 122:231–290
66. Liu Y, Yu K, Hu H, Belouettar S, Potier-Ferry M (2012) A Fourier-related double scale analysis on instability phenomena of sandwich beams. Int J Solids Struct 49:3077–3088
67. Yu K, Hu H, Chen S, Belouettar S, Potier-Ferry M (2013) Multi-scale techniques to analyze instabilities in sandwich structures. Compos Struct 96:751–762
68. Koiter WT (1945) On the Stability of Elastic Equilibrium. Dissertation, Delft. NASA Technical Translation F-10. 833, Clearinghouse, US Dept. of Commerce/Nat. Bur. of Standards N67–25033, 1967
69. Budiansky B (1974) Theory of buckling and post-buckling behaviour of elastic structures. Adv Appl Mech 14:1–65
70. Cochelin B, Damil N, Potier-Ferry M (2007) Méthode Aymptotique Numérique. Hermes-Lavoisier, Paris
71. Jacques N, Potier-Ferry M (2005) On mode localization in tensile plate bukling. Comptes Rendus Mecanique 333:804–809
72. Nayfeh AH (1973) Perturbation Methods. John Wiley and Sons, New York
73. Sanchez-Palencia E (1980) Non-Homogeneous Media and Vibration Theory. In: Lecture Notes in Physics, vol 127. Springer-Verlag, Heidelberg

9

Coupling local and non-local damage evolutions with the Thick Level Set model

Nicolas Moës[1*], Claude Stolz[1,2] and Nicolas Chevaugeon[1]

*Correspondence: nicolas.moes@ec-nantes.fr
[1] Ecole Centrale de Nantes, GeM Institute, UMR CNRS 6183, 1 Rue de la Noe, 44321 Nantes, France
Full list of author information is available at the end of the article

Abstract

Background: The Thick Level Set model (TLS) is a recent method to delocalize local constitutive models suffering spurious localization. It has two major advantages compared to other delocalization methods. The first one is that the transition from localization to fracture is taken into account in the model. The second one is that the delocalization only acts when and where needed. In other words, the TLS has no effect when the local model is stable. The former advantage was already detailed in several papers (IJNME 86:358-380, 2011, CMAME 233:11-27, 2012, IJF 174:49-60, 2012). This paper concentrates on the latter advantage.

Methods: The TLS delocalization approach is formulated as a bound on the damage gradient. The non-local zone is defined as the zone where the bound is met whereas the local zone is defined as the zone where it is not met. The boundary (localization front) between the local and non-local zone is the main unknown in the problem.

Results: Based on the new model, a 1D pull-out test is solved both analytically and numerically. Different regimes are observed in the solution as the loading progresses: fully elastic, local damage, coupled local/non-local damage and, finally, purely non-local damage.

Conclusions: The new model introduces delocalization as an inequality allowing local damage to develop in zones whereas non-local damage may develop in other zones. This reduces dramatically the cost of implementation of such models compared to fully non-local models.

Keywords: Damage; Delocalization; Non-local damage models; Level set; TLS

Background

Although the scope of TLS application is much wider, we consider in this paper the fracture of quasi-brittle structures under quasi-static loading and under small deformation assumption. The loading is proportional to a scalar parameter. The material is modelled by a time-independent elasto-damage constitutive model with scalar damage. Due to quasi-static analysis, the loading parameter must be controlled especially when bifurcation occurs.

The TLS model was introduced in several papers [1,2] and [3,4]. It lies between continuum damage mechanics and fracture mechanics. Indeed, crack opening is allowed across fully damaged zones (see [2] for instance). The fully damaged zone is located by a level set. Let us note that the description above is different from a diffuse vision of the crack in which crack opening is not explicitly modeled as in the phase-field approach [5-7] or the variational approach to fracture [8,9]. We are rather in the vein of transition from damage to fracture as in [10]. However, the TLS will not be in need of a cohesive zone to perform the transition. The model can be considered as a continuous transition from damage to fracture.

The main idea of the TLS for quasi-brittle fracture is to bound the spatial gradient of the damage variable d, thus avoiding spurious localization. One imposes that the spatial damage distribution satisfies at all time

$$\|\nabla d\| \leq f(d) \text{ on } \Omega \tag{1}$$

where Ω is the domain of interest. The choice of the function $f(d)$ will be discussed in what follows. As damage evolves, one eventually wants to locate the crack, i.e. the zone for which $d = 1$. However, finding the iso-contour $d = 1$ for a quantity d than cannot go beyond 1 is a tedious operation. This is where the level set ingredient comes into play. Variable d is expressed in terms of a level set ϕ as depicted in Figure 1. This relation introduces a length scale l_c. Finding the zone $d = 1$, is now well-posed since the level set ϕ is not strictly limited to l_c but may go beyond. With the use of the surrogate variable ϕ, condition (1) may be rewritten as

$$\begin{cases} \|\nabla\phi\| \leq 1 \\ d = d(\phi) \end{cases} \tag{2}$$

where $f(d)$ in (1) is related to $d(\phi)$ by $f(d) = d'(\phi(d))$ (the prime indicating the derivative of d with respect to ϕ). The function $d(\phi)$ is called the damage shape function and is the main ingredient of the TLS. Equation (2) above indicates that ϕ is a distance function in the zone where the constraint is active (we name this zone the localization zone). The evolution of a distance function has been analyzed and updating algorithm proposed in [11]. In the localization zone, the evolution of ϕ is non-local, indeed

$$\|\nabla\phi\| = 1 \Rightarrow \nabla\dot{\phi} \cdot \nabla\phi = 0 \tag{3}$$

The rate of change of ϕ is thus uniform on any segment aligned with $\nabla\phi$ and the rate of d is given by $\dot{d} = d'(\phi)\dot{\phi}$. Such segments over which $\dot{\phi}$ is uniform are depicted in Figure 2. In the local zone, the evolution of ϕ stems from the evolution of d and the relation $d = d(\phi)$.

The delocalization (1) used in the TLS is different from existing delocalization techniques. Indeed, it directly uses the norm of the damage gradient. It is thus a Hamilton-Jacobi type equation. On the contrary, damage gradient models [12-14] yield Laplacian damage type equation rising the question of proper boundary conditions.

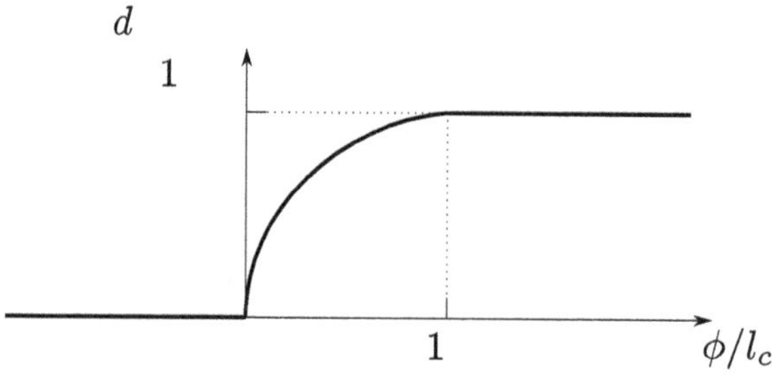

Figure 1 An example of damage shape function.

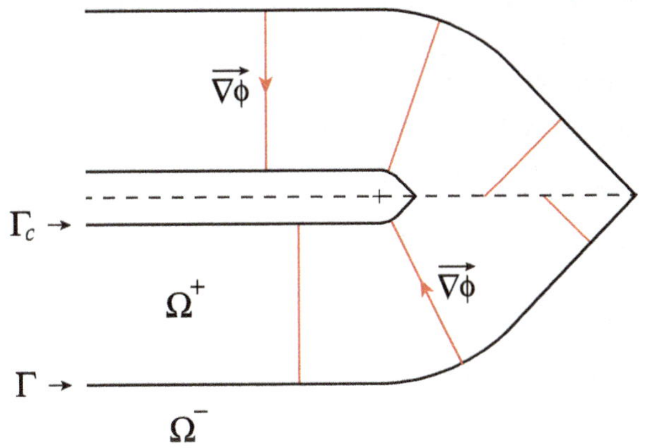

Figure 2 Local (Ω^-) and localization (Ω^+) domains as well as cracks faces (Γ_c). Inside Ω^+, we have $\|\nabla\phi\| = 1$. It can be noted that $\nabla\phi$ (and thus ∇d) is discontinuous along the dashed line which is the so-called skeleton of the distance function.

The TLS shares some similarities with the so-called non-local integral approach [15,16] in which weighted averages are performed over segments (1D), disks (2D) and spheres (3D) of fixed size. In the TLS approach, however, weighted averages are always performed on segments (Figure 2) whatever the dimension of the body and over a length which is not fixed in time but evolves from zero to a maximum length l_c. Finally, note that as l_c is the minimal distance between a point where $d = 0$ and a fully damaged point, $d = 1$, it plays the role of the fracture process zone size.

After this quick introduction of the TLS, we get to the objective of the paper. In previous TLS paper, the delocalization condition (2) was considered as an equality on the whole domain. It meant that d was zero on the domain except in zone where the gradient norm was fixed. The short-coming of this view was that uniform or smooth damage field (because of damage hardening for instance) could not be modeled prior to localization. The inequality analyzed in this paper allows a combination of local and non-local evolutions. In the literature, the possibility to combine both local and non-local approach is seldom discussed with the exception of the so-called morphing numerical technique [17,18].

The paper is organized as follows. The TLS concept with the inequality constraint discussed above are detailed in the first section. Next, the TLS boundary value problem is set up and a dissipation analysis is carried out. A 1D pull-out is solved semi-analytically to show the main feature of the TLS solution. This 1D test is then solved numerically with the TLS to observe the influence of the parameters choice in the model. A conclusion and perspectives end the paper.

Methods

We consider a solid body occupying a domain Ω. The external surface $\partial\Omega$ is composed of two parts $\partial\Omega_u$ and $\partial\Omega_T$ on which the displacements $\lambda \boldsymbol{u}^{\mathrm{d}}$ and the loading $\lambda \boldsymbol{T}^{\mathrm{d}}$ are prescribed, respectively. The parameter λ is a loading parameter.

Small strains and displacements are assumed as well as quasi-static evolution. The current state is characterized by the displacement field $\boldsymbol{u}$, from which the strain field

$\epsilon = \frac{1}{2}(\nabla \boldsymbol{u} + \nabla \boldsymbol{u}^T)$ is derived. The current state is also characterized by an internal scalar variable, the damage denoted d. In this paper, we will not consider other internal variables.

Regarding the material model, we consider a free energy $\psi(\epsilon, d)$ from which the stress tensor $\boldsymbol{\sigma}$ and local energy release rate Y may be derived

$$\boldsymbol{\sigma} = \frac{\partial \psi(\epsilon, d)}{\partial \epsilon}, \quad Y = -\frac{\partial \psi(\epsilon, d)}{\partial d} \tag{4}$$

The potential ψ is assumed for now at least convex with respect to ϵ. The need for other properties will be discussed later.

Regarding the time-independent damage evolution, we consider a function y depending on damage and strain history (through e) such that

$$\dot{d} \geq 0, \quad y(e, d) - Y_c \leq 0, \quad (y(e, d) - Y_c)\dot{d} = 0 \tag{5}$$

where

$$e = e(\epsilon(\tau), \tau \leq t) \tag{6}$$

and Y_c is some threshold. We believe the above formalism encompasses most of the damage models in the literature. To be even more general, one may consider two relations of the kind (5): one for damage in tension and a second one for damage in compression and then combine these damages into d. One has a so-called associated damage model when the y variable is Y. In this case, damage evolution is expressed in terms of the dissipation potential $\varphi^*(Y)$ which is the indicator function of $Y - Y_c \leq 0$:

$$\dot{d} \in \frac{\partial \varphi^*(Y)}{\partial Y} \tag{7}$$

Such model was already considered in [1] for dissymmetric tension-compression evolution. We emphasize the fact that the TLS description is not restricted to associated damage models.

What we have described so far is a purely local damage model. This type of model is known to suffer spurious localizations meaning that the damage gradient may become infinite. The main idea of the TLS approach is to bound damage gradient as expressed in (1). In the TLS model, damage is allowed to go to 1 (but not beyond of course). The location of a crack (or fully degraded zones like in comminution problems) is defined by the set of points for which $d = 1$. Numerically speaking, finding the set of points for which $d = 1$ knowing that d may not go beyond 1 is not very practical. This is why the TLS expresses damage in terms of a surrogate variable ϕ whose values are not limited as depicted in Figure 1. We assume the following regularity on $d(\phi)$

$$\begin{cases} d(\phi) \in C^0(]-\infty, +\infty[) \text{ and monotonically increasing} \\ d(\phi) = 0 \text{ if } \phi \leq 0 \\ d(\phi) = 1 \text{ if } \phi \geq l_c \\ d(\phi) \in C^1(]0, l_c[) \end{cases} \tag{8}$$

Finding the subdomain where $d = 1$ is equivalent to find the subdomain whose boundary is the iso-contour $\phi = l_c$.

In terms of the surrogate variable, ϕ, condition (1) reads

$$\|\nabla \phi\| \leq 1 \tag{9}$$

provided $f(d)$ is given by

$$f(d) = d'(\phi(d)) \tag{10}$$

For instance, if d is linear with respect to ϕ, the gradient of damage will be bounded by a constant

$$d = \phi/l_c, \phi \in [0, l_c] \Longrightarrow \|\nabla d\| \leq \frac{1}{l_c} \tag{11}$$

whereas for more complex function $d(\phi)$, the bound depends on the level of damage. For instance, for the profile shown in Figure 1, we have

$$d = 1 - (1 - (\phi/l_c))^2 \Longrightarrow \|\nabla d\| \leq \frac{2}{l_c}\sqrt{1-d} \tag{12}$$

For a general power law with $n \geq 1$, we obtain

$$d = 1 - (1 - (\phi/l_c))^n \Longrightarrow \|\nabla d\| \leq \frac{n}{l_c}(1-d)^{1-1/n} \tag{13}$$

Whether local or non-local constitutive model should be used at a point $\boldsymbol{x}$ is based on condition (9).

$$\|\nabla\phi(\boldsymbol{x})\| < 1 \Rightarrow \text{Local constitutive model at } \boldsymbol{x} \tag{14}$$

$$\|\nabla\phi(\boldsymbol{x})\| = 1 \Rightarrow \text{Non-Local constitutive model at } \boldsymbol{x} \tag{15}$$

$$\|\nabla\phi(\boldsymbol{x})\| > 1 \quad \text{forbidden} \tag{16}$$

The first condition is the major novelty of this paper, compared to previous paper on the TLS. At any time t, the domain may thus be decomposed into three non-overlapping zones : a local zone Ω^-, a non-local zone Ω^+ and a fully damaged zone Ω_c

$$\overline{\Omega} = \overline{\Omega_c} \cup \overline{\Omega^+} \cup \overline{\Omega^-} \tag{17}$$

$$\Omega^- = \{\boldsymbol{x} \in \Omega : \|\nabla\phi(\boldsymbol{x})\| < 1, \phi(\boldsymbol{x}) < l_c\} \tag{18}$$

$$\Omega^+ = \{\boldsymbol{x} \in \Omega : \|\nabla\phi(\boldsymbol{x})\| = 1, \phi(\boldsymbol{x}) < l_c\} \tag{19}$$

$$\Omega_c = \{\boldsymbol{x} \in \Omega : \phi(\boldsymbol{x}) \geq l_c\} \tag{20}$$

We define also the boundary Γ_c of the fully damaged zone and the interface Γ between the local and non-local zones.

$$\Gamma_c = \partial\Omega_c, \quad \Gamma = \overline{\Omega^+} \cap \overline{\Omega^-} = \partial\Omega^+ \cap \partial\Omega^- \tag{21}$$

The boundary Γ_c defines the crack faces. Figure 2 shows a typical scenario of a crack appearing inside the localization zone.

Note that the volume measure of Ω_c may be zero. This information is part of the solution process. We expect different shapes of Ω_c in comminution and brittle crack propagation.

Eikonal equation

Condition, $\|\nabla\phi(\boldsymbol{x})\| = 1$ is a non-linear first-order partial differential equation. It is called an eikonal equation and belongs to the Hamilton-Jacobi equation family. Among the possible solution satisfying $\|\nabla\phi(\boldsymbol{x})\| = 1$, we will pick the one corresponding to the vanishing viscosity solution [19]. It is characterized by

$$\phi(\boldsymbol{x}) = \min_{\boldsymbol{y}\in\Gamma}(\phi(\boldsymbol{y}) + d(\boldsymbol{x}, \boldsymbol{y})), \boldsymbol{x} \in \Omega^+ \tag{22}$$

where $d(\boldsymbol{x}, \boldsymbol{y})$ is the length of the shortest path connecting $\boldsymbol{x}$ and $\boldsymbol{y}$ inside Ω^+. The value of ϕ at $\boldsymbol{x} \in \Omega^+$ can be thought as the minimal fare to go from Γ to $\boldsymbol{x}$. The fare being the sum of the initial fare $\phi(\boldsymbol{y})$ plus the mileage from $\boldsymbol{y}$ to $\boldsymbol{x}$. Damage on Ω^+ is thus fully determined from values on Γ. A 1D example of ϕ satisfying the eikonal on a segment $[b, d]$ is given in Figure 3.

The fact that damage is related to a variable satisfying the eikonal equation, the cornerstone of the level set technology [11], explains why the damage model is coined Thick Level Set. In the non-local zone, damage is modeled over a thick layer in terms of level sets.

Damage evolution

In the local zone, Ω^-, damage evolution is local and given by (5). In the non-local zone, Ω^+, damage rate is related to $\dot{\phi}$ by

$$\dot{d} = d'(\phi)\dot{\phi} \tag{23}$$

where $\dot{\phi}$ is uniform on segments aligned with $\nabla\phi$, see Equation (3). We denote this space as $\mathcal{A}$:

$$\dot{\phi} \in \mathcal{A} = \left\{a(\boldsymbol{x}) \in L^2(\Omega^+) : \nabla a \cdot \nabla\phi = 0\right\} \tag{24}$$

Non-local damage evolution boils down to decomposing Ω^+ into a set of independent segments and finding a value $\dot{\phi}$ over each of them.

As in [2], we suggest to introduce averaged quantities, $\bar{y}, \bar{\dot{d}}$ over each segments. This may be expressed by a projection operation.

$$\bar{y} \in \mathcal{A} : \int_{\Omega^+} \bar{y} d' a \, \mathrm{d}\omega = \int_{\Omega^+} y d' a \, \mathrm{d}\omega, \quad \forall a \in \mathcal{A} \tag{25}$$

$$\bar{\dot{d}} \in \mathcal{A} : \int_{\Omega^+} \bar{\dot{d}} a \, \mathrm{d}\omega = \int_{\Omega^+} \dot{d} a \, \mathrm{d}\omega, \quad \forall a \in \mathcal{A} \tag{26}$$

We note that the averages satisfy the following property

$$\int_{\Omega^+} \bar{y}\bar{\dot{d}} \, \mathrm{d}\omega = \int_{\Omega^+} y\dot{d} \, \mathrm{d}\omega \tag{27}$$

The above indicates that duality is preserved through the averaging technique. This is not often the case in delocalization techniques as discussed in [20].

The local constitutive model, (5), is then expressed in terms of the non-local quantities

$$\bar{\dot{d}} \geq 0, \quad \bar{y} - Y_c \leq 0, \quad (\bar{y} - Y_c)\,\bar{\dot{d}} = 0 \tag{28}$$

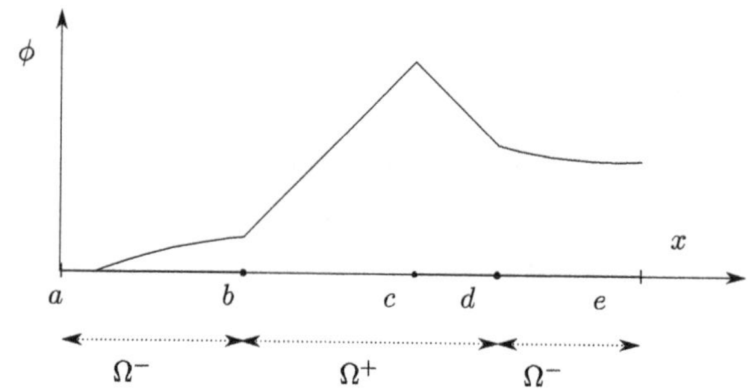

Figure 3 Distribution of ϕ on a 1D domain: $\Omega = [a, e]$, $\Omega^+ = [b, d]$, $\Omega^- = [a, b] \cup [d, e]$, $\Gamma = \{b, d\}$. Slopes at 45 degrees on Ω^+ indicate that ϕ behaves as a distance function ($\|\nabla\phi\| = 1$) whereas $\|\nabla\phi\| < 1$ on Ω^-. Point c is the skeleton of the distance function.

where we have assumed Y_c uniform (if not it needs to be averaged by formula (25)). Finally, we write the relation giving $\dot{\phi}$ in terms of $\dot{\overline{d}}$:

$$\dot{\overline{d}} = \overline{d'}\dot{\phi}, \quad \overline{d'} \in \mathcal{A} : \int_{\Omega^+} \overline{d'} a \, \mathrm{d}\omega = \int_{\Omega^+} d' a \, \mathrm{d}\omega, \quad \forall a \in \mathcal{A} \tag{29}$$

To end this section we illustrate the average formula on the 1D example depicted in Figure 3. Averages are given by

$$\text{On } [b, c] : \overline{y} = \frac{\int_b^c y d'(\phi) \, \mathrm{d}x}{\int_b^c d'(\phi) \, \mathrm{d}x}, \quad \dot{\overline{d}}(\boldsymbol{x}) = \frac{\int_b^c \dot{d} \, \mathrm{d}x}{\int_b^c \mathrm{d}x} \tag{30}$$

$$\text{On } [c, d] : \overline{y} = \frac{\int_c^d y d'(\phi) \, \mathrm{d}x}{\int_c^d d'(\phi) \, \mathrm{d}x}, \quad \dot{\overline{d}}(\boldsymbol{x}) = \frac{\int_c^d \dot{d} \, \mathrm{d}x}{\int_c^d \mathrm{d}x} \tag{31}$$

TLS boundary value problem

We are now able to define the boundary value problem. The set of admissible displacements is given by

$$\mathcal{U} = \left\{ \boldsymbol{u} \in C^0(\Omega \setminus \Omega_c), \int_{\Omega \setminus \Omega_c} \psi(\boldsymbol{\epsilon}(\boldsymbol{u}), d(\phi)) \, \mathrm{d}\omega < +\infty, \boldsymbol{u} = \lambda \boldsymbol{u}^{\mathrm{d}} \text{ on } \partial\Omega_u \right\} \tag{32}$$

The fact that the fully damage zones are removed from the domain is important. It allows the displacement to be discontinuous across Γ_c. Regarding the regularity of the displacement, we request that the energy, i.e. integral of ψ over $\Omega \setminus \Omega_c$, is finite. This space is not simply H^1 as in elasticity since the stiffness is possibly vanishing on Γ_c boundary [21].

Regarding the ϕ variable, it is required to be continuous over Ω and belong to the set $\mathcal{K}$. The admissible set for ϕ is denoted K.

$$\mathcal{K} = \left\{ \phi \in C^0(\Omega) : \|\nabla\phi(\boldsymbol{x})\| = 1, \boldsymbol{x} \in (\Omega^+ \cup \Omega_c), \|\nabla\phi(\boldsymbol{x})\| < 1, \boldsymbol{x} \in \Omega^- \right\} \tag{33}$$

The continuity requirement on ϕ leads to a Hadamard compatibility condition on the moving boundary Γ. Let us define the jump of a quantity f across Γ by

$$[f]_\Gamma(\boldsymbol{x}, t) = f^+(\boldsymbol{x}, t) - f^-(\boldsymbol{x}, t) \tag{34}$$

The exponent $-/+$ placed on some quantities f defined at $\boldsymbol{x}$ on Γ has the following meaning

$$f^{\pm}(\boldsymbol{x}, t) = \lim_{h \to 0^+} f(\boldsymbol{x} \pm h\boldsymbol{n}(\boldsymbol{x}, t), t) \tag{35}$$

where $\boldsymbol{n}$ is the outward normal to Ω^+. With these notations we have

$$[\dot{\phi}]_\Gamma + [\nabla\phi]_\Gamma \cdot \boldsymbol{n} \, v_{\mathrm{n}} = 0 \tag{36}$$

where v_{n} is the normal velocity of Γ counted positively along $\boldsymbol{n}$. This gives the respective evolution of domains Ω^+ and Ω^-.

Potential energy of the domain is given by :

$$\begin{cases} \boldsymbol{u} \in \mathcal{U}, \ \phi \in K \\ E^{\mathrm{pot}}(\boldsymbol{u}, \phi, \lambda) = \int_{\Omega^-} \psi(\boldsymbol{\epsilon}(\boldsymbol{u}), d(\phi)) \, \mathrm{d}\omega + \int_{\Omega^+} \psi(\boldsymbol{\epsilon}(\boldsymbol{u}), d(\phi)) \, \mathrm{d}\omega \\ \qquad - \int_{\partial\Omega_T} \lambda \boldsymbol{T}^{\mathrm{d}} \cdot \boldsymbol{u} \, \mathrm{d}a \end{cases} \tag{37}$$

Note that the same free energy expression, ψ, is used over Ω^+ and Ω^-. In what follows, $\boldsymbol{n}$ is the outward normal vector to Ω on $\partial\Omega$ and to Ω^+ on $\partial\Omega^+$. The set of admissible displacements variations is denoted as $\mathcal{U}_0$. It has the same definition as $\mathcal{U}$ except that $\boldsymbol{u}$ is set to zero on $\partial\Omega_u$.

Assuming, that at time t, the spatial distribution of ϕ of the two volumes Ω^+ and Ω^- is known, the displacement field $\boldsymbol{u}$ is the field that solves the stationarity of the potential energy:

$$E^{\mathrm{pot}}_{,\boldsymbol{u}}\delta\boldsymbol{u} = 0, \quad \forall\delta\boldsymbol{u} \in \mathcal{U}_0 \tag{38}$$

This means;

$$\int_{\Omega^-} \boldsymbol{\sigma} : \boldsymbol{\epsilon}(\delta\boldsymbol{u})\,\mathrm{d}\omega + \int_{\Omega^+} \boldsymbol{\sigma} : \boldsymbol{\epsilon}(\delta\boldsymbol{u})\,\mathrm{d}\omega - \int_{\partial\Omega_T} \lambda\boldsymbol{T}^{\mathrm{d}} \cdot \delta\boldsymbol{u}\,\mathrm{d}a = 0, \quad \forall\delta\boldsymbol{u} \in \mathcal{U}_0 \tag{39}$$

For simplicity, we assume that the boundary Γ_{c} is traction free (no contact on crack faces). The equilibrium (39) yields the following local equations

$$\mathrm{div}\boldsymbol{\sigma} = \boldsymbol{0} \text{ over } \Omega \setminus \Omega_{\mathrm{c}} \tag{40}$$

$$[\boldsymbol{\sigma}]_\Gamma \cdot \boldsymbol{n} = \boldsymbol{0} \text{ on } \Gamma \tag{41}$$

$$\boldsymbol{\sigma} \cdot \boldsymbol{n} = \lambda\boldsymbol{T}^{\mathrm{d}} \text{ on } \partial\Omega_T \tag{42}$$

$$\boldsymbol{\sigma} \cdot \boldsymbol{n} = \boldsymbol{0} \text{ on } \Gamma_{\mathrm{c}} \tag{43}$$

We stress the fact that Γ_{c} denotes the boundary of the fully damage zone and thus in case of a crack Γ_{c} indicates *both* crack lips. To complete the set of equations to be solved for a known damage distribution, we add the stress definition and kinematic relations

$$\boldsymbol{\sigma} = \frac{\partial\psi(\boldsymbol{\epsilon}, d)}{\partial\boldsymbol{\epsilon}} \text{ over } \Omega \setminus \Omega_{\mathrm{c}} \tag{44}$$

$$\boldsymbol{\epsilon} = \frac{1}{2}\left(\nabla\boldsymbol{u} + \nabla\boldsymbol{u}^T\right) \text{ over } \Omega \setminus \Omega_{\mathrm{c}} \tag{45}$$

$$[\boldsymbol{u}]_\Gamma = \boldsymbol{0} \text{ on } \Gamma \tag{46}$$

$$\boldsymbol{u} = \lambda\boldsymbol{u}^{\mathrm{d}} \text{ on } \partial\Omega_u \tag{47}$$

Finally, we need to add damage evolution equations in the local zone (5) and non-local zone (28).

Dissipation analysis and fields regularity

The goal of this section is to analyze the expression of the dissipation as well as looking at the fields regularity across the boundary Γ.

Taking into account the conservation law for the total energy during the evolution of the system, the total dissipation associated with the loading rate $\dot{\lambda}$ is:

$$D = \int_{\partial\Omega_T} \lambda\boldsymbol{T}^{\mathrm{d}} \cdot \dot{\boldsymbol{u}}\,\mathrm{d}a + \int_{\partial\Omega_u} \dot{\lambda}\boldsymbol{u}^{\mathrm{d}} \cdot \boldsymbol{\sigma} \cdot \boldsymbol{n}\,\mathrm{d}a \tag{48}$$

$$-\frac{\mathrm{d}}{\mathrm{d}t}\left(\int_{\Omega^-} \psi(\boldsymbol{\epsilon}, d(\phi))\,\mathrm{d}\omega + \int_{\Omega^+} \psi(\boldsymbol{\epsilon}, d(\phi))\,\mathrm{d}\omega\right) \tag{49}$$

In the above, we did not consider the energy inside Ω_{c} because it is assumed to be zero. Indeed, no compression is considered in this zone (see (43)).

Using Leibniz formula for the time derivative of moving domains as well as the relation:

$$\dot{\psi} = \boldsymbol{\sigma} : \boldsymbol{\epsilon}(\dot{\boldsymbol{u}}) - Y\dot{d} \tag{50}$$

we obtain

$$D = \int_{\partial\Omega_T} \lambda \boldsymbol{T}^{\mathrm{d}} \cdot \dot{\boldsymbol{u}}\, \mathrm{d}a + \int_{\partial\Omega_T} \dot{\lambda} \boldsymbol{u}^{\mathrm{d}} \cdot \boldsymbol{\sigma} \cdot \boldsymbol{n}\, \mathrm{d}a - \int_{\Omega^-} \boldsymbol{\sigma} : \boldsymbol{\epsilon}(\dot{\boldsymbol{u}})\, \mathrm{d}\omega - \int_{\Omega^+} \boldsymbol{\sigma} : \boldsymbol{\epsilon}(\dot{\boldsymbol{u}})\, \mathrm{d}\omega$$
$$+ \int_{\Gamma} [\psi]_{\Gamma}\, v_{\mathrm{n}}\, \mathrm{d}a - \int_{\Gamma_{\mathrm{c}}} \psi\, v_{\mathrm{n}}\, \mathrm{d}a + \int_{\Omega^-} Y\dot{d}\, \mathrm{d}\omega + \int_{\Omega^+} Y\dot{d}\, \mathrm{d}\omega$$

Integrating the domain integral by parts in the second line above and using the equations characterizing the equilibrium state, we get

$$D = \int_{\Gamma} [\psi]_{\Gamma}\, v_{\mathrm{n}} + \boldsymbol{n} \cdot \boldsymbol{\sigma} \cdot [\dot{\boldsymbol{u}}]_{\Gamma}\, \mathrm{d}a - \int_{\Gamma_{\mathrm{c}}} \psi\, v_{\mathrm{n}}\, \mathrm{d}a + \int_{\Omega^-} Y\dot{d}\, \mathrm{d}\omega + \int_{\Omega^+} Y\dot{d}\, \mathrm{d}\omega \tag{51}$$

During the propagation of the interface, perfect contact is assumed on Γ, that is the displacement jump across Γ must be zero at all time. As a consequence, the derivative along the moving interface of the displacement jump must be zero, [22], yielding the so called first Hadamard compatibility condition between the front velocity v_{n} and the jump in material velocities $[\dot{\boldsymbol{u}}]_{\Gamma}$:

$$[\dot{\boldsymbol{u}}]_{\Gamma} + [\nabla \boldsymbol{u}]_{\Gamma} \cdot \boldsymbol{n}\, v_{\mathrm{n}} = 0 \tag{52}$$

Equation (51), now becomes

$$D = \int_{\Gamma} \boldsymbol{n} \cdot [\boldsymbol{P}]_{\Gamma} \cdot \boldsymbol{n}\, v_{\mathrm{n}}\, \mathrm{d}a - \int_{\Gamma_{\mathrm{c}}} \psi\, v_{\mathrm{n}}\, \mathrm{d}a + \int_{\Omega^-} Y\dot{d}\, \mathrm{d}\omega + \int_{\Omega^+} Y\dot{d}\, \mathrm{d}\omega \tag{53}$$

where $\boldsymbol{P}$ is the Eshelby tensor

$$\boldsymbol{P} = \psi \boldsymbol{I} - \boldsymbol{\sigma} \cdot \nabla \boldsymbol{u} \tag{54}$$

The first term is the dissipation created by the interface propagation. We show now that due to damage continuity on Γ this term is zero.

Since normal stress and displacement are continuous across Γ, the product of the jump in stress and strain across Γ is zero, [23,24]:

$$[\boldsymbol{\sigma}]_{\Gamma} : [\boldsymbol{\epsilon}]_{\Gamma} = 0 \tag{55}$$

Let $\psi_{\mathrm{d}}(\boldsymbol{\epsilon})$ be the density of free energy for a given value of damage and let $\psi_{\mathrm{d}}^*(\boldsymbol{\sigma})$ be its dual by the Legendre-Fenchel transform. Since the couples $(\boldsymbol{\epsilon}^+, \boldsymbol{\sigma}^+)$ and $(\boldsymbol{\epsilon}^-, \boldsymbol{\sigma}^-)$ do satisfy the constitutive model (4), we have

$$\psi_{\mathrm{d}}(\boldsymbol{\epsilon}^+) + \psi_{\mathrm{d}}^*(\boldsymbol{\sigma}^+) - \boldsymbol{\sigma}^+ : \boldsymbol{\epsilon}^+ = 0, \quad \psi_{\mathrm{d}}(\boldsymbol{\epsilon}^-) + \psi_{\mathrm{d}}^*(\boldsymbol{\sigma}^-) - \boldsymbol{\sigma}^- : \boldsymbol{\epsilon}^- = 0 \tag{56}$$

Summing the two relations above and using (55), we have

$$\left(\psi_{\mathrm{d}}(\boldsymbol{\epsilon}^+) + \psi_{\mathrm{d}}^*(\boldsymbol{\sigma}^-) - \boldsymbol{\sigma}^- : \boldsymbol{\epsilon}^+\right) + \left(\psi_{\mathrm{d}}(\boldsymbol{\epsilon}^-) + \psi_{\mathrm{d}}^*(\boldsymbol{\sigma}^+) - \boldsymbol{\sigma}^+ : \boldsymbol{\epsilon}^-\right) = 0 \tag{57}$$

Since both terms above are greater or equal to zero (classical property of convex analysis, see [25]), we have

$$\psi_{\mathrm{d}}(\boldsymbol{\epsilon}^+) + \psi_{\mathrm{d}}^*(\boldsymbol{\sigma}^-) - \boldsymbol{\sigma}^- : \boldsymbol{\epsilon}^+ = 0, \quad \psi_{\mathrm{d}}(\boldsymbol{\epsilon}^-) + \psi_{\mathrm{d}}^*(\boldsymbol{\sigma}^+) - \boldsymbol{\sigma}^+ : \boldsymbol{\epsilon}^- = 0 \tag{58}$$

This implies that the couples $(\boldsymbol{\epsilon}^+, \boldsymbol{\sigma}^-)$ and $(\boldsymbol{\epsilon}^-, \boldsymbol{\sigma}^+)$ do satisfy the constitutive model. Assuming that the convex potential $\psi_{\mathrm{d}}(\boldsymbol{\epsilon})$ is such that the stress associated to any strain is unique, we have

$$[\boldsymbol{\sigma}]_{\Gamma} = [\boldsymbol{\epsilon}]_{\Gamma} = \boldsymbol{0} \tag{59}$$

The continuity of the strain and displacement across Γ leads to the continuity of the displacement gradient

$$[\nabla \boldsymbol{u}]_\Gamma = \boldsymbol{0} \tag{60}$$

leading finally to the continuity of the Eshelby tensor.

$$[\boldsymbol{P}]_\Gamma = \boldsymbol{0} \tag{61}$$

Dissipation is thus reduced to

$$D = \int_{\Omega^-} Y\dot{d}\, \mathrm{d}\omega + \int_{\Omega^+} Y\dot{d}\, \mathrm{d}\omega - \int_{\Gamma_c} \psi\, v_n\, da \tag{62}$$

The dissipation must be positive. For classical models in which Y is positive, this implies that damage may only grow. Damage growth will create a growth of the fully damaged zone Ω_c (and thus a negative velocity v_n). The last term in (62) is thus automatically positive. Whether this term is zero or not depends on the regularity of ψ on the boundary Γ_c. This regularity must be assessed from the non-local constitutive model condition: $\overline{y} - Y_c \leq 0$.

Note that dissipation may also be written

$$D = \int_{\Omega^-} Y\dot{d}\, \mathrm{d}\omega + \int_{\Omega^+} \overline{Y}\dot{\overline{d}}\, \mathrm{d}\omega - \int_{\Gamma_c} \psi\, v_n\, da \tag{63}$$

where $\overline{Y}$ is defined by (25) (y replaced by Y). The above expression exhibits the duality between $\overline{Y}$ and $\dot{\overline{d}}$ in the localization zone.

Results

We consider a 1D axisymmetric fiber pull-out depicted in Figure 4. The fiber of radius r_i is considered rigid and infinitely long. It is pulled out of a clamped circular domain of radius $r_e = r_i + L$. The only non-zero stress component is the shear stress τ satisfying the following equilibrium conditions

$$(\tau r)_{,r} = 0 \Rightarrow \tau(r) = \frac{\tau(r_i) r_i}{r} \tag{64}$$

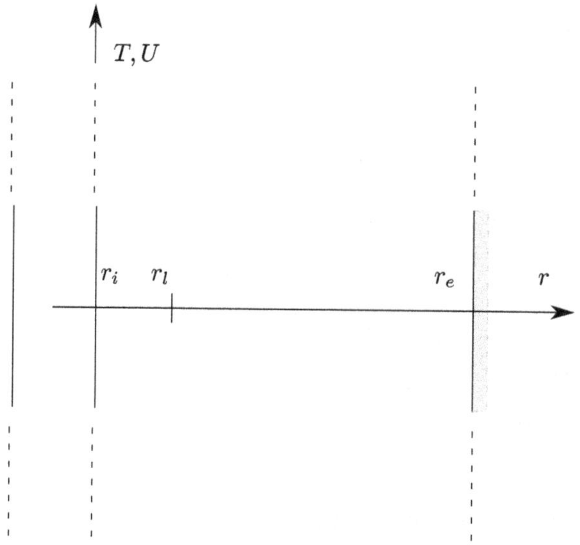

Figure 4 Pull-out of an infinite fiber of radius r_i from a tube of radius r_e. Radius r_l indicates the (evolving) extent of the non-local damage zone.

The only non-zero strain is the shear strain, derivative of the displacement along the fiber direction

$$\gamma = u_{,r} \tag{65}$$

We consider the following free energy density involving some hardening function $h(d)$, satisfying $h(1) = 0$. The shear stiffness is denoted μ and Y_c is also a material parameter.

$$\psi(\gamma, d) = \frac{1}{2}\mu(1-d)\gamma^2 + Y_c h(d) \tag{66}$$

So, state laws read

$$\tau = \mu(1-d)\gamma, \quad Y = \frac{1}{2}\mu\gamma^2 - Y_c h' \tag{67}$$

The local evolution model is given by

$$\dot{d} \geq 0, \quad Y - Y_c \leq 0, \quad (Y - Y_c)\dot{d} = 0 \tag{68}$$

The condition $Y = Y_c$ reduces to

$$\frac{\tau}{\tau_c} = \underbrace{(1-d)\sqrt{h'+1}}_{g(d)}, \quad \tau_c = \sqrt{2\mu Y_c} \tag{69}$$

Let us now be more precise on the type of function $g(d)$ we will be considering. Basically, we are interested by C^1 positive concave functions with a maximum value at some damage $d_c < 1$:

$$g(d) \in C^1([0,1]) : g'' < 0, \; g(0) = 1, \; g(1) = 0, \; g'(d_c) = 0 \tag{70}$$

We shall use in what follows

$$g(d) = (1-d)\exp\left(\frac{d}{1-d_c}\right) \tag{71}$$

The corresponding stress strain curve is given in Figure 5. We will now search for the complete solution linking the (non-dimensional) shear stress T needed to move by a (non-dimensional) displacement U the fiber:

$$T = \frac{\tau(r_i)}{\tau_c}, \quad U = \frac{u(r_i)\mu}{r_i \tau_c} \tag{72}$$

Four regimes will be observed. They are depicted in Figure 6: elastic, local damage, local and non-local damage and finally purely non-local damage. The first two regimes may

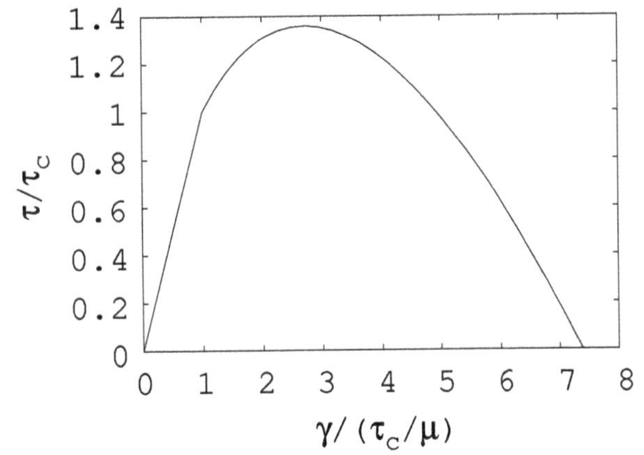

Figure 5 Local constitutive model: stress versus rising strain (case $d_c = 0.5$).

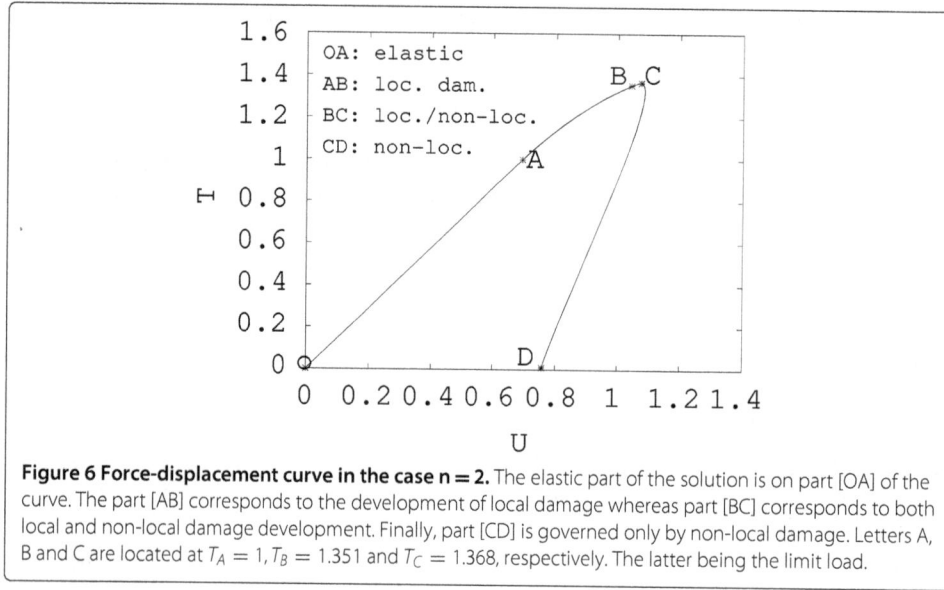

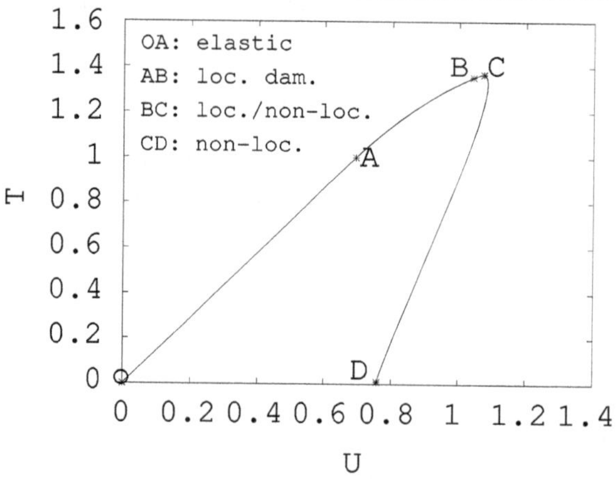

Figure 6 Force-displacement curve in the case n = 2. The elastic part of the solution is on part [OA] of the curve. The part [AB] corresponds to the development of local damage whereas part [BC] corresponds to both local and non-local damage development. Finally, part [CD] is governed only by non-local damage. Letters A, B and C are located at $T_A = 1, T_B = 1.351$ and $T_C = 1.368$, respectively. The latter being the limit load.

be solved analytically whereas the two last one may not. We however pursue as much as possible the analytical path. Next section is devoted to a numerical solver.

Pure elastic regime: $T \in [0, T_A = 1]$

The displacement solution is given by

$$u(r) = T\frac{\tau_c}{\mu}\int_r^{r_e}\frac{r_i}{r}\,\mathrm{d}r = T\frac{\tau_c}{\mu}r_i\log\left(\frac{r_e}{r}\right) \tag{73}$$

We thus have a linear relationship between the stress and displacement

$$T = U\left(\log\left(\frac{r_e}{r_i}\right)\right)^{-1} \tag{74}$$

Local damage regime: $T \in [1, T_B]$

When T reaches 1 local damage starts around the fiber. Its distribution is obtained by combining (64) and (69)

$$T = g(d)\frac{r}{r_i} \tag{75}$$

This distribution of local damage is acceptable provided the condition below holds true

$$\|\nabla\phi\| \leq 1 \text{ i.e. } \|\nabla d\| \leq d'(\phi(d)) \tag{76}$$

The norm of the damage gradient is maximum at $r = r_i$ and of value

$$\left|\frac{\mathrm{d}d}{\mathrm{d}r}\right| = \frac{1}{r_i}\frac{g(d_i)}{g'(d_i)} \tag{77}$$

where $d_i = d(r_i)$. Considering a general power law damage profile (13), the condition (76) is

$$\left|\frac{g(d_i)}{g'(d_i)}\right| \leq \frac{r_i}{l_c}n(1-d_i)^{1-1/n} \tag{78}$$

Let us denote by d_i^B the smallest value of damage for which the condition above is violated and T_B the corresponding loading. Due to the fact that $g'(d_c) = 0$, it is clear that d_i^B will be slightly lower than d_c. We note that as the material length gets bigger with respect to r_i, non-locality (violation of (78)) will step in for smaller and smaller damage d_i. Considering the choice (71), we get the condition

$$\left|\frac{(1-d_i)(1-d_c)}{d_c-d_i}\right| \leq \frac{r_i}{l_c} n(1-d_i)^{1-1/n} \tag{79}$$

For instance if $n = 1$, we get

$$d_i^B = \frac{d_c - (1-d_c)(l_c/r_i)}{1-(1-d_c)(l_c/r_i)} \tag{80}$$

As a numerical application, with $d_c = 0.5$ and $l_c/r_i = 0.1$, we get $d_i^B = 0.47$.

Combined local and non-local damage regime: $T \in [T_B, T_C]$

For a loading higher than T_B, non-local damage will develop close to the fiber. Let $[r_i, r_l]$ be the current extension of the non-local damage zone in which damage ranges from d_i to d_l following:

$$r(d, d_i) = r_i + \phi(d_i) - \phi(d) \tag{81}$$

The condition for the non-local zone to grow is $\overline{Y} = Y_c$, i.e.

$$\int_{r_i}^{r_l} (Y - Y_c) \frac{\mathrm{d}d}{\mathrm{d}\phi} r \, \mathrm{d}r = 0 \tag{82}$$

Using, (81), we may rewrite it as

$$\int_{d_l}^{d_i} (Y - Y_c) r(d, d_i) \, \mathrm{d}d = 0 \tag{83}$$

Now, using (69), we get

$$T = \left(\frac{\int_{d_l}^{d_i} (1-d)^{-2} g^2(d) (r(d,d_i)/r_i) \, \mathrm{d}d}{\int_{d_l}^{d_i} (1-d)^{-2} (r(d,d_i)/r_i)^{-1} \, \mathrm{d}d} \right)^{1/2} \tag{84}$$

Since loading is rising, so does local d_l damage at $r = r_l$, following

$$T = \frac{r_l}{r_i} g(d_l) \tag{85}$$

Given T, system (84)-(85) returns unknowns d_i and d_l as well as the extent of the non-local zone $r_l = r(d_l, d_i)$. We note that for $T = T_B$, we have $d_l = d_i$ and $r_l = r_i$. Let T_C be the loading above which the system has no solution.

Non-local damage regime: T decreases from T_C to 0

There is no solution of the problem for a loading higher than T_C. When the loading decreases below T_C, there is of course a possible elastic solution. Another possible solution is the further development of the non-local damage zone (while damage in the local zone no longer evolves since loading is decreasing). The system of equations to solve still involves (84)

$$T = \left(\frac{\int_{d_l}^{d_i} (1-d)^{-2} g^2(d) (r(d,d_i)/r_i) \, \mathrm{d}d}{\int_{d_l}^{d_i} (1-d)^{-2} (r(d,d_i)/r_i)^{-1} \, \mathrm{d}d} \right)^{1/2} \tag{86}$$

Equation (85) is now different and reads

$$T_C = \frac{r_l}{r_i} f(d_l) \tag{87}$$

Indeed the damage at r_l did not change from its value at load T_C because the load has been decreasing afterwards.

Analysis of the displacement of the fiber

The displacement of the fiber is given by

$$U = T\int_{r_i}^{r_e} \frac{1}{(1-d)r}\,\mathrm{d}r \tag{88}$$

As the damage around the fiber goes to 1, the integrand goes to infinity. But, at the same time the loading goes to zero. Let us study the limit of the fiber displacement for the loading going to zero. The loading is given by (84) recalled below

$$T = \left(\frac{\int_{d_l}^{d_i}(1-d)^{-2}g^2(d)(r(d,d_i)/r_i)\,\mathrm{d}d}{\int_{d_l}^{d_i}(1-d)^{-2}(r(d,d_i)/r_i)^{-1}\,\mathrm{d}d}\right)^{1/2} = \left(\frac{\int_{d_l}^{d_i}N(d,d_i)\,\mathrm{d}d}{\int_{d_l}^{d_i}D(d,d_i)\,\mathrm{d}d}\right)^{1/2} \tag{89}$$

Due to the property of $g(d)$, (70), we have

$$g(d) = O(1-d) \text{ as } d \to 1 \tag{90}$$

So

$$0 < N(d,d_i) < +\infty, \forall d, d_i \in [0,1] \tag{91}$$

Finally, we have

$$T = O\left(\sqrt{1-d_i}\right) \text{ as } d_i \to 1 \tag{92}$$

Note that this property does not depend on the choice of $d(\phi)$. Going back to the displacement expression, (88), we have

$$U = T\int_{d_l}^{d_i}(1-d)^{-1}(r(d,d_i))^{-1}\frac{\mathrm{d}r(d,d_i)}{\mathrm{d}d}\,\mathrm{d}d + CT \tag{93}$$

where C is a finite constant and

$$r(d,d_i) = r_i + \phi(d_i) - \phi(d) = r_i + l_c\left((1-d)^{1/n} - (1-d_i)^{1/n}\right) \tag{94}$$

assuming a power law asymptotic behavior of $\phi(d)$ as d goes to 1:

$$\phi(d) = l_c(1-(1-d)^{1/n}) \tag{95}$$

Finally, we get

$$U = T\,O\left((1-d_i)^{1/n-1} + C\right) = O\left((1-d_i)^{1/n-1/2}\right) \text{ as } d_i \to 1 \tag{96}$$

We conclude that there exists three regimes of delocalization.

$$\lim_{d_i\to 1} U = 0 \quad \text{if } n < 2 \tag{97}$$

$$0 < \lim_{d_i\to 1} U < +\infty \quad \text{if } n = 2 \tag{98}$$

$$\lim_{d_i\to 1} U = +\infty \quad \text{if } n > 2 \tag{99}$$

When $n < 2$, the fiber displacement must be zero for total failure. When $n = 2$, there exists a limit value of fiber displacement before total failure and when $n > 2$, it takes an infinite displacement before total failure. It is interesting to note that these three regimes also exist in gradient damage models [20].

Numerical solve

Last section gave some insight on the different regimes in the solution. In order to plot the solution, we detail here a 1D numerical solver. This code is rather ad hoc for 1D problem, since we force the advance of the Γ boundary and find the corresponding loading and fields. General 2D and 3D solvers will be detailed in a forthcoming paper. We search for the solution at a set of discrete times. Consider the solution known at time t_n, the solution at time t_{n+1} must satisfy the following equations.

Kinematics and equilibrium on $]r_i, r_e[$.

$$u^{n+1} \in \mathcal{U} = \left\{u \in H^1(]r_i, r_e[) : u(r_e) = 0\right\} \tag{100}$$

$$\gamma\left(u^{n+1}\right) = u^{n+1}_{,r} \tag{101}$$

$$\int_{r_i}^{r_e} \tau^{n+1}\ \gamma(u^*)\ r\ \mathrm{d}r = \tau_c T^{n+1} u^*(r_i) r_i, \quad \forall u^* \in \mathcal{U} \tag{102}$$

State laws and $d(\phi)$ relation on $]r_i, r_e[$.

$$\tau^{n+1} = \tau\left(\gamma\left(u^{n+1}\right), d^{n+1}\right) = \left(1 - d^{n+1}\right) \mu\gamma\left(u^{n+1}\right) \tag{103}$$

$$Y^{n+1} = Y\left(\gamma\left(u^{n+1}\right), d^{n+1}\right) = \frac{1}{2}\mu\gamma\left(u^{n+1}\right)^2 - Y_c h'\left(d^{n+1}\right) \tag{104}$$

$$d^{n+1} = d\left(\phi^{n+1}\right) \tag{105}$$

Non-local evolution law on $]r_i, r_l^{n+1}[$.

$$a = \int_{r_i}^{r_l^{n+1}} \left(Y^{n+1} - Y_c\right) d'\left(\phi^{n+1}\right) r\ \mathrm{d}r \leq 0, \tag{106}$$

$$b = \phi^{n+1}(r_i) - \phi^n(r_i) \geq 0, \ ab = 0 \tag{107}$$

$$\phi^{n+1} = \phi^{n+1}\left(r_l^{n+1}\right) + r_l^{n+1} - r, \quad \text{on }]r_i, r_l^{n+1}[\tag{108}$$

Local evolution law on $]r_l^{n+1}, r_e[$.

$$Y^{n+1} - Y_c \leq 0, \ \phi^{n+1} - \phi^n \geq 0, \ \left(Y^{n+1} - Y_c\right)\left(\phi^{n+1} - \phi^n\right) = 0 \tag{109}$$

Regarding space discretization, the segment $]r_i, r_e[$ is discretized with a set of finite elements. Initially, the non-local zone is empty and we proceed with a classical Newton-Raphson scheme depicted in the solver flowchart without non-local zone.

Solver flowchart without non-local zone

1. initialization: $u^0 = d^0 = T^0 = 0$
2. elastic step: find the load step for which damage starts
3. load step $T^{n+1} = T^n + \Delta T$
4. iterations initialization $k = 0$

5. solve linear system (110) to find Δu
6. after the first iteration adapt the load step so that the maximum damage increment is d_{inc}
7. update using (111)-(118)
8. if residual $\leq$ tol, go to 9 else go to 5
9. if $\|\nabla\phi^{n+1}\| \leq 1$, go to 3, else go to solver flowchart with non-local zone

The linear problem at each iteration reads: find $\Delta u \in \mathcal{U}$ such that:

$$\int_{r_i}^{r_e} H^k \gamma(\Delta u)\gamma(u^*) r \, \mathrm{d}r = \tau_c T^{n+1} u^*(r_i) r_i - \int_{r_i}^{r_e} \tau^k \gamma(u^*) r \, \mathrm{d}r, \forall u^* \in \mathcal{U} \tag{110}$$

where the right hand side is the residual at iteration k. Once the displacement correction is obtained, the local update of the fields is computed from

$$u^{k+1} = u^k + \Delta u \tag{111}$$

$$\tau^{k+1} = \tau\left(\gamma\left(u^{k+1}\right), d^{k+1}\right), \; Y^{k+1} = Y\left(\gamma\left(u^{k+1}\right), d^{k+1}\right) \tag{112}$$

$$Y^{k+1} - Y_c \leq 0, d^{k+1} - d^n \geq 0, \left(Y^k - Y_c\right)\left(d^k - d^n\right) = 0 \tag{113}$$

$$\phi^{k+1} = \phi\left(d^{k+1}\right) \tag{114}$$

whereas tangent operators are obtained by

$$H^{k+1} = H_{\gamma\gamma}^{k+1} - \eta^k H_{\gamma d}^{k+1} \left(H_{dd}^{k+1}\right)^{-1} H_{d\gamma}^{k+1} \tag{115}$$

$$\eta^{k+1} = 1, \text{ if } d^{k+1} - d^k > 0 \text{ and } 0 \text{ otherwise} \tag{116}$$

$$H_{\gamma\gamma}^{k+1} = \frac{\partial \tau}{\partial \gamma}|_{k+1}, \; H_{\gamma d}^{k+1} = \frac{\partial \tau}{\partial d}|_{k+1}, \tag{117}$$

$$H_{d\gamma}^{k+1} = -\frac{\partial Y}{\partial \gamma}|_{k+1}, H_{dd}^{k+1} = -\frac{\partial Y}{\partial d}|_{k+1} \tag{118}$$

At the end of each load step, the gradient of the level set is computed. If it is below 1 everywhere the next load step is applied. If not, a non-local zone is placed and the solver flowchart with non-local zone is used.

Solver flowchart with non-local zone

1. initialization: $r_l^0 = r_i$
2. increase non-local zone: $r_l^{n+1} = r_l^n + \Delta r_l$
3. iterations initialization: $k = 0$
4. linear solve: solve (119) to find $\Delta u, \Delta T, \Delta\phi$
5. load update: $T^{k+1} = T^k + \Delta T$
6. update in local zone (111)-(118), and non-local zone (120)-(124)
7. if residual $\leq$ tol, go to 8, else go to 4
8. if domain not fully broken ($d(r_i) < 1$), go to 2, else go to 9
9. end

The extent of the non-local zone is imposed and one tries to find a continuous displacement and damage field satisfying the problem.

The linear symmetric problem to be solved at each iteration when the non-local zone is not empty is to find $\Delta u \in \mathcal{U},\ \Delta\phi \in \mathcal{A}, \Delta T \in R$ such that

$$\begin{aligned}
&\int_{r_i}^{r_l^{n+1}} \left(H^k_{\gamma\gamma}\gamma(\Delta u) + H^k_{\gamma d} d'\left(\phi^k\right)\Delta\phi\right)\gamma(u^*) r\,\mathrm{d}r + \int_{r_l^{n+1}}^{r_e} H^k\gamma(\Delta u)\gamma(u^*) r\,\mathrm{d}r - \tau_c \Delta T u^*(r_i) = \\
&\qquad \tau_c T^k u^*(r_i) r_i - \int_{r_i}^{r_e} \tau^k \gamma(u^*) r\,\mathrm{d}r, \forall u^* \in \mathcal{U} \\
&\int_{r_i}^{r_l^{n+1}} \left(H^k_{d\gamma} d'\left(\phi^k\right)\gamma(\Delta u) + \left(H^k_{dd} d'^2\left(\phi^k\right) + \left(Y_c - Y^k\right) d''\left(\phi^k\right)\right)\Delta\phi\right)\phi^* r\,\mathrm{d}r = \\
&\qquad \int_{r_i}^{r_l^{n+1}} \left(Y^k - Y_c\right) d'\left(\phi^k\right)\phi^* r\,\mathrm{d}r, \quad \forall\phi^* \in \mathcal{A} \\
&\qquad - \Delta\phi + \eta^k \frac{\left(H^k_{dd}\right)^{-1} H^k_{d\gamma}\gamma(\Delta u)}{d'(\phi^k)\,|_{r_l^{n+1,+}}} = \left(\phi^k - \phi\left(d^k\right)\right)|_{r_l^{n+1,+}}
\end{aligned} \tag{119}$$

where η^k is evaluated following (116). The update in the local zone follows (111)-(118) whereas in the non-local zone we have

$$u^{k+1} = u^k + \Delta u,\ \phi^{k+1} = \phi^k + \Delta\phi, \tag{120}$$

$$d^{k+1} = d\left(\phi^{k+1}\right) \tag{121}$$

$$\tau^{k+1} = \tau\left(\gamma\left(u^{k+1}\right), d^{k+1}\right),\ Y^{k+1} = Y\left(\gamma\left(u^{k+1}\right), d^{k+1}\right) \tag{122}$$

$$H^{k+1}_{\gamma\gamma} = \frac{\partial\tau}{\partial\gamma}|_{k+1},\ H^{k+1}_{\gamma d} = \frac{\partial\tau}{\partial d}|_{k+1}, \tag{123}$$

$$H^{k+1}_{d\gamma} = -\frac{\partial Y}{\partial\gamma}|_{k+1}, H^{k+1}_{dd} = -\frac{\partial Y}{\partial d}|_{k+1} \tag{124}$$

Is is interesting to note the difference between the two solver flowcharts. When the non-local zone is empty, the linear solve deals only with displacement increments and the local update deals with the damage variable. On the contrary, when the non-local zone is not empty, the linear solve involves both displacement and damage (or more precisely the surrogate ϕ variable) increment in the non-local zone (local zone being treated as before).

The mesh is built so that it is much finer in the localization zone. Node j is located at a position $x(j)$ given by

$$x(j) = \frac{(r_i + ((j-1)/N)(r_e - r_i))^2}{r_e - r_i} + r_i, \quad j = 1, \ldots, N+1 \tag{125}$$

where N is the number of elements considered. Results will be shown for the following mechanical parameters:

$$r_i = 0.1m, \quad r_e = 0.2m, \quad l_c = 0.02m, \quad \frac{\tau_c}{\mu} = 10^{-4}, \quad d_c = 0.5 \tag{126}$$

and numerical parameters

$$N = 200, \quad d_{\text{inc}} = 0.02 \tag{127}$$

Regarding parameter Δr_l, the non-local zone is advanced by one element at a time or smaller when damage gets close to 1 at r_i. This is done in order to capture the full load-displacement curve. The formula used in the simulation is

$$\Delta r_l = \max\left(\min\left(h^*, \left(l_c - \phi^n(r_i)\right)/2\right), 1.e^{-12} * l_c\right) \tag{128}$$

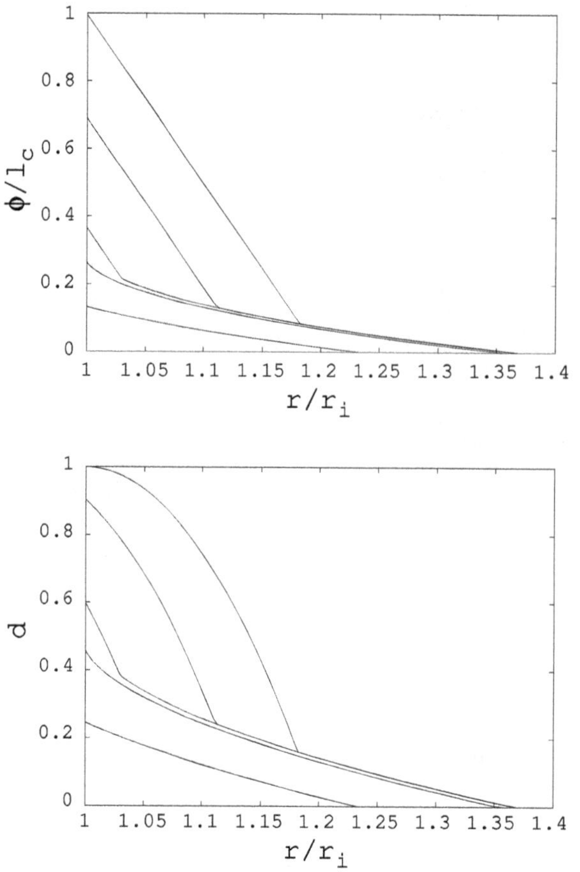

Figure 7 Distributions of ϕ/l_c (top figure) and damage (bottom figure) for different loadings. Details on the curves in each plot from bottom to top: Bottom curve corresponds to $T = 1.232$, damage evolution is purely local. The next curve is for $T = 1.351$. It corresponds to the load at which $\|\nabla\phi\| = 1$ at $r = r_i$ and non-locality steps in. Next curve is for $T = 1.368$, it is the limit load. The load then decreases and damage evolution is purely non-local. Last two curves are for $T = 1.012$ and $T = 0.02$, respectively. The latter case depicts the profile at complete decohesion of the fiber.

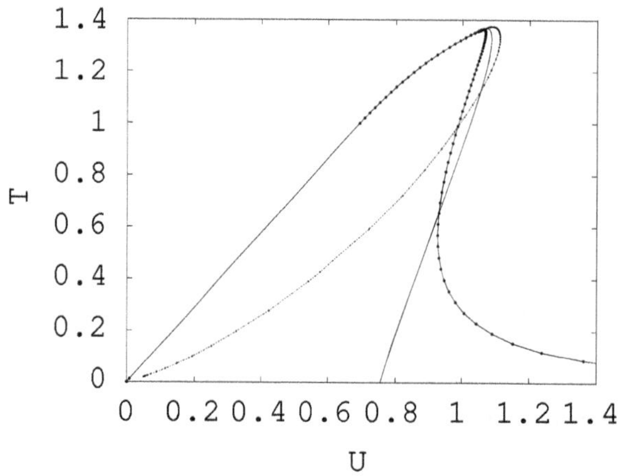

Figure 8 Force-displacement curves for n = 1 (small dots), n = 2 (solid line), n = 3 (big dots).

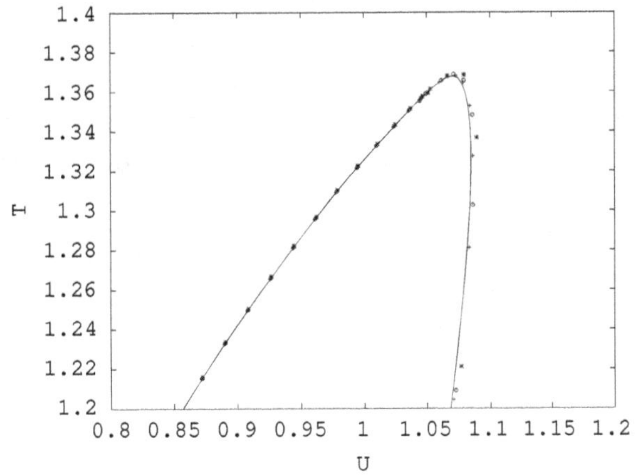

Figure 9 Force-displacement curves (zoom) for n = 2 and d_{inc} = 0.02 with different mesh sizes: N = 200 (solid line), N = 20 stars, N = 30 circles, N = 40 plus sign.

where h^* is the size of the element adjacent to the non-local zone at time step n. The initial ($n = 0$) non-local zone needs to be more than one-element for convergence. Between 5 and 10 elements are used.

As a final remark on the solver flowchart with the non-local zone, we noticed that in non-local zone update, it was more efficient (reduced number of iterations) to take $\Delta\phi$ as the one ensuring damage continuity rather that picking the one coming from step 3.

Discussion

In Figure 6, the force-displacement curve in the case n = 2 is shown. The figure indicates the different regime of the solution (pure elastic, local damage, coupled and pure non-local damage). Note that snap-back is taken into account automatically since the loading is not imposed but an unknown in the numerical scheme. Profiles of ϕ and damage along the radius at different loads are depicted in Figure 7.

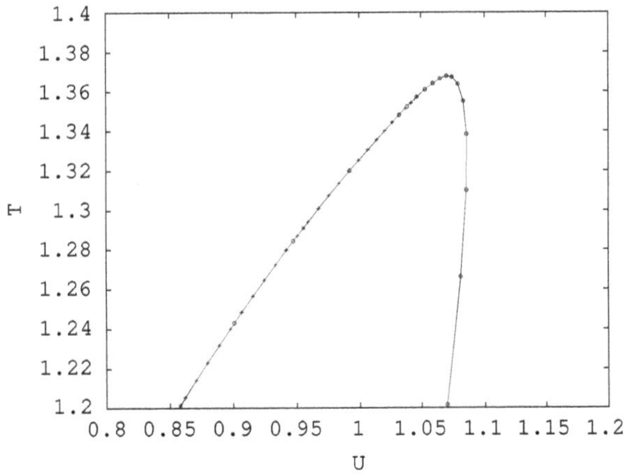

Figure 10 Force-displacement curves (zoom) for n = 2 and N = 50 with different values of d_{inc}: d_{inc} = 0.02 (solid line), d_{inc} = 0.1 stars, d_{inc} = 0.05 circles, d_{inc} = 0.01 plus sign.

Figure 8 shows the influence of the delocalization parameter n. Plots confirm the analytical limit results (97). As long as damage is purely local, all curves are superposed. As non-locality steps in, the delocalization parameters n plays a role.

Finally, in order to show the insensitivity of the model with respect to the discretization parameters N, d_{inc}, we show Figure 9 the influence of the choice of the N parameter (for the case $n = 2$ and $d_{\text{inc}} = 0.02$). In Figure 10, we show the influence of d_{inc} (for the case $n = 2$ and $N = 50$). Note that as expected, parameter d_{inc} has only an influence when damage is purely local (rising part of the curve). For both figures, a zoom was used. Otherwise, curves cannot be distinguished.

Conclusions

The Thick Level Set damage model allows coupling local damage evolution in some part of the domain to a non-local damage evolution in the localization zone. Damage gradient is bounded. The bound is reached in the non-local zone (localization zone) and not reached in the local one. The localization zone boundary is the main unknown in the model. It evolves ensuring damage continuity. A semi-analytical 1D solution has been developed showing different regimes in the solution (elastic, local damage, coupled local and non-local damage and finally pure non-local damage). The solution was plotted using a numerical scheme. This numerical scheme is ad hoc for the 1D problem considered. The corresponding numerical implementation for 2D and 3D cases will be the subject of a forthcoming publication.

Competing interests

The authors declare that they have no competing interests.

Authors' contributions

All authors contributed to the main ideas in the paper: the way to couple local and non-local evolutions of damage. NM came up with the analytical solution. NM and NC designed the 1D code to plot the results. All authors read and approved the final manuscript.

Acknowledgements

The support of the ERC Advanced Grant XLS no 291102 is greatfully acknowledged. Professor Antonio Huerta is also acknowledged for his advice.

Author details

[1]Ecole Centrale de Nantes, GeM Institute, UMR CNRS 6183, 1 Rue de la Noe, 44321 Nantes, France. [2]Lamsid, EDF-CEA-CNRS UMR 2832, Avenue du Général de Gaulle, 92141 Clamart, France.

References

1. Moës N, Stolz C, Bernard P-E, Chevaugeon N (2011) A level set based model for damage growth: the thick level set approach. Int J Numer Meth Eng 86:358–380
2. Bernard P-E, Moës N, Chevaugeon N (2012) Damage growth modeling using the Thick Level Set (TLS) approach: efficient discretization for quasi-static loadings. Comput Meth Appl Mech Eng 233–236:11–27
3. Stolz C, Moës N (2012) A new model of damage: a moving thick layer approach. Int J Fract 174:49–60
4. Stolz C, Moës N (2012) On the rate boundary value problem for damage modelization by Thick Level Set. In: ACOME 2012 Proceeding. Ho-Chi-Minh, Viet Nam. http://hal.archives-ouvertes.fr/hal-00725635
5. Karma A, Kessler D, Levine H (2001) Phase-field model of mode III dynamic fracture. Phys Rev Lett 87(4):045501
6. Miehe C, Welschinger F, Hofacker M (2010) Thermodynamically consistent phase-field models of fracture: Variational principles and multi-field FE implementations. Int J Numer Meth Eng 83(10):1273-1311
7. Spatschek R, Brener E, Karma A (2011) Phase field modeling of crack propagation. Phil Mag 91(1):75–95
8. Francfort GA, Marigo J-J (1998) Revisiting brittle fracture as an energy minimization problem. J Mech Phys Solid 46:1319–1412
9. Bourdin B, Francfort GA, Marigo J-J (2008) The Variational Approach to Fracture Vol. 91. pp 5–148, http://link.springer.com/10.1007/s10659-007-9107-3
10. Comi C, Mariani S, Perego U (2007) An extended FE strategy for transition from continuum damage to mode I cohesive crack propagation. Int. J. Numer. Anal. Meth. Geomech 31:213–238

11. Sethian JA (1999) Level set methods and fast marching methods: evolving interfaces in computational geometry, fluid mechanics, computer vision and material science. Cambridge University Press, UK
12. Maugin GA (1990) Internal variables and dissipative structures. J Non-Equilibrium Therm 15:173–192
13. Frémond M, Nedjar B (1996) Damage, gradient of damage and principle of virtual power. Int J Solid Struct 33(8):1083-1103
14. Comi C (1999) Computational modelling of gradient-enhanced damage in quasi-brittle materials. Mech Cohesive-Frictional Mater 36(April 1997):17–36
15. Pijaudier-Cabot G, Bazant ZP (1987) Nonlocal dalmage theory. J Eng Mech ASCE 113:1512–1533
16. Bazant ZP, Jirasek M (2002) Nonlocal integral formulations of plasticity and damage: survey of progress. J Eng Mech 128(November):1119–1149
17. Lubineau G, Azdoud Y, Han F, Rey C, Askari A (2012) A morphing strategy to couple non-local to local continuum mechanics. J Mech Phys Solid 60(6):1088–1102
18. Azdoud Y, Han F, Lubineau G (2013) A Morphing framework to couple non-local and local anisotropic continua. Int J Solid Struct 50(9):1332–1341
19. Lions P-L (1982) Generalized solutions of Hamilton-Jacobi equations. Pitman Advanced Publishing Program, Boston
20. Lorentz E, Godard V (2011) Gradient damage models: toward full-scale computations. Comput Meth Appl Mech Eng 200(21–22):1927–1944
21. Chung-Min L, Rubinstein J (2006) Elliptic equations with diffusion coefficient vanishing at the boundary: theoretical and computational aspects. Quaterly Appl Math 64:725–747
22. Pradeilles-Duval RM, Stolz C (1995) Mechanical transformations and discontinuities along a moving surface. J Mech Phys Solid 43(1):91–121
23. Hill R (1986) Energy-momentum tensors in elastostatics: some reflections on the general theory. J. Mech. Phys. Solids 34(3):305–317
24. Stolz C (2010) On micro-macro transition in non-linear mechanics. Materials 3(1):296–317
25. Rockafellar RT (1970) Convex analysis. Princeton University Press, USA

Permissions

All chapters in this book were first published in AMSES, by Springer; hereby published with permission under the Creative Commons Attribution License or equivalent. Every chapter published in this book has been scrutinized by our experts. Their significance has been extensively debated. The topics covered herein carry significant findings which will fuel the growth of the discipline. They may even be implemented as practical applications or may be referred to as a beginning point for another development.

The contributors of this book come from diverse backgrounds, making this book a truly international effort. This book will bring forth new frontiers with its revolutionizing research information and detailed analysis of the nascent developments around the world.

We would like to thank all the contributing authors for lending their expertise to make the book truly unique. They have played a crucial role in the development of this book. Without their invaluable contributions this book wouldn't have been possible. They have made vital efforts to compile up to date information on the varied aspects of this subject to make this book a valuable addition to the collection of many professionals and students.

This book was conceptualized with the vision of imparting up-to-date information and advanced data in this field. To ensure the same, a matchless editorial board was set up. Every individual on the board went through rigorous rounds of assessment to prove their worth. After which they invested a large part of their time researching and compiling the most relevant data for our readers.

The editorial board has been involved in producing this book since its inception. They have spent rigorous hours researching and exploring the diverse topics which have resulted in the successful publishing of this book. They have passed on their knowledge of decades through this book. To expedite this challenging task, the publisher supported the team at every step. A small team of assistant editors was also appointed to further simplify the editing procedure and attain best results for the readers.

Apart from the editorial board, the designing team has also invested a significant amount of their time in understanding the subject and creating the most relevant covers. They scrutinized every image to scout for the most suitable representation of the subject and create an appropriate cover for the book.

The publishing team has been an ardent support to the editorial, designing and production team. Their endless efforts to recruit the best for this project, has resulted in the accomplishment of this book. They are a veteran in the field of academics and their pool of knowledge is as vast as their experience in printing. Their expertise and guidance has proved useful at every step. Their uncompromising quality standards have made this book an exceptional effort. Their encouragement from time to time has been an inspiration for everyone.

The publisher and the editorial board hope that this book will prove to be a valuable piece of knowledge for researchers, students, practitioners and scholars across the globe.

List of Contributors

Wolfgang Boiger
Department of Mathematics, Humboldt-Universität zu Berlin, Unter den Linden 6, 10099, Berlin, Germany
Department of Computational Science and Engineering, Yonsei University, Unter den Linden 6, 120-749, Seoul, Korea

Carsten Carstensen
Department of Mathematics, Humboldt-Universität zu Berlin, Unter den Linden 6, 10099, Berlin, Germany
Department of Computational Science and Engineering, Yonsei University, Unter den Linden 6, 120-749, Seoul, Korea

Carl Sandstöm

Fredrik Larsson
Department of Applied Mechanics, Chalmers University of Technology, Hörsalsvägen 7, 412 96 Göteborg, Sweden

Kenneth Runesson
Department of Applied Mechanics, Chalmers University of Technology, Hörsalsvägen 7, 412 96 Göteborg, Sweden

Nissrine Akkari
University of La Rochelle, LaSIE - Laboratory of the Engineering Sciences for the Environment, Avenue Michel Crépeau, 17042 La Rochelle Cedex 1, France
Lebanese University, LaMA - Laboratory of Mathematics and Applications, B.P 826 Tripoli, Lebanon

Aziz Hamdouni
University of La Rochelle, LaSIE - Laboratory of the Engineering Sciences for the Environment, Avenue Michel Crépeau, 17042 La Rochelle Cedex 1, France

Erwan Liberge
University of La Rochelle, LaSIE - Laboratory of the Engineering Sciences for the Environment, Avenue Michel Crépeau, 17042 La Rochelle Cedex 1, France

Mustapha Jazar
Lebanese University, LaMA - Laboratory of Mathematics and Applications, B.P 826 Tripoli, Lebanon

Sandrine Germain

Philipp Landkammer
University of Erlangen-Nuremberg, Chair of Applied Mechanics, Egerlandstrasse 5, 91058 Erlangen, Germany

Paul Steinmann
University of Erlangen-Nuremberg, Chair of Applied Mechanics, Egerlandstrasse 5, 91058 Erlangen, Germany

Tristan Carrier Baudouin
Université catholique de Louvain, Institute of Mechanics, Materials and Civil Engineering, Bâtiment Euler, Avenue Georges Lemaître 4, Louvain-la-Neuve 1348, Belgium

Jean-François Remacle
Université catholique de Louvain, Institute of Mechanics, Materials and Civil Engineering, Bâtiment Euler, Avenue Georges Lemaître 4, Louvain-la-Neuve 1348, Belgium

Emilie Marchandise
Université catholique de Louvain, Institute of Mechanics, Materials and Civil Engineering, Bâtiment Euler, Avenue Georges Lemaître 4, Louvain-la-Neuve 1348, Belgium

François Henrotte
Université catholique de Louvain, Institute of Mechanics, Materials and Civil Engineering, Bâtiment Euler, Avenue Georges Lemaître 4, Louvain-la-Neuve 1348, Belgium

Christophe Geuzaine
Université de Liège, Dept. of Electrical Engineering and Computer Science, Montefiore Institute, Bâtiment B28, Sart-Tilman, Liège 4000, Belgium

Matthew L Rossi
Nuclear Regulatory Commission, Rockville, MD 20852, USA
Los Alamos National Laboratory, Los Alamos, NM 87545, USA

Christopher D Taylor
Fontana Corrosion Center, The Ohio State University, Columbus, OH 43210, USA.
Los Alamos National Laboratory, Los Alamos, NM 87545, USA

Adri CT van Duin
Mechanical and Nuclear Engineering, Pennsylvania State University, University Park, PA 16802, USA

Grégory Legrain

Nicolas Moës
LUNAM Université, GeM, UMR CNRS 6183, Ecole Centrale de Nantes, 1 rue de la Noë, 44321 Nantes, France

Noureddine Damil
Laboratoire d'Ingénierie et Matériaux LIMAT, Faculté des Sciences Ben M'Sik, Université Hassan II Mohammedia Casablanca, Sidi Othman, Casablanca, Maroc.

Michel Potier-Ferry
LEM3, Laboratoire d'Etudes des Microstructures et de Mécanique des Matériaux, UMR CNRS 7239, Université de Lorraine, Ile du Saulcy, 57045, Metz Cedex 01, France
Laboratory of Excellence on Design of Alloy Metals for low-mAss Structures (DAMAS), Université de Lorraine, Lorraine, France

Heng Hu
School of Civil Engineering, Wuhan University, 8 South Road of East Lake, 430072, Wuhan, PR China

Nicolas Moës
Ecole Centrale de Nantes, GeM Institute, UMR CNRS 6183, 1 Rue de la Noe, 44321 Nantes, France

Claude Stolz
Ecole Centrale de Nantes, GeM Institute, UMR CNRS 6183, 1 Rue de la Noe, 44321 Nantes, France
Lamsid, EDF-CEA-CNRS UMR 2832, Avenue du Général de Gaulle, 92141 Clamart, France

Nicolas Chevaugeon
Ecole Centrale de Nantes, GeM Institute, UMR CNRS 6183, 1 Rue de la Noe, 44321 Nantes, France